METHODS IN MOLECULAR BIOLOGY

Series Editor
John M. Walker
School of Life and Medical Sciences
University of Hertfordshire
Hatfield, Hertfordshire, UK

For further volumes:
http://www.springer.com/series/7651

For over 35 years, biological scientists have come to rely on the research protocols and methodologies in the critically acclaimed Methods in Molecular Biology series. The series was the first to introduce the step-by-step protocols approach that has become the standard in all biomedical protocol publishing. Each protocol is provided in readily-reproducible step-by step fashion, opening with an introductory overview, a list of the materials and reagents needed to complete the experiment, and followed by a detailed procedure that is supported with a helpful notes section offering tips and tricks of the trade as well as troubleshooting advice. These hallmark features were introduced by series editor Dr. John Walker and constitute the key ingredient in each and every volume of the Methods in Molecular Biology series. Tested and trusted, comprehensive and reliable, all protocols from the series are indexed in PubMed.

Manuscript Instructions https://drive.google.com/file/d/1K4Az5vpAFVzYadJ3jKx2-QSo7Dqa_SNA/view?usp=sharing

Evolutionary Genomics

Methods and Protocols

Edited by

Gustavo Caetano-Anollés

Department of Crop Sciences, University of Illinois, Urbana, IL, USA

Humana Press

Editor
Gustavo Caetano-Anollés
Department of Crop Sciences
University of Illinois
Urbana, IL, USA

ISSN 1064-3745 ISSN 1940-6029 (electronic)
Methods in Molecular Biology
ISBN 978-1-0716-4835-3 ISBN 978-1-0716-4836-0 (eBook)
https://doi.org/10.1007/978-1-0716-4836-0

This Humana imprint is published by the registered company Springer Science+Business Media, LLC, part of Springer Nature.
The registered company address is: 1 New York Plaza, New York, NY 10004, U.S.A.

Preface

What if we could rewind the tape of life's molecular history? Over the past few decades, evolutionary genomics has granted us an unprecedented power—to decode the hidden narratives written in DNA, reconstruct ancient genetic innovations, and even predict how genomes might evolve in response to tomorrow's challenges. From the origins of complex traits to the real-time evolution of pandemic viruses, this field has reshaped our understanding of life's blueprint. Yet, as sequencing technologies outpace our ability to interpret their revelations, we stand at a critical juncture. Will we merely accumulate data, or can we forge new frameworks to truly understand evolution in action?

Evolutionary genomics has already transformed our understanding of how life evolves at the molecular level, offering unprecedented insights into the genetic and structural mechanisms driving diversity, adaptation, and innovation. By integrating genomic data with evolutionary theory, researchers have uncovered processes that shape genomes over time, from the emergence of novel traits to the dynamics of speciation and adaptation. The rapid advancements in sequencing technologies and computational tools have opened new frontiers in this discipline, yet they also present significant challenges in data interpretation, integration, and scalability. This volume seeks to address some of these challenges by presenting cutting-edge methodologies and conceptual frameworks that push the boundaries of evolutionary genomics.

A foundational aspect of evolutionary genomics is retrodiction—the reconstruction of prior states and evolutionary histories. This requires a critical evaluation of assumptions and methodologies to ensure that inferences are both robust and biologically meaningful. By reflecting on the philosophical underpinnings of evolutionary inference, this volume highlights the importance of rigor in combining hypothesis-driven approaches with data-driven discoveries. This perspective informs the practical methodologies discussed throughout the book, which tackle complex questions about genome evolution, structural variation, and molecular adaptation.

Reconstructing evolutionary relationships is central to evolutionary genomics. Recent methodological advances allow researchers to resolve conflicting signals caused by processes such as horizontal gene transfer, recombination, or incomplete lineage sorting. These approaches include tools for inferring phylogenetic networks that capture reticulate evolution, scalable algorithms for placing fragmentary or metagenomic data into phylogenies, and innovative techniques for whole-genome comparisons using language-inspired models. Together, these methods provide powerful ways to infer evolutionary relationships across diverse taxa and genomic scales.

While sequence-based analyses remain a cornerstone of evolutionary genomics, increasing attention is being paid to the role of structure in understanding molecular evolution. Protein function and allosteric regulation are shaped by biophysical constraints that influence how proteins evolve over time. Structural phylogenomics offers a window into these processes by integrating molecular dynamics simulations with phylogenetic analysis to uncover deep evolutionary patterns. Similarly, studies on RNA secondary structures reveal how conserved structural motifs encode evolutionary histories and functional innovations. Efforts to reconstruct ancient molecular architectures, such as those present in the Last Universal Common

Ancestor (LUCA), further demonstrate how structural comparisons can illuminate life's earliest evolutionary events.

The study of viral evolution has taken on new urgency in light of recent global pandemics. Evolutionary genomics provides tools for analyzing genetic diversity within viral populations, tracking geographic and temporal changes in viral genomes, and predicting adaptive trajectories based on proteome structure. These approaches are critical for understanding how viruses evolve under selective pressures such as host immune responses or antiviral treatments. By combining genomic data with structural modeling and spatiotemporal metadata, researchers can gain actionable insights into viral evolution that inform public health strategies.

The vast scale of genomic datasets necessitates innovative computational frameworks capable of processing complex information efficiently. Advances in methods for detecting structural variants provide robust pipelines for identifying genomic changes associated with adaptation or disease. Techniques for analyzing error-prone genomes, such as those of RNA viruses, ensure accurate characterization of genetic diversity even under challenging conditions. Furthermore, integrating network theory with genomic data allows researchers to trace molecular chronologies onto biological networks, offering a systems-level perspective on evolutionary processes.

Evolutionary genomics is rapidly developing into a predictive science, capable of addressing foundational questions about life's history while tackling pressing global challenges such as biodiversity conservation and pandemic preparedness. By advancing our ability to reconstruct evolutionary histories, analyze genomic variation, and predict future trajectories, I hope that the chapters in this volume serve as an essential toolkit for exploring the mechanisms underlying genome evolution across scales—from individual molecules to entire ecosystems—and, in doing so, inspire new discoveries that deepen our understanding of life's molecular blueprint while charting a path toward a sustainable future.

Urbana, IL, USA ***Gustavo Caetano-Anollés***

Contents

Contributors

ADRIÁN CRUZ-GONZÁLEZ • *Facultad de Ciencias, Universidad Nacional Autónoma de México, CDMX, Mexico City, Mexico*

ANA ISABEL DE ÁVILA • *Centro de Biología Molecular "Severo Ochoa" (CSIC-UAM), Campus de Cantoblanco, Madrid, Spain*

ANTHONY M. POOLE • *School of Biological Sciences, The University of Auckland, Auckland, New Zealand*

ANTONIO LAZCANO • *El Colegio Nacional, Donceles 104, Centro Histórico, Mexico City, Mexico; Facultad de Ciencias, UNAM, Cd. Universitaria, Mexico City, Mexico*

ARSHAN NASIR • *Department of Crop Sciences, University of Illinois at Urbana-Champaign, Urbana, IL, USA*

ARTURO BECERRA • *Facultad de Ciencias, UNAM, Cd. Universitaria, Mexico City, Mexico*

ASHAR J. MALIK • *Australian Centre for Ecogenomics, The University of Queensland, Brisbane, QLD, Australia; Computational Biology and Clinical Informatics, Baker Heart and Diabetes Institute, Melbourne, VIC, Australia; School of Chemistry and Molecular Biosciences, The University of Queensland, Brisbane, QLD, Australia*

BRENDA MARTÍNEZ-GONZÁLEZ • *Centro de Biología Molecular "Severo Ochoa" (CSIC-UAM), Campus de Cantoblanco, Madrid, Spain; Department of Clinical Microbiology, Instituto de Investigación Sanitaria-Fundación Jiménez Díaz University Hospital, Universidad Autónoma de Madrid (IIS-FJD UAM), Madrid, Spain*

BYUNG-JU KIM • *Department of Chemistry and Center for Computational Biology, University of California, Berkeley, CA, USA; Department of Dermatology, Chobanian & Avedisian School of Medicine, Boston University, Boston, MA, USA*

CARLOS GARCÍA-CRESPO • *Centro de Biología Molecular "Severo Ochoa" (CSIC-UAM), Campus de Cantoblanco, Madrid, Spain*

CAROLINE PUENTE LELIEVRE • *School of Biological Sciences, The University of Auckland, Auckland, New Zealand*

CELIA PERALES • *Centro de Biología Molecular "Severo Ochoa" (CSIC-UAM), Campus de Cantoblanco, Madrid, Spain; Department of Clinical Microbiology, Instituto de Investigación Sanitaria-Fundación Jiménez Díaz University Hospital, Universidad Autónoma de Madrid (IIS-FJD UAM), Madrid, Spain*

CHENGZE SHEN • *Siebel School of Computing and Data Science, University of Illinois Urbana-Champaign, Urbana, IL, USA*

CORAL C. G. LUNA • *Facultad de Ciencias, UNAM, Cd. Universitaria, Mexico City, Mexico; Posgrado en Ciencias Biológicas, UNAM, Cd. Universitaria, Mexico City, Mexico*

DEREK CAETANO-ANOLLÉS • *Black Canyon Consulting LLC, Fairfax, VA, USA; Data Sciences Platform, Broad Institute of MIT & Harvard, Cambridge, MA, USA; National Center for Biotechnology Information, National Institutes of Health, Bethesda, MD, USA*

DÉSIRÉE B. LANGER • *School of Biological Sciences, The University of Auckland, Auckland, New Zealand*

ELEANOR WEDELL • *Siebel School of Computing and Data Science, University of Illinois Urbana-Champaign, Urbana, IL, USA*

ELLIOTT SOBER • *Philosophy Department, Stanford University, Stanford, CA, USA; Philosophy Department, University of Wisconsin-Madison, Madison, WI, USA*
ERIC BAPTESTE • *Department of Computational, Quantitative and Synthetic Biology (CQSB), UMR7238, Sorbonne Université, CNRS, IBPS, Paris, France*
ESTEBAN DOMINGO • *Centro de Biología Molecular "Severo Ochoa" (CSIC-UAM), Campus de Cantoblanco, Madrid, Spain*
FENGJIE SUN • *Department of Biological Sciences, School of Science and Technology, Georgia Gwinnett College, Lawrenceville, GA, USA*
FRANÇOIS-JOSEPH LAPOINTE • *Département de Sciences Biologiques, Complexe Des Sciences, Université de Montréal, Montréal, QC, Canada*
GUSTAVO CAETANO-ANOLLÉS • *Department of Crop Sciences, University of Illinois at Urbana-Champaign, Urbana, IL, USA*
HUGO BONNEFOUS • *Department of Computational, Quantitative and Synthetic Biology (CQSB), UMR7238, Sorbonne Université, CNRS, IBPS, Paris, France*
IGNACIO GADEA • *Centre for Biomedical Network Research on Infectious Diseases (CIBERINFEC), Madrid, Spain; Department of Clinical Microbiology, Instituto de Investigación Sanitaria-Fundación Jiménez Díaz University Hospital, Universidad Autónoma de Madrid (IIS-FJD UAM), Madrid, Spain*
IGOR N. BEREZOVSKY • *Bioinformatics Institute, Agency for Science, Technology and Research (A*STAR), Singapore, Singapore; Department of Biological Sciences (DBS), National University of Singapore (NUS), Singapore, Singapore*
ISABEL GALLEGO • *Centro de Biología Molecular "Severo Ochoa" (CSIC-UAM), Campus de Cantoblanco, Madrid, Spain*
JAEJIN CHOI • *Department of Chemistry and Center for Computational Biology, University of California, Berkeley, CA, USA*
JANE R. ALLISON • *School of Biological Sciences, The University of Auckland, Auckland, New Zealand*
JOSÉ ALBERTO CAMPILLO-BALDERAS • *Facultad de Ciencias, Universidad Nacional Autónoma de México, CDMX, Mexico City, Mexico*
KYUIN HWANG • *Division of Life Sciences, Korea Polar Research Institute, Incheon, Republic of Korea*
KYUNG MO KIM • *Division of Life Sciences, Korea Polar Research Institute, Incheon, Republic of Korea*
LUAY NAKHLEH • *Department of BioSciences, Rice University, Houston, TX, USA; Department of Computer Science, Rice University, Houston, TX, USA*
M. FAYEZ AZIZ • *Department of Crop Sciences, University of Illinois at Urbana-Champaign, Urbana, IL, USA*
MARÍA EUGENIA SORIA • *Centro de Biología Molecular "Severo Ochoa" (CSIC-UAM), Campus de Cantoblanco, Madrid, Spain; Department of Clinical Microbiology, Instituto de Investigación Sanitaria-Fundación Jiménez Díaz University Hospital, Universidad Autónoma de Madrid (IIS-FJD UAM), Madrid, Spain*
MATTHEW A. B. BAKER • *School of Biotechnology and Biomolecular Sciences at the University of New South Wales, Sydney, NSW, Australia*
MEHRDAD TAMIJI • *Department of Computer Science, Rice University, Houston, TX, USA*
MUHAMMAD ASIF ALI • *Department of Crop Sciences, University of Illinois at Urbana-Champaign, Urbana, IL, USA*

Nicholas J. Matzke • *School of Biological Sciences, The University of Auckland, Auckland, New Zealand*
Nicolae Sapoval • *Department of Computer Science, Rice University, Houston, TX, USA*
Pablo Mínguez • *Bioinformatics Unit, Instituto de Investigación Sanitaria-Fundación Jiménez Díaz University Hospital, Universidad Autónoma de Madrid (IIS-FJD UAM), Madrid, Spain; Centre for Biomedical Network Research on Rare Diseases (CIBERER), Instituto de Salud Carlos III, Madrid, Spain; Department of Genetics & Genomics, Instituto de Investigación Sanitaria-Fundación Jiménez Díaz University Hospital, Universidad Autónoma de Madrid (IIS-FJD UAM), Madrid, Spain*
Paul Klemm • *Center for Synthetic Microbiology (SYNMIKRO), Philipps-University Marburg, Marburg, Germany*
Philippe Lopez • *Department of Computational, Quantitative and Synthetic Biology (CQSB), UMR7238, Sorbonne Université, CNRS, IBPS, Paris, France*
Pilar Somovilla • *Centro de Biología Molecular "Severo Ochoa" (CSIC-UAM), Campus de Cantoblanco, Madrid, Spain; Departamento de Biología Molecular, Universidad Autónoma de Madrid, Campus de Cantoblanco, Madrid, Spain*
Ricardo Hernández-Morales • *Facultad de Ciencias, Universidad Nacional Autónoma de México, CDMX, Mexico City, Mexico*
Rodrigo Jácome • *Facultad de Ciencias, Universidad Nacional Autónoma de México, CDMX, Mexico City, Mexico*
Sung-Hou Kim • *Department of Chemistry and Center for Computational Biology, University of California, Berkeley, CA, USA*
Tandy Warnow • *Siebel School of Computing and Data Science, University of Illinois Urbana-Champaign, Urbana, IL, USA*
Tre Tomaszewski • *School of Information Sciences, University of Illinois at Urbana-Champaign, Urbana, IL, USA*
Ward C. Wheeler • *Division of Invertebrate Zoology, American Museum of Natural History, New York, NY, USA*
Yu Zhang • *Department of Computational, Quantitative and Synthetic Biology (CQSB), UMR7238, Sorbonne Université, CNRS, IBPS, Paris, France*

Check for updates

Chapter 1

Retrodiction in Evolutionary Genomics: A Philosophical Perspective

Elliott Sober

Abstract

I begin with an intuitively plausible principle: if proposition H retrodicts or predicts that proposition D is true, then Pr(D|H) > ½. I then consider the difference between forward-directed and backward-directed conditional probabilities. Until the 1980s, population genetics theory was dominated by the former; the situation changed with the appearance of coalescent theory, which deploys the latter. A forward-directed probability model cannot retrodict, and a backward-directed model cannot predict. A Bayesian model will do both if it deploys both types of conditional probability. I then describe ideas from model selection theory in statistics that throw doubt on the idea that true models of a process will always be more accurate predictors or retrodictors than false models of that process. I then discuss theorems in information theory that show how the passage of time degrades information in a causal chain. The rate of information loss can be reduced and even cancelled by the branching process in phylogenetic trees. The separation of within-lineage from across-lineage assessments of information loss is an instance of Simpson's paradox.

Keywords AIC, Bayesianism, Evolution, Forward/backward directed probabilities, Frequentism, Information theory, Instrumentalism, Likelihood, Model selection, Prediction, Retrodiction, Simpson's paradox

1 What Hypotheses Predict and Retrodict

Here is an intuitively plausible principle concerning what it takes for a proposition to predict or retrodict:

(RP) If hypothesis H describes (truly or falsely) the present state of a system and it retrodicts or predicts that proposition D is true, then $Pr_M(D|H) > ½$.

D is a dictum, with prediction being one kind and retrodiction another. The "M" subscript on the probability operator is a reminder: when H describes the state of a system at one time and D describes the state of that system at another, a model of the relationship between those two states needs to be postulated for H to confer a probability on D. The RP constraint entails that hypotheses that assign a

Gustavo Caetano-Anollés (ed.), *Evolutionary Genomics*, Methods in Molecular Biology, vol. 2981,
https://doi.org/10.1007/978-1-0716-4836-0_1,

probability to D that is lower than ½ can't predict or retrodict D. The same is true of hypotheses that assign no probability at all to D.

One ramification of RP becomes clear when you move beyond the dichotomy of proposition D and its negation, not D. Suppose there are n discrete states that might obtain at some past (or future) time, that H assigns each of them a probability, and that these probabilities sum to 1. At most one of these states can have a probability greater than ½, which means that H can predict or retrodict at most one of them. Additional possibilities arise, however, by taking disjunctions of states into account. H may not be able to retrodict/predict s_1 and it may not be able to retrodict/predict s_2, but this leaves it open that H might retrodict/predict a disjunction – s_1 or s_2. A similar idea applies to continuous variables. Suppose tomorrow's temperature in Madison will be between 0 and 20°C and that each point value in that region has a probability of zero. That space of possibilities can be divided into a finite number of exclusive and exhaustive regions where at most one of those regions has a probability greater than ½; each region is an infinite disjunction with each disjunct having a positive probability density. As before, a disjunction of regions can have a probability that is greater than ½. RP also has implications about conjunctions. It says that even if H predicts (or retrodicts) D_1 and also predicts (or retrodicts) D_2, H might not predict (or retrodict) their conjunction, D_1 and D_2.

2 Forward-Directed and Backward-Directed Probabilities

The central models developed in population genetics before the 1980s involved forward-directed conditional probabilities, not backward-directed probabilities. A forward-directed conditional probability has the form Pr(Y is true at time t_2|X is true at time t_1) where $t_1 < t_2$. The probabilities deployed in the models of Fisher [1], Haldane [2], Wright [3], Moran [4], Kimura [5], and others are forward-directed. They facilitate prediction, but not retrodiction. If you know that X is true now, you can use a forward-directed model to predict what will probably happen later, but forward-directed models provide no clue about what (probably) happened earlier. This fixation on forward-directed models subsided somewhat with the invention of coalescent theory by Kingman [6] and others, which is based on backward-directed probabilities. Models of coalescence facilitate retrodiction but not prediction.

The earlier focus on forward-directed models was no accident; it went hand-in-hand with the anti-Bayesian statistical practice that then dominated the field. Anti-Bayesianism encompasses a diverse family of procedures (e.g., significance tests, Neyman–Pearson hypothesis tests, confidence intervals, likelihood-ratio tests). The standard name for this family is "frequentism" and its first commandment is *thou shalt not assign probabilities to theories*. The governing idea was that probabilities apply only to possible outcomes of chance processes, where O's being an outcome of chance process C presupposes that O happened after C got rolling. For example, a one-locus two allele model of heterozygote superiority makes a prediction about how allelic frequencies will change, but the observed subsequent allelic frequency does not confer a probability on the population's earlier instantiating that model. Why heterozygote superiority itself appears as a genotype/fitness relationship is a different question, whose answer would call for a different model, which also deploys forward-directed probabilities. This suggests that it isn't just the frequentist framework that led population genetics to be built by constructing models that had no backward-directed probabilities. The idea that evolution is a process that moves from earlier states to later states naturally leads to the thought that models of that process should show how earlier states transition to later states; that transition could be probabilistic or deterministic.

The division of labor between probabilistic models that are forward-directed and models that are backward-directed is not a feature of Bayesian models, because Bayes's theorem knits the two together into a single formula:

$$\text{(BAYES)}\ \frac{\Pr(\text{Y is true at time } t_2 \mid \text{X is true at time } t_1)}{\Pr(\text{Y is true at time } t_2)} = \frac{\Pr(\text{X is true at time } t_1 \mid \text{Y is true at time } t_2)}{\Pr(\text{X is true at time } t_1)}$$

The value of each of the four probabilities in this equation can be computed from the values of the other three, but not from the values of just one of them or two. This means that a Bayesian model can make predictions and retrodictions that conform to the RP constraint.

3 What Makes a Hypothesis a Good Predictor or Retrodictor?

Hypotheses sometimes make predictions that are utterly useless to people who want their predictions to be accurate. For example, consider the hypothesis that the moon is made of green cheese; this hypothesis makes a prediction about what astronauts will find when they land on the moon, but this hypothesis rightly played no role in NASA's planning.

This example suggests a thesis—that false hypotheses are always less accurate than true ones in the predictions and retrodictions they make. A surprising upshot of work in the part of statistics called model selection theory shows that this generalization is wrong.

Idealized models (whose simplifications render the model false) are often predictively accurate, and true models often do a poor job of making accurate predictions.

These two points are illustrated by the following example. Suppose you bicycle in the countryside outside Madison in August and stop for a breather between two large fields of corn. You are interested in two models of how the mean heights in the two fields are related:

$$\text{(Null)}\ \ m_1 - m_2 = 0$$

$$\text{(DIFF) There exists a height h such that } |m_1 - m_2| = h$$

You then select 100 corn plants at random from each field and measure how tall each is. You find that the sample means are $s_1 = 64$ inches and $s_2 = 60$ inches. You want to decide which of these models will do a better job of predicting what you will observe if you draw a new random sample of 100 plants from each of the two fields.

It is clear what NULL predicts about the new samples you'll draw, but what does DIFF predict? The answer seems to be *nothing*; DIFF is close to being a tautology. However, you can use DIFF to construct a fitted model by replacing the adjustable parameter h in DIFF with h's maximum likelihood estimate:

$$f(\text{DIFF})|m_1 - m_2| = 4 \text{ inches}$$

Now, NULL and DIFF each make a prediction about the new data set. NULL does this directly, whereas DIFF does this indirectly, via f(DIFF).

It might seem that f(DIFF) will make a more accurate prediction because it fits the old data better than NULL does. However, the lived experience of scientists has taught them to be wary of this naïve empiricism. They have learned that a model can be made to fit the old data perfectly by making it sufficiently complex, but complex models often do a poor job of predicting new data drawn from the same underlying reality. They also have learned that simpler models often fit the old data less well than models that are more complex, but those simpler models often do a better job of predicting new data. In such cases, the complex model is said to have overfitted the data.

This lived experience is not a brute fact, thanks to statisticians working in model selection theory who have produced mathematical reasons for expecting simpler models to sometimes (but not always!) be more predictively accurate than models that are more complex. A landmark in this enterprise was Akaike's [7] formulation of a model selection criterion, now called AIC, the Akaike Information Criterion, where a model M's AIC value is

$$\mathrm{AIC(M)} = \log(\mathrm{pr[observations} \mid \mathrm{f(M)]}) - \mathrm{k}.$$

Akaike proved that AIC(M) provides an unbiased estimate of M's predictive accuracy (*see* **Note 1**). Here, k is the number of adjustable parameters in model M, and f(M) is the result of fitting M to the observations. Notice that AIC(M) penalizes models for their complexity, where complexity is measured by counting adjustable parameters. In the present example, DIFF has one adjustable parameter (h), but Null has no adjustable parameters. It is no surprise that DIFF fits the data better than NULL does. However, AIC tells you that the proper question to ask is whether DIFF fits the data *sufficiently* better than NULL, so that DIFF thereby "overcomes" the fact that it is more complex. The right question is whether AIC(DIFF) − AIC(NULL) > 1 (*see* **Note 2**).

I think you know, before you look at any data from the two fields of corn, that NULL is false and that DIFF is true. In this example, a model known to be false may have a higher estimated predictive accuracy than a model known to be true. This opens the door to an instrumentalist thesis about the goal of scientific inference. Instrumentalism says that the goal is predictive accuracy; realism says that the goal is truth.

NULL is a special case of DIFF in that NULL can be deduced from DIFF by setting h = 0. In statistical parlance, NULL is "nested" in DIFF. Although AIC and other model selection criteria (e.g., cross validation) are open to the possibility that NULL may be "better" than DIFF, it is important to recognize that NULL can't be better than DIFF in the sense of "better" that matters to Bayesians. This is because NULL cannot have a higher probability than DIFF, regardless of what evidence you have. This is a mathematical fact, a consequence of the axioms of probability. Nested models provide a clear illustration of how Bayesianism and Frequentism conflict.

4 Are Anti-Bayesians Entitled to Use a Forward-Directed Model to Make Retrodictions?

Anti-Bayesians often use models to make retrodictions from present data. For example, consider the use of maximum likelihood estimation to estimate the character state of the most recent common ancestor (MRCA) of several current species. Consider a binary character whose two states are 0 and 1. Suppose you observe that each current species in your data set is in state 0. Which assignment of character state to their MRCA maximizes the probability of the observations?

All extant models of phylogenetic inference assume that the character states of ancestors and descendants are correlated (Lemey et al. [8]), which in the case of a binary character means that

(COR) For each branch in a phylogenetic tree,
Pr (the branch ends in state 0 | the branch begins in state 0)>
Pr (the branch ends in state 0 | the branch begins in state 1).

This inequality holds so long as the time between beginning and end is finite. Notice that COR describes forward-directed probabilities. Given COR and the observation that all leaf species are in state 0, the maximum likelihood estimate of the character state of their MRCA is that it too is in state 0.

So far so good, but can we take the further step of saying that the observed character states of the leaf species, along with the assumption of correlation, *retrodicts* that the MRCA was in state 0? To do so would violate the RP principle. True, the observed character state of current species and the COR assumption together indicate that MRCA=0 has a higher *likelihood* than MRCA=1. But likelihood is not probability. No conclusion follows as to whether MRCA=0 is more probable than MRCA=1.

This conclusion is not undercut by the fact that correlation is a symmetrical relation, which means that COR is equivalent to

For each branch in a phylogenetic tree,
Pr (the branch begins in state 0 | the branch ends in state 0)>
Pr (the branch begins in state 0 | the branch ends in state 1)

Here we have two backward-directed probabilities, but no conclusion can be drawn as to whether the probability that the branch began in state 0 is greater than the probability that it began in state 1.

Does this mean that frequentists should reject the RP criterion? The general approach to inference that frequentists take suggests that they will. The usual line is that scientists should make predictions by using hypotheses that were obtained by applying inferential procedures that have good "operating characteristics." In the case of maximum likelihood estimation, frequentists will point to the fact that maximum likelihood estimators are consistent, unbiased, and (sometimes) admissible. Beyond that, they don't care about the value of the likelihood in a particular case, and they also have no interest in the probability of hypotheses about the state of the MRCA

Here's another example: consider what the method of maximum likelihood says about inferring how contemporary Humans, Chimpanzees, and Gorillas are genealogically related. There are three possible bifurcating trees – (HC)G, H(CG), and (HG)C. After choosing a model of the evolutionary process at work in the branches of the three phylogenetic trees, you use the observed characteristics of the three species to figure out which of these tree hypotheses confers the highest probability on the data. With large data sets, the typical situation is that all three trees say that the data are very improbable.

However, this does not rule out the possibility that the three pairwise log-likelihood ratios are all very far from 0. Frequentist biologists may be inclined to describe what they are doing here by saying that they use present data to retrodict the topology of the branching process that unfolded in the past. This statement falls afoul of what the RP criterion demands of a retrodiction. Frequentists can be expected to respond—"so much the worse for RP."

This frequentist reaction to RP is reminiscent of what Neyman and Pearson [9] said about evidence:

> No test based upon the theory of probability can by itself provide any valuable evidence of the truth or falsehood of [an] hypothesis ... But we may look upon the purpose of tests from another viewpoint. Without hoping to know whether each separate hypothesis is true or false, we may search for rules that govern our behavior with regard to them, in following which we ensure that, in the long run of experience, we shall not be too often wrong (p. 291).

Frequentists focus on the long-run performance of rules for evaluating hypotheses; Bayesians focus on hypotheses one by one and try to determine their probabilities.

There is a third statistical philosophy that is "intermediate" between Frequentism and Bayesianism. I call it Likelihoodism. It says that the goal of inference is to see whether the evidence favors one hypothesis over another in the following sense:

The Qualitative Law of Likelihood: Observation O favors hypothesis H_1 over hypothesis H_2 if and only if Pr $(O|H_1) > Pr(O|H_2)$.

The Quantitative Law of Likelihood: If O favors H_1 over H_2, the strength of that favoring is given by the likelihood ratio $\frac{Pr(O|H_1)}{Pr(O|H_2)}$.

Likelihoodists do not forsake the idea that observations can provide evidence about the truth of hypotheses, but they have no truck with hypotheses' prior and posterior probabilities (Hacking [10], Edwards [11], Royall [12]; see Sober [13], [14], and [15] for discussion). Likelihoodists can accept the RP constraint and say that prediction and retrodiction is not their job, or they can join frequentists and reject the RP constraint.

5 What Does Information Theory Say About Retrodiction?

Claude Shannon's [16] concept of mutual information is a workhorse in information theory. It quantifies the amount of information that observing the state of one variable provides about the state of another. Our interest here is the case in which one variable represents the present state of the system and the other represents its past. Consider a system that has the following properties:

(i) **Discrete states and times**: At any time, the system is in one of n possible states (s_1, s_2, …, s_n), and evolves in discrete time steps.

(ii) **Irreducibility**: For any two states, each is accessible from the other, either directly or by the system's moving through intermediate states.

(iii) **The Markov property**: For any two times $t_1 < t_2$, the state of the system at t_1 screens-off the system's history prior to t_1 from the state at t_2. That is, for any i and j, Pr (system is in state j at t_2|system is state i at t_1) = Pr(system is in state j at t_2 | system is state i at t_1 & system's history prior to t_1).

These three assumptions lead to an important result:

The Markov Chain Convergence Theorem: If a system's evolution satisfies assumptions (i), (ii), and (iii), then I(Past; Present) approaches zero as the time separating past and present approaches infinity (Cover and Thomas [17]).

Here, I (X; Y) is the mutual information linking the two variables. When the variables are discrete, the formula for this quantity is:

$$I(X;Y) = \sum_{y \in Y} \sum_{x \in X} p(x,y) \log \left[\frac{p(x,y)}{p(x)p(y)} \right]$$

Here, $p(x, y)$ is the probability of the conjunction ($X=x$ & $Y=y$), and $p(x)$ and $p(y)$ are the probabilities of conjuncts ($X=x$) and ($Y=y$), respectively. The logarithm is base 2. The mutual information is zero when X and Y are independent; otherwise, it is positive. Mutual information is symmetrical: I (X;Y) = I (Y;X). The decay of mutual information described by the theorem occurs exponentially fast (Sober and Steel [18], Proposition 6). There is no requirement in the convergence theorem that transition probabilities are constant through time.

Mutual information is a Bayesian construct, since it requires you to consider the unconditional probabilities of hypotheses about the state of each variable. Indeed, the quotient for specific values of X and Y in the definition of mutual information is equivalent to the logarithm of $\frac{\Pr(Y=y \mid X=x)}{\Pr(Y=y)}$; Bayesians philosophers call this the ratio measure of degree of confirmation. Note that this quotient stands on one side of the BAYES equation given above.

The convergence theorem does not ensure a monotonic decline in information as the temporal separation of past from present increases, but monotonicity is guaranteed by a second result:

The data processing inequality: In a causal chain from a distal cause D to a proximate cause P to an effect E, if P screens-off D from E (meaning that Pr(E=z|P=y)=Pr(E=z|P=y & D=x), for all x, y, z), then I (E; D) ≤ I (E; P) and I (E; D) ≤ I (P; D) (Cover and Thomas [17]).

D, P, and E are variables. For a discrete-state process, the two inequalities are strict if P probabilistically depends on D and on E but is perfectly correlated with neither. The data processing inequality does not require that the process linking D to P and the process linking P to E are of the same type.

It may seem that the two theorems about information loss provide a blanket guarantee that present evidence provides more information about what happened in the recent past than it does for events that are farther back in time, but this is an overstatement. Both theorems describe the relationship of events *within* a single causal chain, but nothing follows about relationships *between* chains. For example, consider two chains that are independent of each other, where each runs from a distal cause to a proximate cause and then to an effect. Chain 1 is $D_1 \rightarrow P_1 \rightarrow E_1$ and Chain 2 is $D_2 \rightarrow P_2 \rightarrow E_2$. Suppose that E_1 and E_2 are variables that apply to the current state of species 1 and species 2, respectively, and that D_1 occurred much earlier than D_2. The theorems say nothing about whether $I(D_1; E_1) < I(D_2; E_2)$.

How are the above results about mutual information related to the task of retrodiction where retrodiction is assumed to satisfy the RP constraint? One difference is that mutual information is about two variables, whereas RP is about two propositions, each of which describes a single state of a variable. Another gap is that mutual information is symmetrical, but the relationship described in the RP principle is not; $\Pr(D|H) > ½$ does not entail that $\Pr(H|D) > ½$. A third is that mutual information is about how much you learn about the state of one variable by observing the state of the other; mutual information involves comparisons of $\Pr(Y=y|X=x)$ with $\Pr(Y=y)$. Retrodiction isn't about a comparison of two probabilities; the RP constraint says only that a single probability must be greater than ½.

In spite of these differences between mutual information and retrodiction, there is nonetheless a simple connection between information loss and retrodiction. The two theorems about information loss together entail that each term of the form $\log[\frac{\Pr(Y=s_j|X=s_i)}{\Pr(Y=s_j)}]$ in the summation that defines mutual information approaches zero as the temporal gap between X and Y widens. This means that $\Pr(Y=s_j|X=s_i)$ gets closer and closer to $\Pr(Y=s_j)$, for each i, j. Note that $\Pr(Y=s_1)$, $\Pr(Y=s_2)$, ..., and $\Pr(Y=s_n)$ form a partition, so at most one of them is greater than ½. That means that with enough time between the two events, at most one of the conditional probabilities that have the form $\Pr(Y=s_j|X=s_i)$ will have a value greater than ½. The upshot is that the prospects for retrodiction in the limit are very modest at best. When $\Pr(Y=s_i) < 0.5$ for all i, the prospect is zero.

6 Branching and Information Loss

Although the passage of time *within lineages* induces information loss (if the process governing that passage conforms to the assumptions of the two theorems described above), the branching that takes place in evolution can reduce the rate of information loss, since it creates *new lineages*. In a symmetrical bifurcating tree with no extinctions, the single ancestor at the root of the tree has 2^n descendants in generation n. The result is that an ancient ancestor may have numerous present day descendants, each of which provides evidence about the ancestor's characteristics. This multiplicity of "witnesses" can compensate for the loss of information that occurs within each lineage in the tree. In a symmetric two-state evolutionary process, it is possible to say exactly how often branching must occur for information loss to be canceled (Evans et al. [19]) (*see* **Note 3**). The mathematics for other processes is more complicated.

Just as the fact of branching affects information loss, so too does the topology of the branching process itself. Consider the following conjecture about the two genealogies depicted in the Fig. 1: observing the four descendants will provide more information about the character state of their most recent common ancestor if they are related to each other by a star phylogeny than would be the case if they were related to each other by a balanced bifurcating tree. The conjecture seems reasonable, since in genealogy (a), the observations are all probabilistically independent of each other, conditional on the state of the root, whereas in genealogy (b), the observations are not all pairwise conditionally independent. If the same symmetric two-state process is at work in all the branches in the two genealogies, the conjecture holds (Evans et al. [19] but it sometimes fails in symmetric processes with more than two states (Sly [20]). With an asymmetric two-state process, the conjecture can fail [15].

A further complexity is introduced when trees are not balanced. For a simple example, consider a bifurcating tree whose left half bifurcates twice as fast as its right. If r is the rate of bifurcation on the right and the time between root and leaves is t, there are 2rt

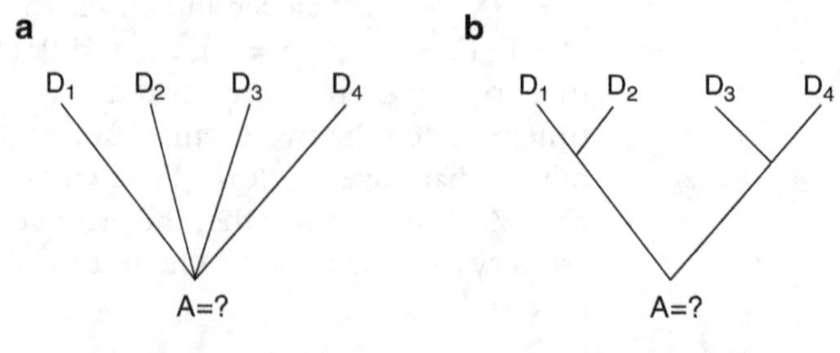

Fig. 1 Does observing the states of the four descendants (D_1, D_2, D_3, D_4) in the star phylogeny (**a**) provide more information about the state of the ancestor (A) than observing the states of the four descendants in the bifurcating tree (**b**)?

generations between root and leaves on the left and rt generations on the right. This means that there are $(2)^{2rt}$ leftist leaves and $(2)^{rt}$ rightists. Because of their greater numbers, the leftist leaves provide more information about the state of the MRCA than the rightist leaves. Since the leftist leaves and the rightist leaves are independent of each other, conditional on the state of the MRCA, the mutual information that they collectively provide about the state of the root is greater than the mutual information supplied by each; this follows from the data-processing inequality.

7 Notes

1. Forster and Sober [21] use the phrase "predictive accuracy" to characterize what AIC estimates, but it is more standard to formulate AIC as a measure of predictive inaccuracy (or distance from the truth), where AIC $(M) = k - \log(\Pr(\text{data}|f(M))$.
2. Model selection theory might better be called model evaluation theory, since criteria like AIC do not require you to put all your eggs in one basket. When evaluating several models with the goal of making a prediction, you might want to take the *n* best-scoring models and average their predictions. Burnham and Anderson [22] recommend using AIC weights to construct this average.
3. This is an instance of Simpson's paradox [23] in that each branch in a tree suffers information loss though the tree as a whole does not.

Acknowledgments

The second half of this paper draws on material from Sober and Steel [18, 24]. My thanks to Gustavo Caetano-Anolles and Mike Steel for useful discussion.

References

1. Fisher RA (1930) The genetical theory of natural selection. Clarendon Press, Oxford
2. Haldane JBS (1932) The causes of evolution. Princeton University Press, New York
3. Wright S (1931) Evolution in mendelian populations. Genetics 16(2):97–159
4. Moran P (1958) Random processes in genetics. Math Proc Camb Philos Soc 54(1):60–71
5. Kimura M (1983) The neutral theory of molecular evolution. Cambridge University Press, Cambridge
6. Kingman JFC (1982) On the genealogy of large populations. J Appl Probab 19:27–43
7. Akaike H (1973) Information theory and an extension of the maximum likelihood principle. In: Petrov B, Csaki F (eds), Second International Symposium on Information Theory. Akademiai Kiado, Budapest, Hungary, 267–281
8. Lemey P, Salemi M, Vandamme AM (2009) The phylogenetic handbook - a practical approach to phylogenetic analysis and hypothesis testing. Cambridge University Press, Cambridge

9. Neyman J, Pearson E (1933) On the problem of the most efficient tests of statistical hypotheses. Phil Trans R Soc A 231:289–337
10. Hacking I (1965) The logic of statistical inference. Cambridge University Press, Cambridge
11. Edwards AWF (1972) Likelihood. Cambridge University Press, Cambridge
12. Royall R (1997) Statistical evidence – a likelihood paradigm. Chapman and Hall, Boca Raton, Florida
13. Sober E (2008) Evidence and evolution – the logic behind the science. Cambridge University Press, Cambridge
14. Sober E (2015) Ockham's razors – a user's manual. Cambridge University Press, Cambridge
15. Sober E (2024) The philosophy of evolutionary theory. Cambridge University Press, Cambridge
16. Shannon C (1948) A mathematical theory of communication. BSTJ 2(4):623–656
17. Cover TM, Thomas JA (2006) Elements of information theory, 2nd ed. Wiley, Hoboken, New Jersey
18. Sober E, Steel M (2014) Time and knowability in evolution. Philos Sci 81(4):558–579
19. Evans W, Kenyon C, Peres Y, Schulman L (2000) Broadcasting on trees and the Ising model. AAP 10(2):410–433
20. Sly A (2011) Reconstruction for the Potts model. Ann Probab 39(4):1365–1406
21. Forster M, Sober E (1994) How to tell when simpler, more unified, or less ad hoc theories will provide more accurate predictions. BJPS 45(1):1–35
22. Burnham K, Anderson D (2002) Model selection and multimodel inference – a practical information-theoretic approach, 2nd ed. Springer-Verlag, New York
23. Simpson EH (1951) The interpretation of interaction in contingency tables. J R Stat Soc B 13(2):238–241
24. Sober E, Steel M (2011) Entropy increase and information loss in Markov models of evolution. Biol Philos 26(2):223–250

Check for updates

Chapter 2

Searching for Phylogenetic Networks

Ward C. Wheeler

Abstract

Phylogenetic networks play an increasingly important role in the comparative analysis of many systems. These include studies of hybridization, linguistic exchange, bacterial and viral evolution, and the origins of zoonotic disease. Computationally, these problems present many NP-hard optimizations requiring heuristic procedures to generate useful results. Here, algorithmic approaches to generate heuristic solutions are discussed and recommendations for real-world use cases are presented for the phylogenetic graph search tool Phylogenetic Graph (PhyG).

Keywords Phylogenetics, Networks, Heuristics, Graphs, Parsimony

1 Introduction

The use of phylogenetic networks as an explanatory model for the evolution of complex biological systems has become commonplace. This is especially seen in species origins through hybridization [1, 2], the origins of viral pathogenesis (e.g., Influenza; [3]; Fig. 1a), zoonotic shifts [4], and historical linguistic analysis [5, 6]. Phylogenetic network scenarios allow for the simultaneous resolution of vertical (i.e., traditional parent to child, tree-like) information transfer and horizontal transfer that occurs between contemporary lineages.

One mechanism of the origin of viral parthenogenesis is reassortment. Using influenza as an example (Fig. 1b), the eight genomic segments (analogous to the chromosomes of eukaryotes) can be mixed among strains that have simultaneously infected a single host. In this process, novel combinations of segments can be created with enhanced infectivity. This is thought to have occurred in the 2009 Swine Flu pandemic [7].

There is a large literature on the mathematics and algorithmics of phylogenetric networks, and multiple types of networks have been described. The most commonly used for evolutionary biological questions is the "soft-wired" network, where the convergent edges

Gustavo Caetano-Anollés (ed.), *Evolutionary Genomics*, Methods in Molecular Biology, vol. 2981,
https://doi.org/10.1007/978-1-0716-4836-0_2,

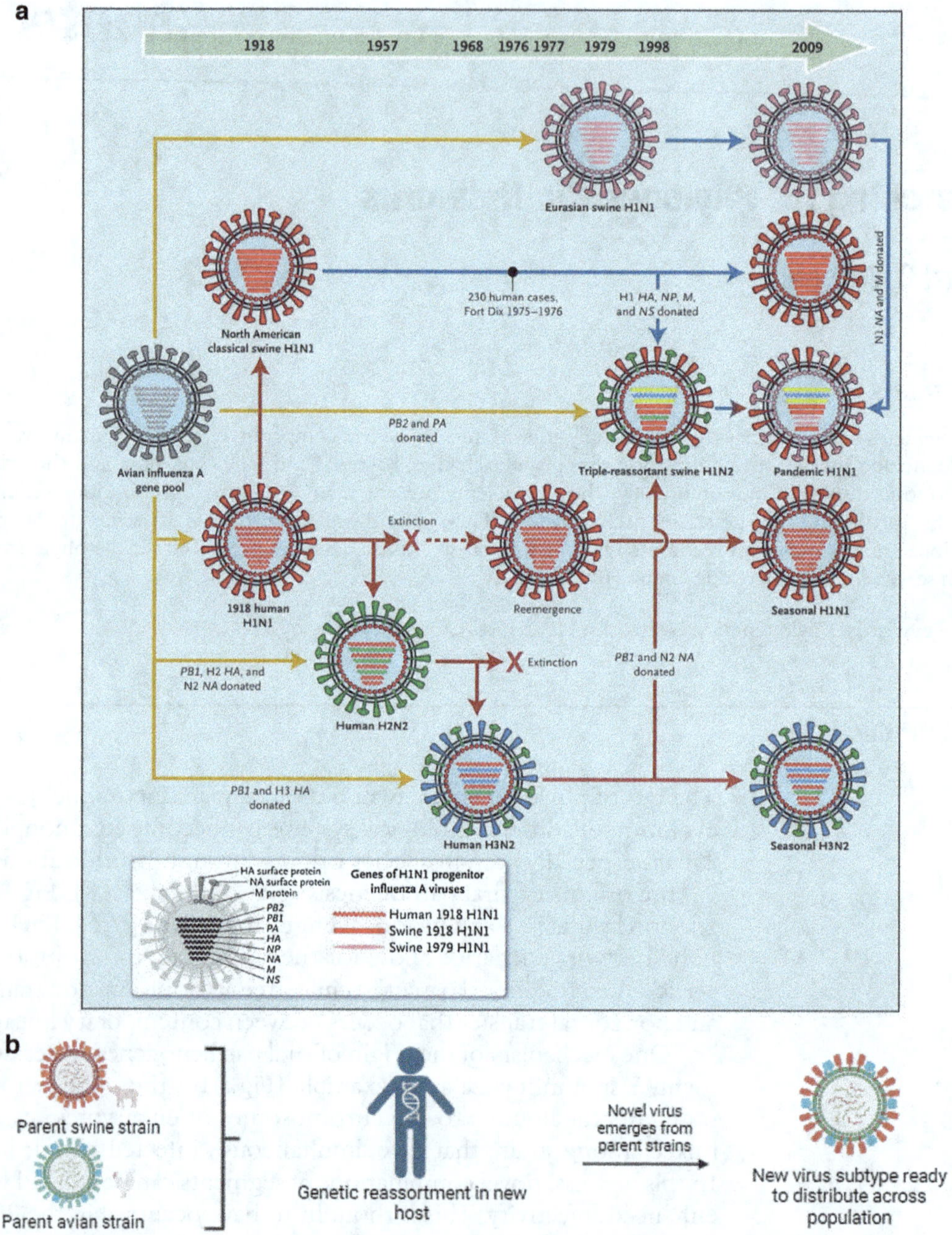

Fig. 1 **(a)**: History of 20thC Influenza showing pandemic-inducing viral reassortment [83]. **(b)**: Origin of new pathogenic viral strains via Antigenic Shift/reassortment within an infected host (biorender.com)

represent alternative parent-child vertex pairs as opposed to "hard-wired" where all parents contribute to each child feature (see below [8]). Systems for zoonotic disease origin as well as viral pathogenesis are usually based on the soft-wired model.

2 Definitions

A phylogenetic graph [9] is a directed acyclic graph ($G = \{V, E\}$) consisting of four types of vertices: a single root vertex, (in-degree 0, out-degree 2), leaf vertices (in-degree 1, out-degree 0), tree vertices (in-degree 1, out-degree 2), and network vertices (in-degree 2, out-degree 1). A tree edge terminates in a tree vertex and a network edge terminates in a network vertex. Edges that lead to leaf vertices are referred to as pendent edges. A tree has no network vertices (Fig. 2).

A "display tree" is created from a network by removing one of each pair of network edges that terminate in the same vertex and removing the resultant in-degree 1, out-degree 1 vertex and "contacting" the two edges into one (Fig. 3). Via this process an

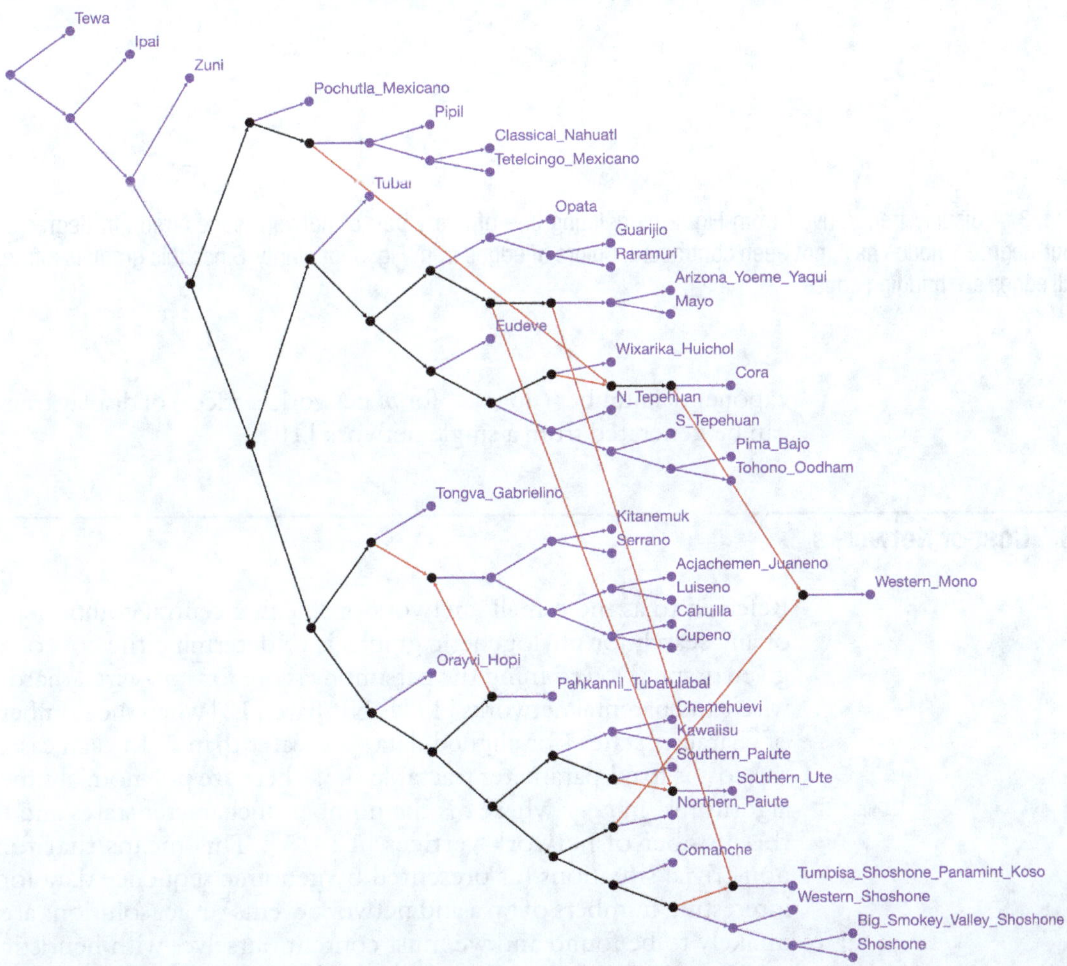

Fig. 2 Phylogenetic network of Uto-Aztecan languages [5]. Network edges are in red, bridging edges in violet, non-bridging edges in black (with black nodes)

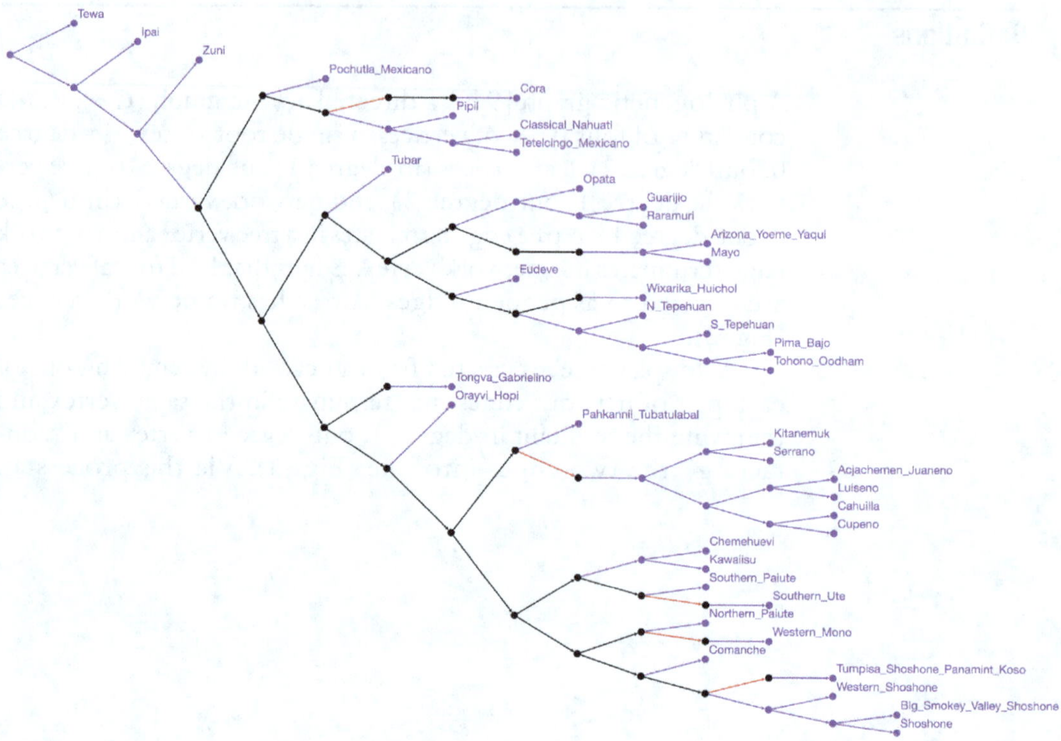

Fig. 3 A display tree derived from Fig. 2 by deleting one of every pair of network edges (red). In-degree 1, out-degree 1 nodes have not been contracted. Colors of edges from Fig. 3 for clarity. Since this graph is a tree, all edges are bridging edges

exponential number (up to 2^m for m network vertices) of display trees can be generated from a single network [10].

3 Cost of Networks

Referred to as the "small" network problem, a central component of any search for phylogenetic graphs is to determine the cost of a given graph. Determining the parsimony score for soft-wired, hard-wired and parental networks [11] is NP-hard [12] when the number of character states (for aligned data) is greater than 2. In each case, the cost is fixed-parameter tractable (i.e., there are polynomial time algorithms) in $c + r$ where c is the number of character states and r the number of network vertices [12–15]. This means that for non-trivial situations (as presented by genomic sequence data for interesting numbers of taxa and network events) exact solutions are unlikely to be found and we must content ourselves with heuristic solutions. This is further complicated when unaligned sequences are analyzed [16], adding a further level of NP-hard optimizations (and their heuristics, e.g., [17, 18]) even for individual display trees.

A second factor in network cost is accounting for the monotonic non-increase in parsimony score with additional network edges. Basically, one could create a "trivial" network that embeds the lowest cost tree for each character (or gene sequence). This is obviously undesirable and several approaches (they can also be used in concert) have been used to deal with this issue. The first is to "bin" the data where each bin contains a number of characters (say a nucleotide sequence) and all characters in that bin must follow the same display tree ("blocks" in PhyG) [19]. This has a logical appeal for genomic data that occur in self-delimiting sets (again such as nucleotide or protein sequences). This, however, only moves the problem up a level and the minimum cost network would be that which included the minimum cost "gene" tree for each locus.

A solution to this seeming arbitrariness is (in the context of parsimony) to add a penalty for each additional network vertex. With such a penalty, tree and network solutions can be competed and a heuristically optimal graph identified. To date, several penalties have been suggested (e.g., [6, 20]) and implemented [21], but none are obviously the best, hence there is no consensual optimality criterion for phylogenetic networks even in the parsimony realm. Algorithmic information theory [22, 23] as applied in Phylogenetic Minimum Descriptive Length (PMDL) [24] offers a more natural (i.e., internally generated) accounting for network costs in terms of added complexity (measured in bits).

The tool used here to demonstrate phylogenetic graph operations, PhyG [21] offers two heuristic methods to determine the underlying network score as well as several penalties that can be employed with parsimony [6, 20], maximum likelihood [25, 26], and information theoretic optimal criteria [24]. The main focus of the discussion here is the "large" network problem or the search for optimal networks within the larger space of possible graphs.

4 Network Search

The total search space of trees is well known to be very large with (for $n \geq 3$ leaf taxa; Eq. (1)) [27].

$$\frac{(2n-4)!}{(n-2)!2^{n-2}} \tag{1}$$

Hence, the search for "best" trees is an NP-hard problem [28–30] requiring heuristic approaches for non-trivial data sets. The search space for networks is even larger [10], expanding the difficulty of identifying heuristically useful graphs.

In this heuristic world, especially where empirical results are desired, there will always be a trade-off between the exhaustiveness of a procedure (hence quality of result) and the time/resources available. An extremely effective procedure that takes thousands of CPU years (even in a world of commodity parallelism) may simply be not usable for a problem (e.g., vaccine design, tenure application) with a shorter time horizon. Modern phylogenetic search procedures use a collection of methods to balance these objectives. In general, methods proceed from initial graph construction followed by local trajectory refinement and then to more global solution searches based in Simulated Annealing [31] and Genetic Algorithm [32]. Recent techniques employing machine learning have further integrated and improved these complex search strategies [33, 34].

4.1 Initial Construction

As with any search process, a starting point is needed. In the phylogenetic graph domain, it is often convenient to begin by constructing initial trees and then later adding potential network edges. There are multiple approaches relying on deterministic and randomized algorithms with varying addition criteria. Furthermore, these initial graphs can be constructed based on overall difference (distance-based methods) or those based on character (e.g., sequence) assignment themselves (reviewed in [35]). In general, distance-based methods are "faster" in a wall-clock sense since they have a lower constant factor (after distance matrix construction) than character-based methods, but have the similar time complexities in terms of graph (i.e., leaf set) size.

All the tree construction procedures (other than pure random trees) discussed here add taxa in turn until all taxa have been added and an initial tree is complete. The rules for which taxa to add and where to add them on the growing tree differentiate the methods yielding $O(n^3)$ time complexity for Wagner methods (both distance and character) [36, 37] (Fig. 4) and Neighbor-Joining [38], while UPGMA (and related) [39] and randomized addition sequence Wager are $O(n^2)$ per build. Randomized Wagner methods are usually performed multiple times (often $>n$), hence will end up as approximately $O(n^3)$ approaches as well (reviewed in ref. [40]). Given the differences among these methods, it can be useful to construct initial graphs with several techniques and create a pool of initial graphs.

4.2 Trajectory Refinement

To improve the initial graph set–and create initial networks, "trajectory" searches are most often employed. The procedures create a trajectory as progressively improved solutions are found and displace the previous, heuristically best result while moving from graph to graph. The course of this trajectory is determined both by the operations performed on the graph and also by any randomization that occurs. Since we are exploring a very large space, small changes in trajectory can result in radically different results.

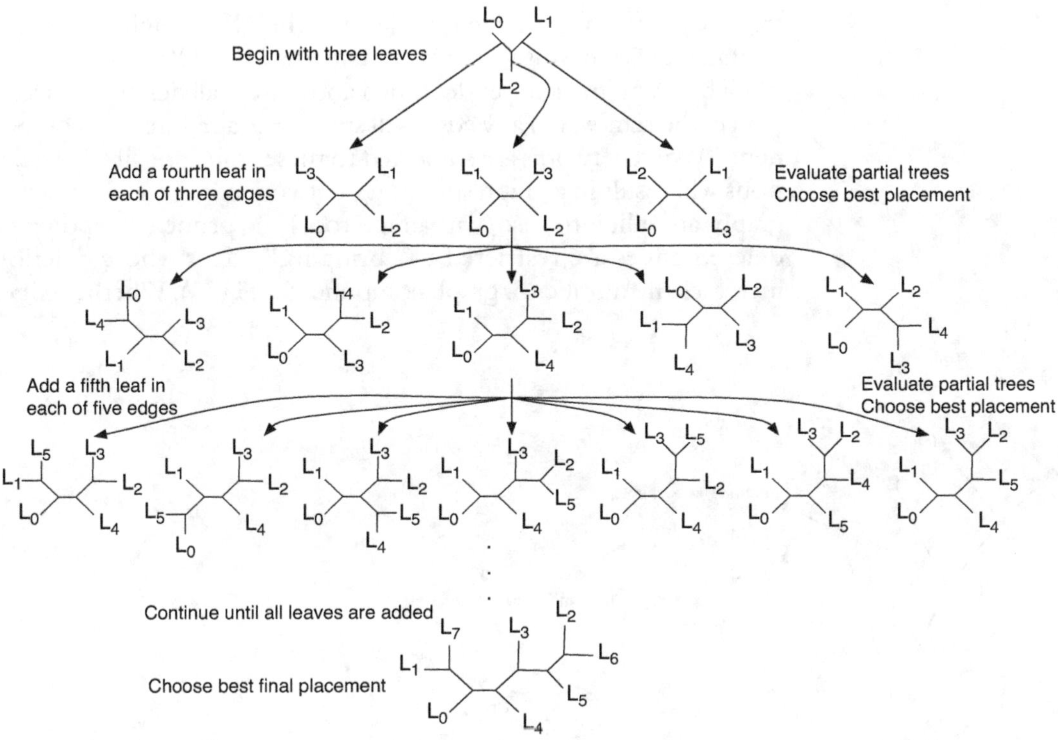

Fig. 4 Wagner method [36, 37] employing sequential addition of terminal taxa until all taxa are added to create an initial tree (after [35])

General Graph Trajectories—Tree rearrangement trajectory methods have been used to improve tree solutions (for any optimality criterion) by defining edits, or changes, to the tree topologies. These methods create neighborhoods of trees that are examined based on a specific edit function. In increasing neighborhood size, the most common of these are Nearest Neighbor Interchange (NNI; [41, 42]), Subtree Pruning and Regrafting (SPR; [43]), and Tree Bisection and Regrafting (TBR; [44–46]). Each rearrangement (or swapping) operation begins with splitting the tree on an edge (resulting in a "base" component including the root and a "pruned" component) and the two pieces are rejoined in different ways to generate new, but related, topologies.

NNI rejoins the tree components in each of the two other positions surrounding the initial edge that was deleted to split the graph. Since there are $n - 3$ edges in the initial tree, there are $2(n - 3)$ trees in the NNI neighborhood that can be examined for increased optimality. SPR rejoins the graph components at all possible edges in the base component, yielding a neighborhood of $2(n - 3)(2n - 7)$ trees [47]. TBR reconnects the base and pruned components via the creation of new edges between each edge in the base graph and each in the pruned component. The neighborhood size depends on the tree shape but overall size is order n^3. The NNI neighborhood is contained within SPR which is contained within TBR (Fig. 5).

The graph trajectories described above were all defined for trees; hence, the removal of any edge will split the graph into two components. When networks are subjected to these edits, not all edge deletions will result in a graph split (e.g., network edges). Hence, when graphs are split into base (with graph root) and pruned components, deleted edges are restricted to "bridging" edges whose deletion increases the number of graph components (Fig. 3). Whether edges

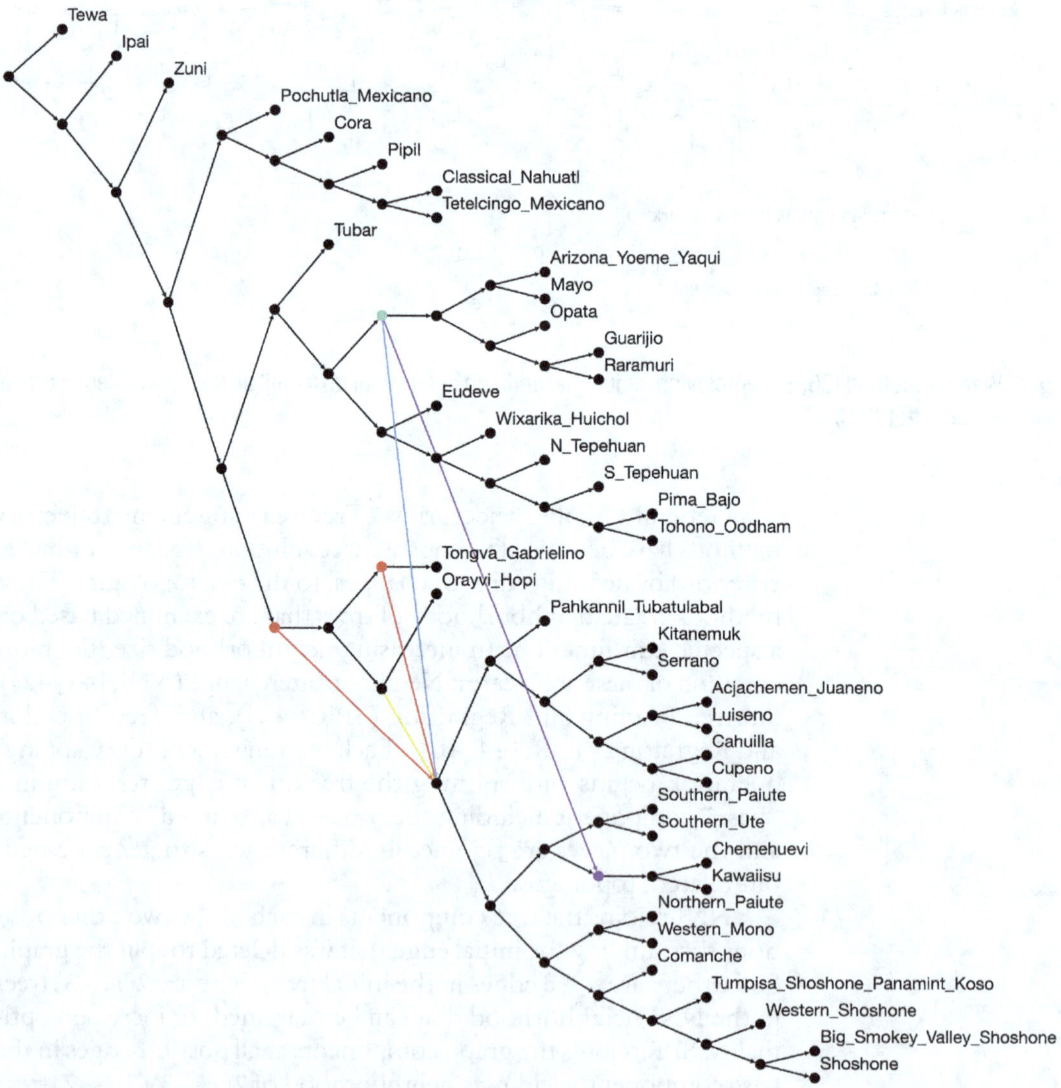

Fig. 5 Branch-swapping refinement. The yellow edge is the original connection and is deleted. The red edges are the two NNI rearrangements, the blue edge is one of many SPR rearrangements that is not an NNI move, and the purple edge is a TBR rearrangement that is neither an SPR nor NNI move

are bridging or not can be determined in linear time via Tarjan's Algorithm [48].

Network Trajectories-—No matter what sort of graph we may have as a starting point (i.e., trees or networks), network trajectory searches will add, delete, and move network edges attempting to find superior graphs. These are the fundamental operations (Fig. 6). Before these are discussed in further, important restrictions on where network edges can occur in a phylogenetic network are required.

Since a network vertex is the terminus of two directed edges, representing a coming together or horizontal exchange of information, the originating vertices of those two edges must be potentially coeval. That is, the partial ordering of the graph (parent older than child, children of same node have same age) constrains the location of new (as in added by a search procedure) network edges. As an example, one could not have a network edge that linked an edge leading to a descendant with one that led to its ancestor. The descendant and ancestral lineages could not (given the graph) have existed at the same time. Such a constraint does not apply to the deletion of an edge, since no time violations could be created. Furthermore, with each additional network edge, new time constraints are created further limiting the location of any possible additional network edge (Fig. 7). Although in principle, there are $2|E|^2$ (given two directions of a network edge) possible edges that could be added to some graph, the time consistency requirement severely limits the actual number of possibilities [9].

As mentioned above with respect to randomization, even deterministic algorithms may yield alternate trajectories when implemented. Since most implementations are "greedy" in that when a "better" graph is found, the search will switch to that new graph, the order in which edges are chosen for graph split and the order of re-connection will affect trajectories. Hence, alternate implementations of the same overall procedure may result in different results. The heuristic solution to this issue, in general, is the use of multiple randomized iterations.

Trajectory searches are effective at identifying local minima, in the sense that these results are surrounded by less optimal solutions. If the optimality space of phylogenetic problems were nicely behaved (i.e., convex) this would be sufficient [49]. Phylogenetic optimization problems are notorious for not being so well behaved; hence, non-trajectory methods have been applied to search for global optima. These include Simulated Annealing and Genetic Algorithm.

4.3 Simulated Annealing

Simulated annealing (SA) is a method allowing temporary acceptance of inferior (i.e., worse optimality score) solutions to escape a local minimum and continue searching for a more global optimum [31, 50–52]. The method is modeled on the annealing of metals where temperature is reduced in steps to allow lower energy (i.e., more ductile, less brittle) materials to be formed. In phylogenetic search,

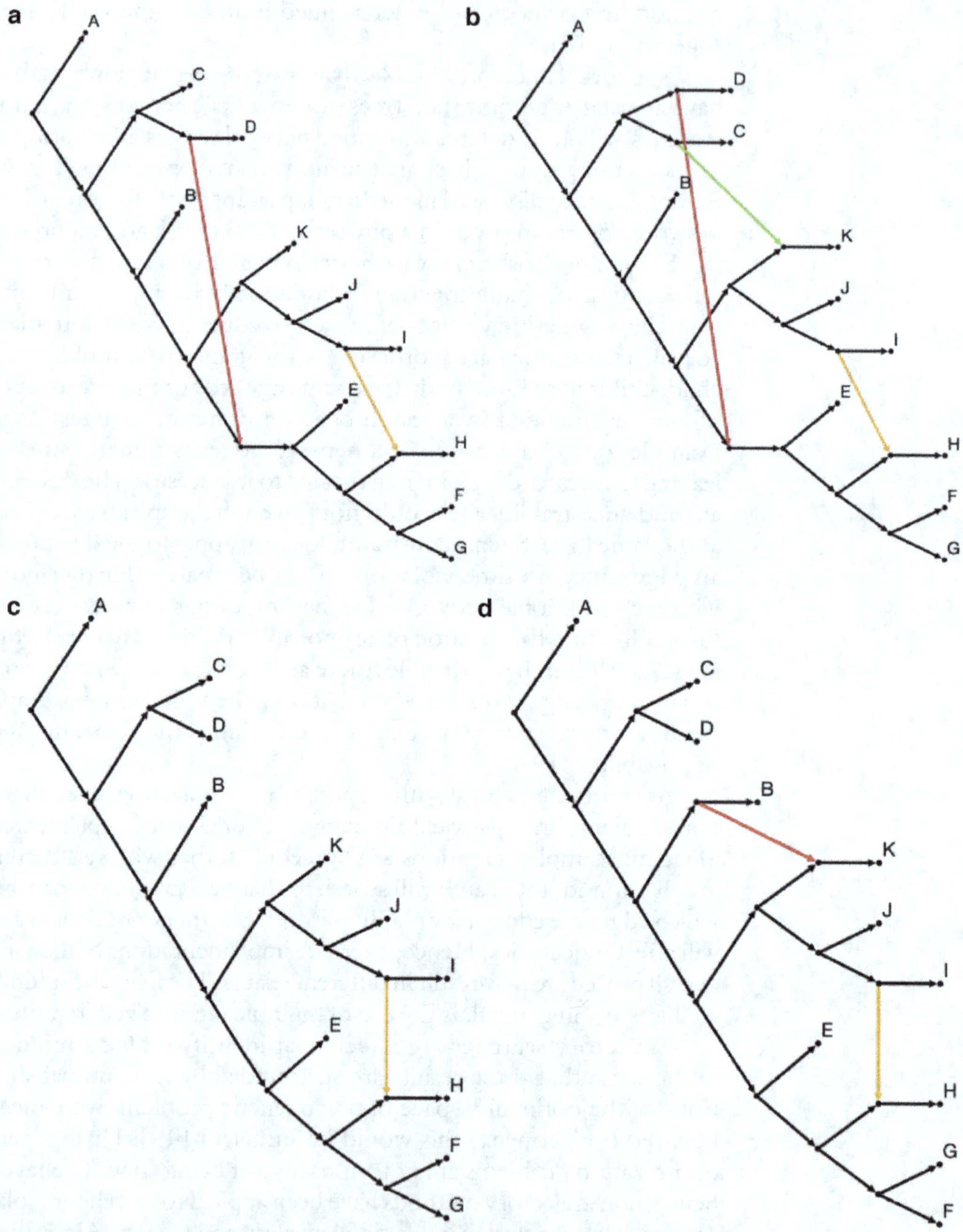

Fig. 6 Network edits. (**a**) Base network highlighting two of four network edges (red and orange). (**b**) Graph after addition of an additional network edge (green). (**c**) Graph after deletion of red network edge. (**d**) Graph after move of red network edge (green)

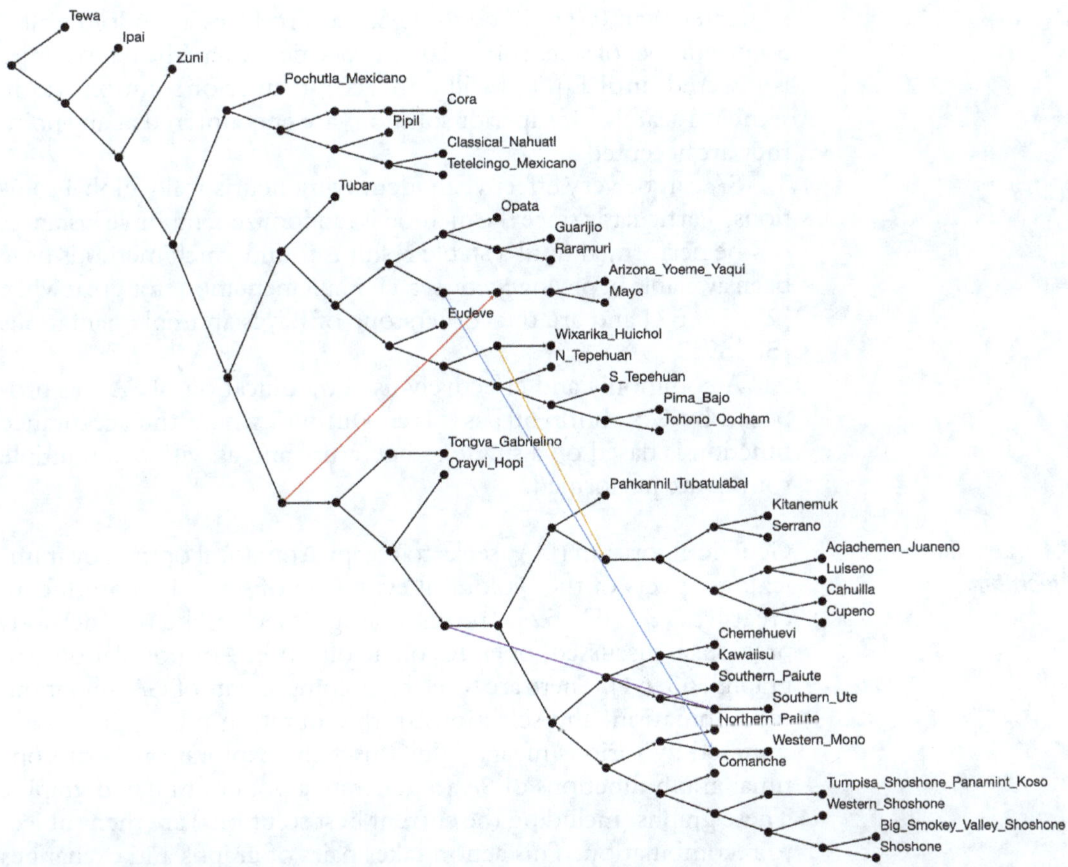

Fig. 7 Network edge time-consistency. The red edge is consistent with the orange and the blue is consistent with the orange. The blue and red edges, however, are inconsistent since the terminal node of red must be later that the starting node of blue and the starting node of red must be earlier than the terminal edge of blue. The purple edge is not time consistent since it requires that its two parent nodes be possibly coeval, but one is an ancestor of the other

analogies to the annealing parameters are defined and as temperatures are cooled the transition from one graph to another moves from a near random walk to one akin to the trajectories discussed above.

Edits are proposed to an existing graph (s) with energy (cost E) that would result in graph s' with energy E' at temperature T. The basic action is that a candidate graph is proposed by some process (akin to graph refinement above) where the probability of accepting the edit is given by Eq. (2).

$$pr(s, s') = \min\left(1,\ e^{(E-E')/T}\right) \tag{2}$$

Trajectory methods would only accept this candidate graph if its cost (i.e., energy) were lower (superior) to the current graph. In SA, however, sub-optimal graphs may be accepted, when the temperature is greater than 0 (at $T = \infty$ the method produces a random walk). Some number of steps (often 10) are specified where the temperature is lowered until Eq. (2) will only accept superior solutions (as in standard search). If superior solutions are encountered at any point they are accepted.

SA can be very effective in identifying heuristically global solutions. Particularly since the method is randomized, multiple instances can be performed until a stable result is found. Such methods have been available in phylogenetic search implementations for great while [21, 53–57] and are the cornerstone of Bayesian implementations [58, 59].

A commonly and effectively used modification of SA was proposed by Goloboff [60] as "Tree Difting" where the acceptance function is based on a single temperature and as with SA, multiple rounds are performed.

4.4 Genetic Algorithm

Genetic algorithm (GA) seeks to escape from local optima by mimicking aspects of the biological evolution of sexually reproducing creatures [32, 61, 62]. In this analogy, GA, unlike the methods previously discussed, operates on a collection—or population—of graphs [63, 64]. There are three basic components of GA: mutation, recombination, and selection. In the mutation phase, graphs are mutated by various means. PhyG uses the exploration of sub-optimal graph functions of SA to generate a pool of mutated graphs. These graphs (including the current best solutions) are then subject to recombination. This action takes pairs of graphs and exchanges non-identical sub-graphs with the same leaf set. It is important to note that these graphs are not "randomized" and even the sub-optimal components are rooted in evidence. After the recombination step, selection occurs based on the graph scores. Some number (population size) of unique graphs are kept including both the current best and near-best graphs to be submitted to the next generation of the algorithm. Multiple generations are usually performed until

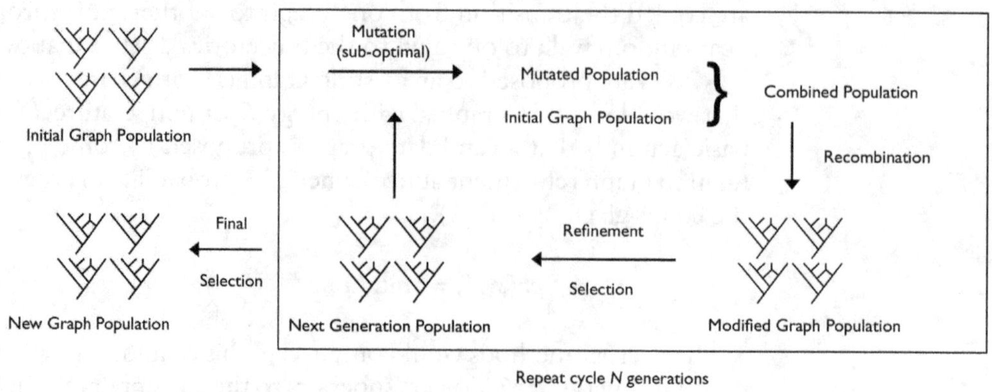

Fig. 8 Genetic algorithm. Mutation, recombination, and refinement/selection are repeated for each generation

some maximum number or lack of improvement (Fig. 8). The graph recombination step has been implemented by Goloboff [60] as "Tree Fusing," focusing on this aspect alone of the GA approach.

5 Search Integration

As can be seen by the descriptions above, there is a large variety of heuristic approaches that can be applied singly and in combination and each one of these algorithms has multiple parameters. Furthermore, methods may well have varying degrees of effectiveness on different data sets, and at different stages of a complex search.

In order to unburden the user from the large number of methodological and paramterization choices the randomized "Search" option was first implemented in POY4 [65]. This meta-algorithm chose optimization procedures at random with appropriate randomized parameters. The method is run for a fixed wall-clock time based on an integer multiple of the wall-clock time of individual trajectory searches (usually TBR). Typically, the search is performed in seriatim, each time doubling the CPU time until no better solutions are found.

This method is effective, but does not allow for the change in algorithmic performance as the search proceeds or for the particular aspects of individual data sets. For example, graph building is very effective (and required) when a search starts, but is less effective later on. Refinement methods such as GA are very productive after initial graphs of good quality are identified, but are notoriously time consuming when used ab initio. Smaller data sets may be satisfactorily approached with refinement methods, but larger ones require multiple rounds of SA and GA. The employment of machine learning techniques via Thompson's Algorithm [33, 34] vastly improves randomized search by revising methodological choice probabilities based on the success (in terms of identifying better solutions) and failure of individual algorithms at the different stages of a search and for the unique aspects of particular data sets. Furthermore, the time required by the search methods (they can differ wildly) is taken into account when revising the probabilities of choosing the next method to apply.

$$\theta'_i = \frac{m}{m+1} \times \theta_i + \frac{1}{m+1} \times (\theta_i + max(0, \delta/t))$$

$$\begin{aligned} \theta'_{\forall j \neq i} &= \frac{m}{m+1} \times \theta_{\forall j \neq i} \\ &+ \frac{1}{m+1} \times \left(\theta_{\forall j \neq i} + \begin{cases} 0, & \text{if } \delta > 0 \\ \dfrac{1}{n-1} & \text{for } n \text{ bandits otherwise.} \end{cases} \right) \end{aligned} \quad (3)$$

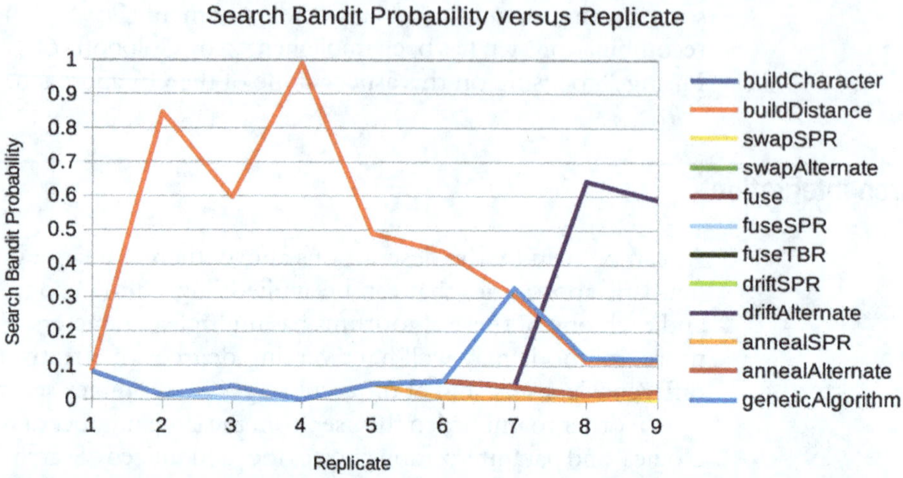

Fig. 9 Thompson search profile from [34]. All search algorithms ("bandits") begin with equal probability. Subsequent replicates show initial increased probability of choice for "build" procedures that then decrease and are superseded by Simulated Annealing ("driftAlternate") and Genetic Algorithm

$$\theta'_i = (1-2^{-m}) \times \theta_i + 2^{-m} \times (\theta_i + max(0, \delta/t))$$

$$\theta'_{\forall j \neq i} = (1-2^{-m}) \times \theta_{\forall j \neq i} + 2^{-m} \times \left(\theta_{\forall j \neq i} + \begin{cases} 0, & \text{if } \delta > 0 \\ \frac{1}{n-1} & \text{for } n \text{ bandits otherwise.} \end{cases} \right) \quad (4)$$

The probability of choosing a search method θ' (or "bandit" in Thompson parlance, initially set to equality) in the next generation is calculated from the probability in the previous generation (θ) the change in graph optimality score (δ), the CPU time consumed by the method, and a memory value (m) to determine how rapidly choice probabilities change between iterations. The update function can employ either linear averaging based on m or exponential smoothing (Eqs. (3) and (4)). Successful results of a single method will increase that method's relative probability in the next generation (θ'_i) to the detriment of the other available methods ($\theta'_{\forall j \neq i}$; Fig. 9).

6 Practical Approaches

When approaching the analysis for a particular data set, the investigator will need to balance the factors of analytical quality, time, and resource consumption. As mentioned above, the quality of a solution (in terms of optimality value) will largely depend on how much time and resource is invested in the analysis. Given the NP-hard nature of many of the optimizations that are performed, there are unlikely to be guarantees as to quality of result, but the investigator can still strive to find the "heuristically best" result. In most cases, this can be defined functionally by a result that is not improved when analytical effort is doubled.

Most of the methods discussed above (e.g., TBR refinement) have a time complexity of $O(n^3)$ with a constant factor determined both by the procedure itself and the number of randomization or generations performed.

6.1 Getting Going

When beginning a search for phylogenetic networks using PhyG, there are two basic approaches: (1) generate an initial tree for subsequent edge addition and network refinement, and (2) generate trees for each data "block" (those data inputs that follow the same display tree) and create an initial network by combining the block trees.

The benefit of the first method is that it is more controlled in number of network edges, hence time complexity. When networks are generated as a start, however, they tend to have larger numbers of network edges representing the alternate patterns in the data. This can cause a large time and memory expense right from the start, and should be followed by network edge deletion before other heuristics are applied.

6.1.1 Starting with a Single Tree

PhyG can construct initial trees using two basic approaches, character and distance. Character builds follow the random addition sequence approach, with cost calculations using the same methods as graph evaluation (optimality criterion). Time complexity quadratic for each addition sequence, and typically as many or more randomized addition sequences (RAS) are performed as the number of taxa resulting in cubic time complexity and a factor for the evaluation of the characters themselves, either linear or greater depending on the type [36]. Distance methods, on the other hand, create an initial pairwise distance matrix [$O(n^2)$] in number of taxa and with the same data factor. After this matrix has been constructed, however, the distance methods are much faster since they do not revisit the character data directly and rely solely on the distances. Several common methods are available including Distance Wager (DW) [37], Neighbor-Joining (NJ) [38], and WPGMA [66], as well as a distance version of RAS builds used for character data. Distance Wagner and Neighbor-Joining are both $O(n^3)$ and WPGMA is $O(n^2)$. For any given data set, DW, NJ, or even WPGMA may yield the best initial results (in terms of tree optimality value when rediagnosed as character data). Large numbers of RAS (>1000) nearly always outperform these deterministic methods. Wall-clock time for the distance methods are vastly shorter than for character builds and yield comparable results (assuming sufficient numbers of randomized builds). This behavior is especially true for large alphabet sequences such as amino acid and linguistic data [40].

6.1.2 Starting with a Network

As mentioned above, initial networks can be created by first generating trees for each data block (by the methods above) and "reconciled" into networks using phylogenetic super-graph techniques [67, 68]. One extremely efficient technique is to employ distance techniques to construct the block trees and then reconcile the trees into graphs using threshold-based edge-union networks (Fig. 10).

As mentioned earlier, the networks constructed in this manner tend to have a large number of network edges (m) and given the $O(2^m)$ display trees that require evaluation to determine the cost of the overall graph, they are time and space-intensive objects to work with.

6.1.3 Network Refinement

Searching for networks after an initial tree build (or input) can follow two paths: (1) try to improve the input tree before adding network edges, or (2) immediately attempting network additions with subsequent refinement. Either (or even both in a randomized search) can be effective, but in general, some initial refinement is desirable (especially in data sets with large amounts of missing data). When networks have been created through tree reconciliation (or input), the best strategy is usually to refine the graph by network edge deletion. In this way, extraneous edges are removed and the time complexity of optimizing a single graph is reduced. This would usually be followed by graph swapping refinement.

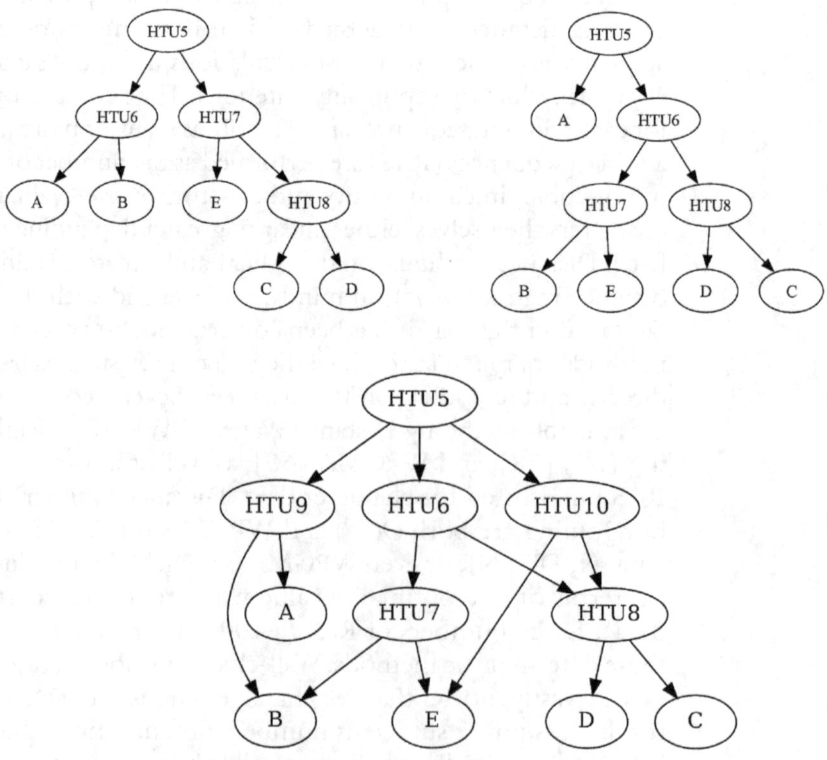

Fig. 10 The top two trees reconciled using Edge-Union-Networks [67, 68]

With either initial strategy, alternation between network refinement and overall graph refinement (e.g., swapping) can be very effective in reducing graph costs. Given the dependent nature of network edge addition (time constraints mentioned above) alternating between edge addition and deletion coupled with network edge moving (keeping overall number the same but shifting the placement network edges) is an important operation. For hardwired networks, moving edges is really the only option for improvement since reducing the number of network edges cannot increase graph cost and adding network edges cannot decrease graph cost.

As mentioned above, branch-swapping refinement operates on bridging edges only. As the number of network edges increases, the number of potential rearrangements (via any method) decreases.

6.1.4 Making It Faster

PhyG has a number of options to control heuristic intensity. Many of these also affect memory footprint with the more effective and time-consumptive options requiring more memory.

There are two main types of options that can be used: (1) those that affect the optimization of a single graph, and (2) those that affect the search among graphs. At this point, it is important to point out the different challenges of optimizing different data types. The parsimony cost of integer-coded and pre-aligned sequence data can be determined exactly for a tree (but not a network, which is still NP-hard) in $O(nm)$ time for N leaf taxa and m characters (anatomical observations or aligned sequence positions). The optimization of unaligned sequence data (i.e., DNA, amino acid, and other [69, 70]) for a single tree is NP-hard [16], hence require heuristic approaches even for a single tree [17, 18, 71] but nearly invariably results in better (i.e., lower cost) solutions [72]. For the cases faced by empirical data sets, the time complexity of these heuristics can approach near linearity [73] but are more commonly $O(n \log m)$ with a worst case of $O(n^2)$ in the simplest implementation as used in [74, 75]. The efficacy of this heuristic was improved in [57, 65] at the cost of a constant factor of 3. The implementation of PhyG further improved the heuristic by allowing individual improvements for each sequence in a multi-sequence data set [21] involving additional factor of the number of unaligned sequences per leaf. With this enhancement, improvements in tree cost ranged from 0.01% to 7.2% in tests of several data sets [5, 76–79]. In order to allow control of this enhanced procedure, PhyG allows the user to turn off the default optimization ("multiTraverse") to use the lower complexity option for sequence characters.

A further, much lower complexity and more approximate option is to create an "Implied Alignment" [80, 81] for a current graph and use that data structure for additional searching ("staticApproximation"). This reduces the tree optimization time complexity to one that is linear in the number of sequence elements. The effect can be dramatic. Since the character correspondences are based on the current graph, however, many patterns are baked in and large departures from the original graph topology are unlikely.

These methods are used in a stochastic fashion in randomized searches and their probability of employment tuned with their efficacy (Thompson sampling above).

The second method of accelerating searches has to do with full evaluation of graphs during searches. PhyG (along with nearly all other phylogenetic search implementations) uses heuristic graph cost calculation methods (often independent of leaf number) to avoid expensive, full graph cost determination. For some data and graph types, these heuristics are approximate when compared to the full (but yet still heuristic) graph costs. A search will examine large numbers of alternate graphs (as mentioned above often $O(n^3)$ for n leaves) and re-evaluate candidate graphs with the more expensive full evaluation. The frequency and fraction of candidate graphs that are reevaluated can be controlled in PhyG by using "better", "bestOnly", "betterN", and "bestAll" to re-evaluate only those candidate graphs with better heuristic costs than the current best graph, the single graph with the best heuristic cost, the best N heuristic cost graphs, or all graphs. The first two options are the most rapid, the final, "bestAll," adds an additional factor of n to the search so is really only useful to evaluate accuracy of the heuristics themselves. In my experience, "betterN" (with N controllable), is the best balance of time complexity and effectiveness.

Restricting the number of simultaneous graphs (PhyG option "keep") for searching to a small number (even 1) will also reduce wall-clock time.

6.1.5 Making It Better

In contrast to the above section, options to yield better graph costs would use multiTraverse. Refinement procedures are made more exhaustive by increasing the "N" of betterN (with PhyG option "graphsSteepest"). A separate, very effective, approach is to add additional randomization to swapping refinement and graph recombination (fuse). Replicate operations can be performed allowing for randomized order of graph actions (splitting and reconnecting), and for graphs chosen to be recombined.

Another useful option is to allow a larger number of graphs for simultaneous consideration (default number of saved graphs is 10).

Finally, the complex strategies of Simulated Annealing and Genetic Algorithm can be used. These methods are best employed when relatively good solutions are already at hand. Graph refinement can also extend the reach of graph recombination when used outside of GA (refinement option "fuse"). GA makes extensive use of randomization in its multi-generational procedure.

6.1.6 Layered Heuristics

When searching new data sets, a layered approach can make effective use of increasingly exhaustive heuristics with reasonable wall-clock time control. Initial heuristics are rapid and coarse, a following stage of intermediate intensity and then a final round of the most time-consumptive procedures. Such "concentric searching" can be efficient and effective, especially for large data sets with thousands of leaves.

6.2 Parallelism

With commodity parallelism broadly available, there is no reason not to make all strategies inherently parallel. Graph problems in general generate many "embarrassingly parallel" opportunities. The challenge, as with all parallel methods, is to scale to large numbers of processing cores as efficiently as possible. Each of the methods discussed above is parallelized in PhyG, and multiple independent search threads (randomized, Thomson-driven) permit the efficient use of processor numbers in the hundreds (potentially thousands for large leaf-size data sets). Currently, PhyG is limited to single machine, multi-core architectures.

6.3 Self-Tuned, Randomized, Timed Search

Many tests of the efficacy of search methods are based on well-known and understood data sets (e.g., "Zilla" [82]). This is, of course, useful but can also result in tuning search parameters to the specifics of the particular data sets used. Randomized search algorithms (e.g., "search" in refs. [57, 65]) were a first attempt to overcome problems presented by unknown data sets in a general manner. This was further improved by adding the self-tuning properties of Thompson's algorithm [33, 34]. These timed searches can operate on their own, or as a later component of a concentric search.

6.3.1 When to Stop?

One of the questions presented to investigators is when to stop? Given the extreme computational complexity of phylogenetic graph problems, one is never guaranteed of a "best" result. One useful criterion is based on doubling of effort. The basic idea is that when search effort is doubled without any improvement, it is less likely (but not guaranteed) that further effort will beneficial. This, of course, presumes that the earlier search iteration was sufficient to generate a result of good quality. Such a criterion is most easily implemented using timed searches. More and longer timed search iterations can be employed to assure a controlled level of effort.

7 Example Strategies

As mentioned above, the size and available analytical time frame of datasets will largely dictate the exhaustiveness of a search. If results are required relatively quickly or data sets are large ($>10^3$ leaves, genomic sequences) only relatively shallow search will be possible. When more analytical resources (including time) are available, deeper searches are possible with concomitantly improved (in terms of optimality) results.

The example strategies of Table 1 show example shallow, intermediate, and deep strategies. Of course, searches are not limited to these scenarios and a concentric search approach could employ all three such strategies in turn.

The performance of these strategies is illustrated by the analysis of Uto-Aztecan linguistic data [5] (Table 2). As one would expected, increased heuristic intensity generally (and absolutely for tree searches) increases the quality (decreases score) of the result. The wall-time cost factor and improvement in score though correlated are not linear. The "deep" search strategy consumed over 40× CPU time compared to the "shallow" with a 0.37% decrease in score. It

Table 1
Example search strategies for lower, intermediate, and longer wall-clock time searches

Shallow	Intermediate	Deep
Distance build	Distance build)	Distance build (rDWag ×1000)
(DWag)	(rDWag ×1000)	(rDWag ×1000)
Keep 1	Keep 3	Keep 10
Static approximation	multiTraverse Off	multiTraverse Off
Swap	Swap	Swap
multiTraverse Off	multiTraverse On	multiTraverse On
Swap	Swap	Swap
	Simulated annealing	Simulated annealing
	(5 replicates)	(20 replicates)
		Genetic algorithm
		(Multiple generations)
		Timed Thompson
		(Multiple replicates Including Character builds)

Table 2
Strategy performance for tree and network searches on Uto-Aztecan language data [5]

	Tree		Network	
Strategy	Score	CPU sec.	Score	CPU sec.
Shallow	10,088	12,325	10,042	11,818,251
Intermediate	10,059	95,042	10,050.13	266,628
Deep	10,051	5,349,772	10,039.70	5,799,772

Network searches were limited to adding 5 network nodes and 10 network edges

should be noted, however, that relatively small differences in the score of phylogenetic graphs can have large topological impacts [72]. The network searches show an alternate pattern. Network search after a "shallow" search took twice as long as the total time for the deep search, but yielded an inferior result. In this case, the use of Thompson-driven searches likely holds the key. The adaptive nature of algorithmic choice was a clear benefit with a notable impact in network score even though it consumed less CPU time.

8 Conclusion

As with much in life, there is no free lunch. There are no guarantees over what must be done and how much must be done to identify a good solution. The solution space is too large, and the heuristic option and parameter space daunting. As researchers, we strive to garner the best quality results we can, given the necessary limits on time and resources. We can, however, address the challenges presented by multiple interrelated computational challenging problems with equally layered heuristic approaches. The methods described here, especially when coupled with machine learning techniques offer easily extended and expanded tools for the analysis of nearly any data set in any problem domain. By progressively increasing the intensity of heuristics, employing self-tuning timed methods, capitalizing on the availability of commodity parallelism and reasonable stopping rules makes possible the analysis of ever increasing data sets and tailoring effort to meet the demands of any network search problem at hand.

9 Software and Data Availability

All source code, documentation, and MacOS and Linux binaries (in GitHub release area) are available at https://github.com/AMNH/PhyG.

Acknowledgment

The author would like to thank the American Museum of Natural History for financial support.

References

1. Abbott R, Albach D, Ansell S, Arntzen JW, Baird SJE, Bierne N et al (2013) Hybridization and speciation. J Evol Biol 26(2):229–246. https://doi.org/10.1111/j.1420-9101.2012.02599.x
2. Pereyra OM, Blotto BL, Baldo D, Chaparro JC, Ron SR, Elias-Costa AJ et al (2021) Evolution in the genus Rhinella: a total evidence phylogenetic analysis of Neotropical true toads (Anura: Bufonidae). Bull Am Mus Nat Hist 447:1–156
3. Janies D, Hill A, Guralnick R, Habib F, Waltari E, Wheeler WC (2007) Genomic analysis and geographic visualization of the spread of avian influenza (H5N1). Syst Biol 56:321–329
4. Machado DJ, Scott R, Janies DA (2021) Fundamental evolution of all Orthocoronavirinae including three deadly lineages descendant from Chiroptera-hosted coronaviruses: SARS-CoV, MERS-CoV and SARS-CoV-2. Cladistics 37(5):461–488. https://doi.org/10.1111/cla.12454
5. Wheeler WC, Whiteley PM (2015) Historical linguistics as a sequence optimization problem: the evolution and biogeography of Uto–Aztecan languages. Cladistics 31(2):113–125
6. Wheeler WC, Washburn AJ (2023) Parsimony optimization of soft-wired phylogenetic networks. Cladistics 39:456–474
7. Michaelis M, Doerr HW, Cinatl J (2009) Novel swine-origin influenza A virus in humans: another pandemic knocking at the door. Med Microbiol Immuno 198:175–183
8. Kannan L, Wheeler WC (2012) Maximum parsimony on phylogenetic networks. Algorithm Mol Biol 7(9):9–19
9. Moret BME, Nakhleh L, Warnow T, Linder CR, Tholse A, Padolina A et al. (2004) Phylogenetic networks: modeling, reconstructibility, and accuracy. IEEE Trans Comput Biol Bioinform 1(1):13–23
10. Linz S, St John K, Semple C (2013) Counting trees in a phylogenetic network is #P-complete. SIAM J Comput 42(4):1768–1776. https://doi.org/10.1137/12089394X
11. Zhu J, Yu Y, Nakhleh L (2016) In the light of deep coalescence: revisiting trees within networks. BMC Bioinform 17(14): 415. https://doi.org/10.1186/s12859-016-1269-1
12. Fischer M, van Iersel L, Kelk S, Scornavacca C (2015) On computing the maximum parsimony score of a phylogenetic network. SIAM J Discrete Math 29:559–585
13. Kannan L, Wheeler WC (2014) Exactly computing the parsimony scores on phylogenetic networks using dynamic programming. J Comput Biol 21(4):303–319
14. Van Iersel L, Jones M, Scornavacca C (2017) Improved maximum parsimony models for phylogenetic networks. Syst Biol 67(3):518–542. https://doi.org/10.1093/sysbio/syx094
15. Scornavacca C, Weller M (2022) Treewidth-based algorithms for the small parsimony problem on networks. Algorithms Mol Biol 17(1):15. https://doi.org/10.1186/s13015-022-00216-w
16. Sankoff DM (1975) Minimal mutation trees of sequences. SIAM J Appl Math 28:35–42
17. Wheeler WC (1996) Optimization alignment: the end of multiple sequence alignment in phylogenetics? Cladistics 12(1):1–9
18. Varón A, Wheeler WC (2012) The tree-alignment problem. BMC Bioinform 13:293
19. Nakhleh L, Jin G, Zhao F, Mellor-Crummey J (2005) Reconstructing phylogenetic networks using parsimony. In: Marstein V (eds) IEEE Computational Systems Bioinformatics Conference (CSB'05). IEEE, pp 93–102
20. Wheeler WC (2005) Phylogenetic network analysis as a parsimony optimization problem. BMC Bioinform 16:296
21. Wheeler WC, Washburn A, Crowley LM (2024) PhylogeneticGraph (PhyG) a new phylogenetic graph search and optimization program. Cladistics 40:97–105
22. Rissanen JJ (1978) Modeling by the shortest data description. Automatica 14:465–471
23. Rissanen J (1989) Stochastic complexity in statistical inquiry. World Scientific https://www.worldscientific.com/doi/abs/10.1142/0822
24. Wheeler WC, Varón A (2025) Phylogenetic minimum description length: an optimality criterion based on algorithmic complexity. Cladistics 41:193–211. https://doi.org/10.1111/cla.12603
25. Than C, Ruths D, Nakhleh L (2008) PhyloNet: a software package for analyzing and reconstructing reticulate evolutionary relationships. BMC Bioinform 9(322):1–16
26. Kong S, Swofford DL, Kubatko LS (2024) Inference of phylogenetic networks from sequence data using composite likelihood. Syst Biol 73:1–17. https://doi.org/10.1093/sysbio/syae054
27. Schröder (1870) Vier combinatoriche probleme. Zeitschrift für Matemmatik und Physik 15:489–503
28. Foulds LR, Graham RL (1982) The Steiner problem in phylogeny is NP-complete. Adv Appl Math 3:43–49
29. Day WHE (1987) Computational complexity of inferring phylogenies from dissimilarity matrices. Bull Math Biol 49:461–467
30. Roch S (2006) A short proof that phylogenetic tree reconstruction by maximum likelihood is hard. Comput Biol Bioinform IEEE/ACM Trans 3:92–94

31. Metropolis NA, Rosenbluth A, Rosenbluth M, Teller A, Teller E (1953) Equation of state calculations by fast computing machine. J Chem Phys 21(6):1087–1092
32. Holland JH (eds) (1975) Adaptation in natural and artificial systems. University of Michigan Press, Ann Arbor Michigan
33. Thompson WR (1933) On the likelihood that one unknown probability exceeds another in view of the evidence of two samples. Biometrika 25(3/4):285–294
34. Wheeler WC (2024) Multi-armed bandits, Thomson sampling and unsupervised machine learning in phylogenetic graph search. Cladistics 40:430–437
35. Wheeler WC (2012) Systematics: a course of lectures. Wiley-Blackwell, Oxford
36. Farris JS (1970) A method for computing Wagner trees. Syst Zool 19:83–92
37. Farris JS (1972) Estimating phylogenetic trees from distance matrices. Amer Nat 106:645–668
38. Saitou N, Nei M (1987) The neighbor-joining method: a new method for reconstructing phylogenetic trees. Mol Biol Evol 4:406–425
39. Michener CD, Sokal RR (1957) A quantitative approach to a problem in classification. Evolution 11:130–162
40. Wheeler WC (2021) Distance Wagner tree refinement as a heuristic approach to character-based initial tree construction. Cladistics 37:829–837. https://doi.org/10.1111/cla.12478
41. Camin JH, Sokal RR (1965) A method for deducing branching sequences in phylogeny. Evolution 19:311–326
42. Robinson DF (1971) Comparison of labeled trees with valency three. J Combinatorial Theory 11:105–119
43. Dayhoff MO (1969) Computer analysis of protein evolution. Sci Am 221(1):86–95
44. Mickevich MF, Farris JS (1980) PHYSYS: phylogenetic analysis system. Stoney Brook, NY
45. Swofford DL (1990) PAUP: phylogenetic analysis using parsimony. Sunderland, MA
46. Farris JS (1988) Hennig86, program and documentation. New York, NY
47. Allen B, Steel M (2001) Subtree transfer operations and their induced metrics on evolutionary trees. Ann Combinatorics 5:1–13
48. Tarjan RE (1974) A note on finding the bridges of a graph. Inf Process Lett 2:160–161
49. Boyd S, Vandenberghe L (2004) Convex optimization. Cambridge University Press, Cambridge, UK
50. Chib S, Greenberg E (1995) Understanding the Metropolis-Hastings algorithm. Am Stat 49:327–335
51. Kirkpatrick S, Gelatt CD, Vecchi MP (1983) Optimization by simulated annealing. Science 220:671–680
52. Cerny V (1985) A thermodynamical approach to the traveling salesman problem: an efficient simulation algorithm. J Optim Theory Appl 45:41–51
53. Felsenstein J (1980) PHYLIP Version 1.0: phylogeny inference package. Seattle, WA
54. Lundy M (1985) Applications of the annealing algorithm to combinatorial problems in statistics. Biometrika 72:191–198
55. Dress A, Krüger M (1987) Parsimonious phylogenetic trees in metric spaces and simulated annealing. Adv App Math 8:8–37
56. Varón A, Vinh LS, Bomash I, Wheeler WC (2008) POY 4.0. American Museum of Natural History. New York, NY
57. Wheeler WC, Lucaroni N, Hong L, Crowley LM, Varón A (2015) POY version 5: Phylogenetic analysis using dynamic homologies under multiple optimality criteria. Cladistics 31:189–19
58. Yang Z, Rannala B (1997) Bayesian inference using DNA sequences: a Markov chain Monte Carlo method. Mol Biol Evol 14:717–724
59. Huelsenbeck JP, Ronquist F (2003) MrBayes: Bayesian inference of phylogeny. Tallahassee, FL
60. Goloboff PA (1999) Analyzing large data sets in reasonable times: solutions for composite optima. Cladistics 15(4):415–428
61. Goldberg DE, Goldberg DE, Deb K, Deb K, Horn J, Horn J (1989) Genetic algorithms. In: Search, optimization, and machine learning. Addison-Wesley, Boston, MA
62. Sastry K, Goldberg D, Kendall G (2005) Search methodologies: introductory tutorials in optimization and decision support methodologies. In: Genetic algorithms. Springer Cambridge, UK. pp 97–125
63. Moilanen A (1999) Searching for most parsimonious trees with simulated evolutionary optimization. Cladistics 15(1):39–50
64. Moilanen A (2001) Simulated evolutionary optimization and local search: introduction and application to tree search. Cladistics 17:S12–S25
65. Varón A, Vinh LS, Wheeler WC (2010) POY version 4: phylogenetic analysis using dynamic homologies. Cladistics 26:72–85
66. Sokal RR, Michener CD (1958) A statistical method for evaluating systematic relationships. Univ Kans Sci Bull 38:1409–1438
67. Miyagi M, Wheeler WC (2019) Reconciling trees using phylogenetic edge union networks. Cladistics 35:688–94
68. Wheeler WC (2022) Phylogenetic supergraphs. Cladistics 38(1):147–158. https://doi.org/10.1111/cla.12485
69. Schulmeister S, Wheeler WC (2004) Comparative and phylogenetic analysis of developmental sequences. Evol Dev 6:50–57
70. Ramírez MJ, Coddington JA, Maddison WP, Midford PE, Prendini L, Miller J et al (2007)

Linking of digital images to phylogenetic data matrices using a morphological ontology. Syst Biol 56(2):283–294. https://doi.org/10.1080/10635150701313848
71. Wheeler WC (2001) Homology and the optimization of DNA sequence data. Cladistics 17:S3–S11
72. Ford E, Wheeler WC (2016) Comparison of heuristic approaches to the generalized tree alignment problem. Cladistics 32(4):452–460. https://doi.org/10.1111/cla.12142
73. Ukkonen E (1985) Finding approximate patterns in strings. J Algor 6:132–137
74. Gladstein DS, Wheeler WC (1997) POY version 2.0. Program and documentation. American Museum of Natural History, New York. http://research.amnh.org/scicomp/projects/poy.php
75. Wheeler WC, Gladstein DS, De Laet J (1996) POY version 3.0. Program and documentation (current version 3.0.11). American Museum of Natural History, New York. http://research.amnh.org/scicomp/projects/poy.php
76. Giribet G, Wheeler WC (1999) The position of arthropods in the animal kingdom: Ecdysozoa, islands, trees and the 'parsimony ratchet'. Mol Phylogenet Evol 10:1–5
77. Giribet G, Wheeler WC (2001) Some unusual small-subunit ribosomal RNA sequences of Metazoans. Am Museum Novitate 3337:1–14
78. Pellens R, D'Haese CA, Bellés X, Piulachs MD, Legendre F, Wheeler WC et al (2007) The evolutionary transition from subsocial to eusocial behaviour in Dictyoptera: phylogenetic evidence for modification of the "shift-in-dependent-care" hypothesis with a new subsocial cockroach. Mol Phylogenet Evol 43(2):616–626. https://doi.org/10.1016/j.ympev.2006.12.017
79. Whiteley PM, Xue M, Wheeler WC (2019) Revising the Bantu tree. Cladistics 35:329–348
80. Wheeler WC (2003) Implied alignment. Cladistics 19:261–268
81. Washburn AJ, Wheeler WC (2020) Efficient implied alignment. BMC Bioinform 21:296
82. Chase MW, Soltis DE, Olmstead RG, Morgan D, Les DH, Mishler BD et al (1993) Phylogenetics of seed plants: an analysis of nucleotide sequences from the plastid gene RBCL. Ann Mol Bot Gard 80:528–580
83. Morens DM, Taubenberger JK, Fauci AS (2009) The persistent legacy of the 1918 influenza virus. N Engl J Med 361(3):225–229

Chapter 3

Phylogenetic Placement Using SCAMPP and Batch-SCAMPP

Eleanor Wedell, Chengze Shen and Tandy Warnow

Abstract

Phylogenetic placement is the problem of adding sequences to an existing phylogenetic tree. While many techniques have been developed for this problem, methods based on optimizing maximum likelihood, such as pplacer and EPA-ng, have been shown to provide the highest accuracy. Unfortunately, these methods are limited to at most moderately large placement trees due to their design. SCAMPP and Batch-SCAMPP are two methods that have been developed to improve the scalability of both pplacer and EPA-ng to very large trees, while maintaining high accuracy. Here, we describe these methods and show how to use them in two applications: metagenomics, including taxon identification and abundance profiling, and incrementally growing large trees. SCAMPP and Batch-SCAMPP are available in open-source form on GitHub and PyPI.

Keywords Phylogenetic placement, Large phylogenies, Pplacer, EPA-ng, Maximum likelihood, Microbiome analysis, Taxon identification, Abundance profiling

1 Introduction

Phylogenetic placement is the problem of adding sequences into an existing phylogenetic tree. A well-known use of phylogenetic placement involves taxon identification or abundance profiling of reads drawn from an environmental sample, especially in a metagenomics analysis. By adding the given sequence into a taxonomic tree, the taxonomic identity of that sequence can be predicted, potentially better than by just looking at its most similar sequence in a database [1–3]. Another use of phylogenetic placement is incremental building of large phylogenies as new sequences are obtained; for example, instead of repeatedly recalculating a tree with each new set of sequences deposited into a public database, the new sequences could be added into an existing, but growing tree [4].

The sequences that are added to the existing phylogeny or taxonomy are referred to as "query sequences" in the literature, to distinguish them from the sequences that are in the placement tree, which may be a maximum likelihood tree for a single gene, a species

Gustavo Caetano-Anollés (ed.), *Evolutionary Genomics*, Methods in Molecular Biology, vol. 2981,
https://doi.org/10.1007/978-1-0716-4836-0_3,

tree estimated in a multilocus analysis [5, 6] (*see* **Note 1**), or a taxonomy [1]. In this paper, we focus on the application where the input sequences are from a single gene and the placement tree is for a single gene with maximum likelihood numeric parameters. This second case is applicable to trees estimated using maximum likelihood methods, such as RAxML [7], or taxonomies that have branch lengths and other numeric parameters estimated under maximum likelihood and with reference to a single gene (i.e., the usage in the TIPP family of methods [1, 2, 8], which is for abundance profiling in metagenomics). This context is, therefore, relevant to both applications described above: metagenomics using marker genes, as in TIPP, as well as in building large phylogenies for single genes.

In the metagenomics application, the query sequences will likely be reads or possibly contigs; and hence, may be short and have high sequencing error; furthermore, there may be many query sequences, possibly in the thousands, that need to be added into the gene tree or taxonomy. For the application of incrementally adding sequences into a gene tree, the sequences are likely to be full-length assembled gene sequences, and the number of these sequences to be added at any one time is smaller. Thus, the two applications are different in terms of the input properties.

In both applications, these query sequences are not necessarily given with an alignment to the sequences in the placement tree. Thus, the first step in a phylogenetic placement analysis may involve computing that alignment. There are many methods to compute alignments on large datasets, with some specifically designed for adding sequences into an existing alignment (*see* **Note 2**). Once the query sequences are added into the multiple sequence alignment for the placement sequences, they can be added into the placement tree using phylogenetic placement methods that use these alignments. However, some phylogenetic placement methods do not require that the query sequences be given in an alignment. One of the most accurate alignment-free methods is AppSpaM [9], but even this technique was shown to be less accurate than some simple alignment-based phylogenetic placement methods [10]. Instead, the most accurate protocols for adding sequences into phylogenies first add the query sequences into a multiple sequence alignment for the placement tree sequences, and then use that alignment in placing the query sequences into the tree.

Phylogenetic placement methods that use alignments vary in approach, with some using distances, such as APPLES 2 [11], some using parsimony, such as UShER [12], and some using maximum likelihood (ML), such as pplacer [13] and EPA-ng [14]. Comparisons between these methods have shown that the distance-based methods are the fastest but they are not as accurate as the ML-based methods, while parsimony methods have high accuracy when rates of evolution are low but are not as accurate as ML-based methods under higher rates [10]. Thus, pplacer and EPA-ng, the two current ML-based phylogenetic placement methods, are the most accurate techniques for phylogenetic placement. However, both pplacer and EPA-ng have limits on the size of the tree they can place into, and attempting to use them on trees beyond where they successfully run either results in out-of-memory errors or increased placement errors.

SCAMPP [15] and Batch-SCAMPP (BSCAMPP henceforth) [10] are two divide-and-conquer approaches that were designed to enable pplacer and EPA-ng to be used on large trees. SCAMPP is designed for use when placing a relatively small number of sequences at a time, while BSCAMPP is designed for use when placing many query sequences (i.e., hundreds to thousands) at once. Both SCAMPP and BSCAMPP can be used with EPA-ng and pplacer, and enable highly accurate phylogenetic placement into large trees (currently established for trees with 180,000 leaves, but likely placing into larger trees is also feasible).

The rest of the paper is organized as follows. Section 2 describes the algorithm design of SCAMPP and BSCAMPP at a high level, and shows how each results in improved scalability for pplacer and EPA-ng. Section 3 describes the software for these methods. Notes are provided in Sect. 4.

2 SCAMPP and BSCAMPP Algorithms

2.1 Common Themes to SCAMPP and BSCAMPP

Phylogenetic placement methods based on maximum likelihood, such as pplacer and EPA-ng, take query sequences and find the best position for the query within a given reference tree. These methods can be used with both protein and nucleotide sequences, although an appropriate sequence evolution model will need to be specified (*see* **Note 3**). The output of pplacer and EPA-ng provides information on alternative positions, as well as the statistical support for the positions (*see* **Note 4**). Maximum likelihood-based phylogenetic placement provides higher accuracy compared to methods that are based on distances or parsimony, but are computationally intensive, and typically are limited to relatively small trees. This is why SCAMPP and BSCAMPP were developed: to help a selected phylogenetic placement method, such as EPA-ng and pplacer, scale to large trees.

To achieve this, SCAMPP and BSCAMPP take the input placement tree and divide the tree into subtrees that are small enough for the phylogenetic placement method to succeed in placing query sequences with high accuracy. The locations of the query sequences within the small subtrees are then used to identify the correct locations in the input placement tree. This last step requires consideration of branch lengths in the placement tree, as well as specifically where the query sequence is placed along the edge in the subtree.

The high-level difference between SCAMPP and BSCAMPP is that SCAMPP places each query sequence independently; this improves accuracy as each query sequence can pick a subtree that is really best suited to it. Consequently, SCAMPP may be placing k query sequences into k different subtrees, which increases the running time.

To reduce the running time, BSCAMPP uses the query sequences to determine a small set of subtrees and then selects for each query sequence the best subtree *in that set*. Afterward, placement of the query sequences into the subtrees operates identically as for SCAMPP.

In what follows, we will use the following notation to define these pipelines:

- SCAMPP(pplacer) (or SCAMPP(p)) when using pplacer within SCAMPP
- SCAMPP(EPA-ng) or SCAMPP(e) when using EPA-ng within SCAMPP
- BSCAMPP(pplacer) or BSCAMPP(p) when using pplacer within BSCAMPP
- BSCAMPP(EPA-ng) or BSCAMPP(e) when using EPA-ng within BSCAMPP

With this understanding, we illustrate the concepts by beginning with the design of SCAMPP.

2.2 SCAMPP Algorithm Design

SCAMPP requires a reference phylogenetic tree T and a multiple sequence alignment A of the reference sequences and query sequences. Each query sequence is placed into the tree independently of the others.

SCAMPP has several algorithmic parameters (*see* **Note 5**), but only two main algorithmic parameters:

- Φ, the phylogenetic placement method (currently, either pplacer or EPA-ng), and
- B, the maximum size of the placement subtree for each query sequence (an integer greater than 2)

To place a single query sequence q into T, SCAMPP uses three stages:

- Stage 1: SCAMPP picks a closest leaf in T for q (based on one of several possible criteria), and then expands out from that closest leaf to select a subtree t of T that has at most B leaves.
- Stage 2: SCAMPP places q into t using the phylogenetic placement method Φ.
- Stage 3: SCAMPP then uses the branch length in the edge of t that includes q to find the corresponding edge of T for q.

Figure 1 illustrates the three stages of SCAMPP for two different query sequences on a toy phylogenetic tree. Note the algorithmic parameter B determines the size of the subtree in Stage 2. If B is larger than the size of the phylogeny, then the resulting output will be that of running EPA-ng or pplacer on the entire phylogenetic tree. The default for the parameter B defining the size of the subtree was set based on our experimental results to B=2000 when using pplacer within SCAMPP and B=10,000 when using EPA-ng within SCAMPP (*see* **Note 6**).

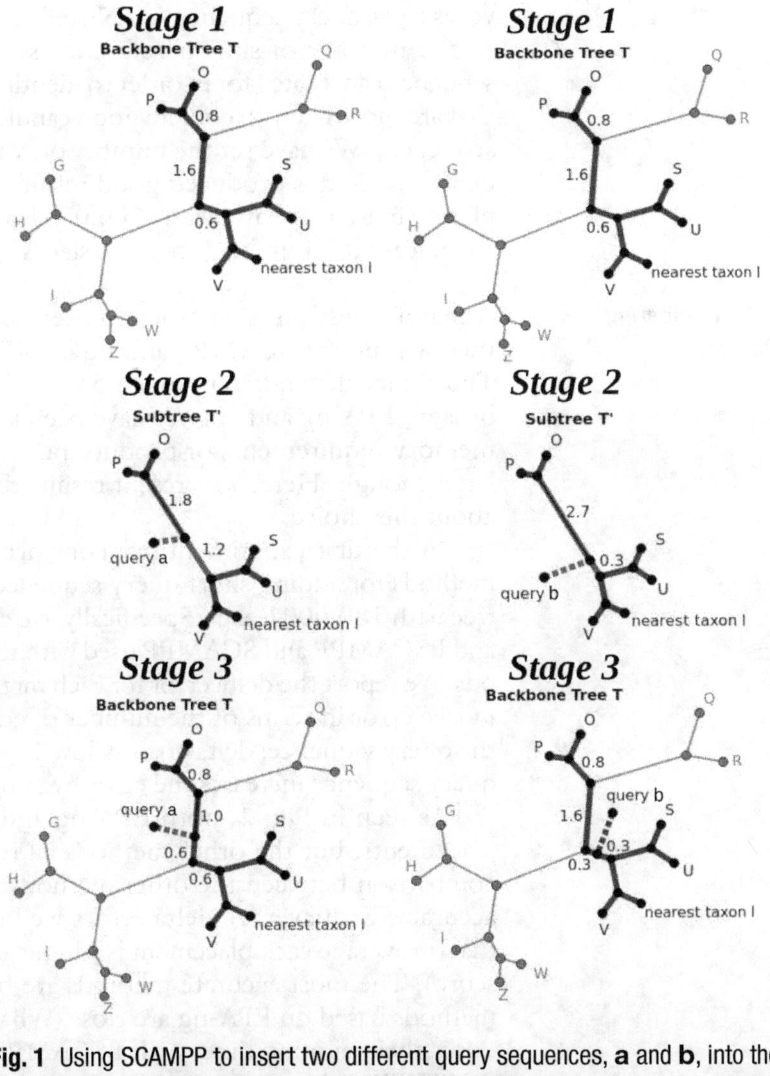

Fig. 1 Using SCAMPP to insert two different query sequences, **a** and **b**, into the same backbone tree. In Stage 1, each query sequence picks the same nearest taxon and extracts the same placement subtree. Stage 2, each query sequence is placed into the same edge, but at a different location along that edge as indicated by the way the length (3.0) of the edge is distributed. In Stage 3, the two query sequences end up in different edges in the backbone tree

2.3 BSCAMPP

As discussed earlier, BSCAMPP and SCAMPP share high-level design features and differ only in that SCAMPP places each query sequence independently, while BSCAMPP places the query sequences into a small number of subtrees. Thus, BSCAMPP must use the information in the query sequences to determine the set of subtrees, before doing any placement.

BSCAMPP, like SCAMPP, requires a subtree size B and an underlying phylogenetic placement method Φ which will be run on subtrees of size B. Because of its design, the default setting for B changes (*see* **Note 6**).

The algorithmic parameter new to BSCAMPP is the number of votes v per query sequence (*see* **Note 7**). This number corresponds to the number of similar reference sequences that each query sequence can "vote" for in order to identify regions of the phylogeny that are more likely to contain good candidate edges for many query sequences. We have set the number of votes per query sequence to default to 5; this produced good results for all settings, including placing in to trees with nearly 200,000 leaves, but the user may wish to increase this number if the tree size is very large.

2.4 Comparing Pipelines

A major question is whether to use pplacer, EPA-ng, or one of these within the SCAMPP and BSCAMPP pipelines (*see* **Note 8**). The choice depends very much on the size of the placement tree, because EPA-ng and pplacer have been known to either fail due to memory requirements or produce poor accuracy once the tree is large enough. Here, we present results that will inform the reader about this choice.

In the first experiment, we compare phylogenetic placement methods for adding short query sequences into an estimated gene tree with 180,000 leaves. Specifically, we compare EPA-ng, pplacer, and BSCAMPP and SCAMPP used with these two placement methods. We report the delta error for each method, which is the increase in tree error in terms of the number of edges, produced by adding the query sequence; delta error below 1 means that on average each query sequence increases the error by at most one edge.

As seen in Fig. 2, both EPA-ng and pplacer failed to return placements, but the other methods all returned placements. The comparison between the other methods shows they all have high accuracy, with average delta errors well below 1 (thus indicating that on average each placement is adding at most 1 edge to the error score). The most accurate methods are based on pplacer, but the methods based on EPA-ng are close. What distinguishes the methods is their running time, with BSCAMPP(e) by far the fastest and SCAMPP(p) the slowest.

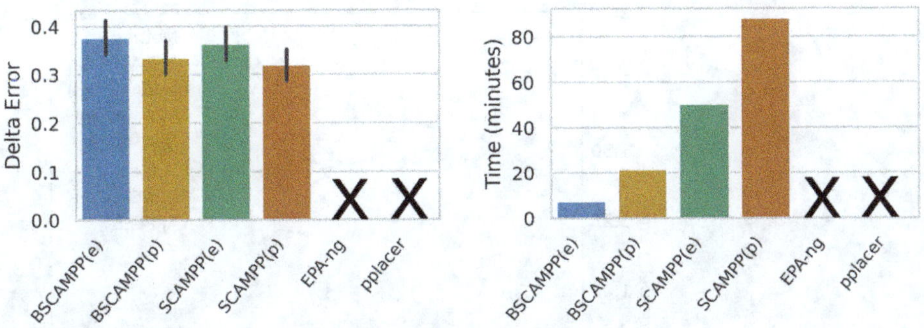

Fig. 2 Average delta error and runtime of four different pipelines placing 2000 fragmentary query sequences into a 180,000-leaf tree of the RNASim dataset, using an estimated (FastTree 2) tree on the true alignment of sequences. We show results using BSCAMPP (EPA-ng or pplacer) and SCAMPP (EPA-ng or pplacer). pplacer [13] and EPA-ng [14] were unable to run on these datasets and are marked with an **X**. BSCAMPP used with EPA-ng is the fastest method, and SCAMPP used with pplacer is the most accurate

In the second experiment, we examine conditions with a smaller tree, containing only 50,000 leaves. This is a tree size that is too large for EPA-ng but not too large for pplacer. In this comparison, each method places 10,000 short query sequences into a tree with 50,000 leaves. We show a comparison between our four pipelines, i.e., SCAMPP and BSCAMPP with EPA-ng or pplacer, the distance-based method APPLES 2, and the parsimony method, UShER (Fig. 3). For each method, we show the runtime on the x-axis and the average delta error on the y-axis. The two most accurate methods are pplacer and SCAMPP(p), i.e., SCAMPP used with pplacer, with an advantage to pplacer. Furthermore, on this dataset, pplacer is somewhat faster than SCAMPP(p). The next most accurate methods are the three likelihood-based methods: BSCAMPP(e), BSCAMPP(p), and SCAMPP(e), and of these, BSCAMPP(e) is the fastest, followed by BSCAMPP(p). UShER is slightly less accurate than these three likelihood-based methods and about the same speed as SCAMPP(e). Finally, APPLES 2 is the fastest but least accurate of the methods we examined. Memory usage is shown by the ball size, indicating that pplacer used the most memory of all methods. In fact, all the methods other than pplacer use under 3 GB of memory, while pplacer uses more than 20 GB of memory.

Thus, Figs. 2 and 3 show the ability of SCAMPP and BSCAMPP to enable EPA-ng and pplacer to scale to large trees, and also the relative impact each has on accuracy and runtime.

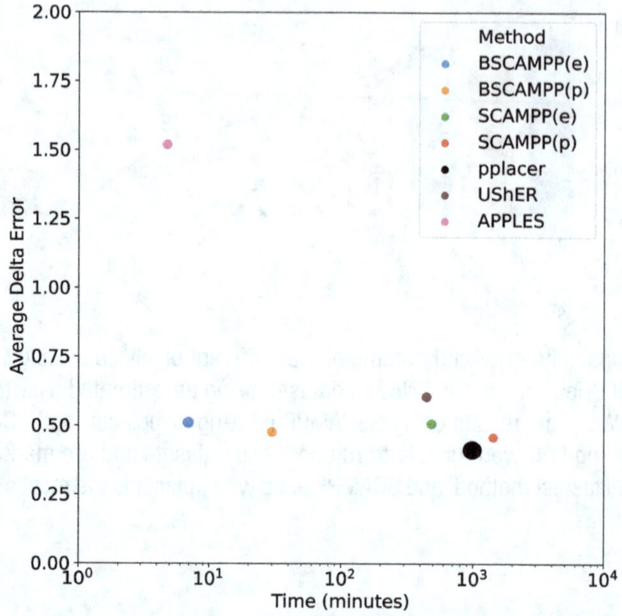

Fig. 3 Average delta error and runtime of different phylogenetic placement methods placing 10,000 fragmentary query sequences into a 50,000-leaf tree of the RNASim dataset, using an estimated (FastTree 2) tree on the true alignment of sequences. Memory usage is indicated by the size of the balls for each method. We show results using BSCAMPP (EPA-ng or pplacer) and SCAMPP (EPA-ng or pplacer), APPLES 2 [11], pplacer [13], and UShER [12]. All methods require less than 3 GB of memory, except for pplacer, which uses more than 20 GB

3 Methods

3.1 Initial Comments on Using the Software

BSCAMPP and SCAMPP have been implemented for use on Linux machines. We are testing BSCAMPP and SCAMPP for use with EPA-ng and pplacer on Macs. We have not developed these methods for use on Windows, and do not plan to.

BSCAMPP and SCAMPP can be downloaded and used in two ways: (1) from the GitHub page or (2) from PyPI (Fig. 4). We recommend using the second way as it is easier and will simultaneously install any required Python packages (*see* Table 1).

The binaries are compiled on a Linux-based machine. The binaries (found in `bscampp/tools/hamming_distance`, folder under the root directory) can be recompiled using `cmake` and `make` with the provided `CMakeLists.txt` file.

3.2 Usage

Here, we provide details of input and output for BSCAMPP and SCAMPP.

Software Input: The user must provide as input a reference alignment and tree, as well as a set of query sequences that will be

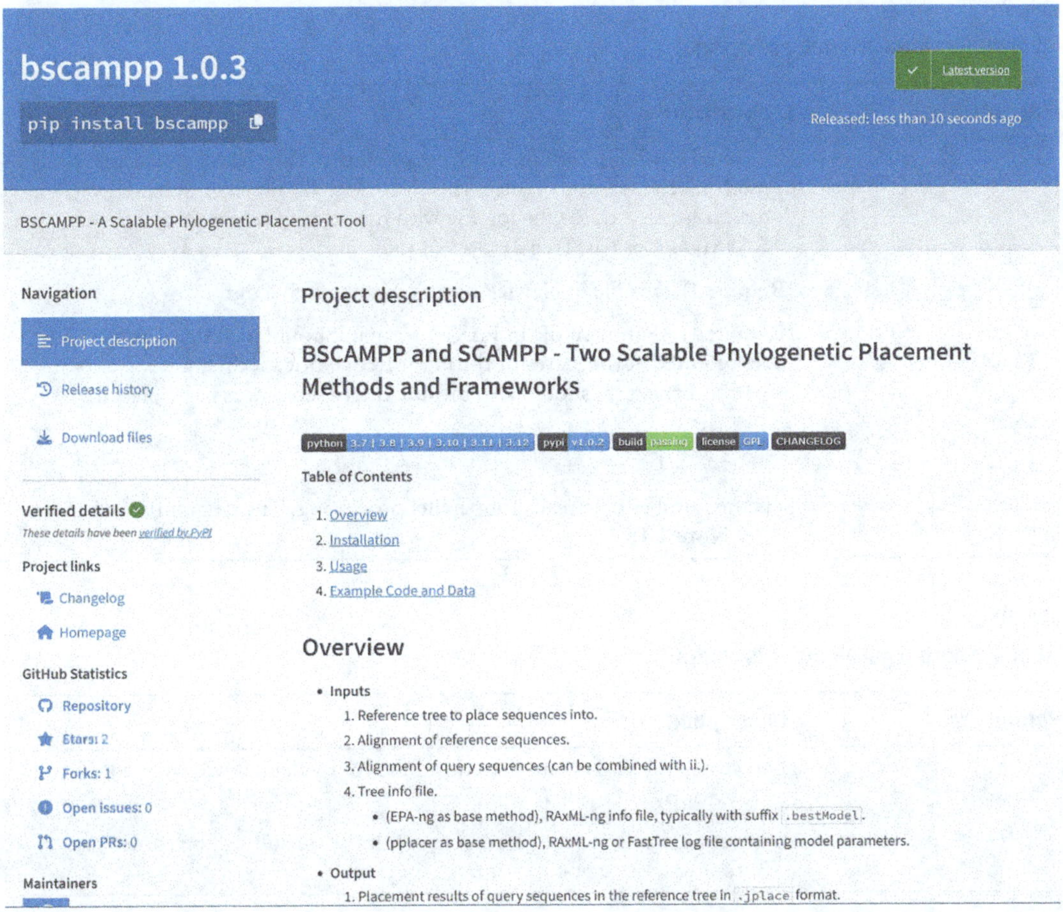

Fig. 4 BSCAMPP v1.0.3 PyPI release page, with both functionalities from BSCAMPP and SCAMPP. BSCAMPP and SCAMPP can be downloaded and installed using `pip install bscampp`

Table 1
Installation instruction for BSCAMPP and SCAMPP

Source	Instruction
`GitHub page`	Clone or download the GitHub repository at https://github.com/ewedell/BSCAMPP.git. Then, in the downloaded folder, install all requirements by running "`pip install -r requirements.txt`" (no quotation marks) in a console. Finally, run BSCAMPP with `run_bscampp.py` and SCAMPP with `run_scampp.py`.
`PyPI (pip)`	Install by running "`pip install bscampp`" (no quotation marks) in a console. Finally, run BSCAMPP with `run_bscampp.py` or `bscampp`, and SCAMPP with `run_scampp.py` or `scampp`.

Table 2
List of supported parameters for input

Parameter	Description
`-i, --info FILE`	(Required) Info file with numeric parameters to run the base placement method. For `epa-ng`, this is the `.bestModel` file from RAxML-ng; for `pplacer`, this is the log file with model parameters from either RAxML-ng or FastTree 2.
`-t, --tree FILE`	(Required) Tree file in Newick format with branch lengths.
`-a, --alignment FILE`	(Required) Alignment file in FASTA format. Should include sequences represented at the leaves of the input tree; if not specifying `-q, --qalignment`, should also include query sequences to be placed.
`-q, --qalignment FILE`	Query alignment file in FASTA format.
`--threads, --num-cpus NUM`	Sets the number of threads for parallel processing. The default uses all cores (*see* **Note 10**).

Table 3
List of supported parameters for output

Parameter	Description
`-d, --outdir FOLDER`	Directory/folder to write the output file to. By default will write to `{scampp,bscampp}_output/`.
`-o, --output NAME`	Output file name for the placement result. By default will be named as `{scampp,bscampp}_result.jplace`

placed into the reference tree. An alignment of the query sequences to the reference sequences must also be provided (*see* **Note 2**). If `epa-ng` is selected for phylogenetic placement, then the user must provide a RAxML-ng best model, if necessary by re-estimating numeric parameters on the reference tree. If `pplacer` is used, then the user must prepare an info file from RAxML-ng or FastTree 2 (*see* **Note 3**). See Table 2 for the list of input parameters.

Software Output: The output of BSCAMPP and SCAMPP is a single .jplace [16] file containing the placement information for each input query sequence. The parameters shown in Table 3 can be set for output control. Logging information running the software will be printed to the console. In some cases, a query sequence will not appear in the output file; when this occurs, it indicates that the pipeline did not find any leaf sequence for that query sequence (*see* **Note 9**).

Algorithm Configuration: Additional parameters can be set to control how BSCAMPP and SCAMPP are run, including the subtree size and the number of votes (for BSCAMPP). We have a `README.md` file with more detailed parameter settings on the GitHub page mentioned earlier in this section.

Examples: The examples are based on the executable installed with PyPI.

1. For a default usage case with EPA-ng as the base placement method, the user can run the following commands for BSCAMPP:

   ```
   $ bscampp -i [ raxml best model ] -t [ reference tree ] \
     -a [ reference alignment ] -q [ query alignment ]
   ```

   SCAMPP uses the same required parameters:

   ```
   $ scampp -i [ raxml best model ] -t [ reference tree ] \
     -a [ reference alignment ] -q [ query alignment ]
   ```

   Note that `[reference alignment]` and `[query alignment]` should have the same alignment length. The user can also combine the two files for `-a` and omit `-q`:

   ```
   $ bscampp -i [ raxml best model ] -t [ reference tree ] \
     -a [ alignment with all sequences ]
   ```

2. To use `pplacer` as the base placement method, add `--placement-method pplacer` to the command:

   ```
   $ bscampp -i [ raxml info file / fasttree log file ] \
     -t [ reference tree ] \
     -a [ alignment with all sequences ] \
     --placement - method pplacer
   ```

3. Use `--num-cpus [NUM]` to change the number of CPUs for multiprocessing. By default, the software uses all available CPU cores (*see* **Note 10**):

   ```
   $ bscampp -i [ raxml best model ] -t [ reference tree ] \
         -a [ alignment with all sequences ] \
         --num - cpus [ NUM ]
   ```

4. Use `-d [FOLDER]` and `-o [NAME]` to control where to write the output file and the file name:

   ```
   $ bscampp -i [ raxml best model ] -t [ reference tree ] \
     -a [ alignment with all sequences ] \
     -d [ output folder ] -o [ output name ]. jplace
   ```

5. Use `--subtreesize [NUM]` to adjust the subtree size that the base placement method will run on (*see* **Note 6**):

   ```
   $ bscampp -i [ raxml best model ] -t [ reference tree ] \
     -a [ alignment with all sequences ] \
     -- subtreesize [ NUM ]
   ```

6. (BSCAMPP only) Use `-V [NUM]` to change the number of votes per query (*see* **Note 7**):

   ```
   $ bscampp -i [ raxml best model ] -t [ reference tree ] \
     -a [ alignment with all sequences ] \
     -V [ NUM ]
   ```

4 Notes

1. The phylogenetic placement methods discussed in this paper can be used for the application, where the sequences at the leaves of the placement tree are derived from a multilocus analysis, and so are concatenations of aligned sequences across different loci. However, there are some challenges in using these methods for this kind of analysis since the sequence lengths will be much longer, and this will increase both memory usage and runtime. However, there is also another concern, which is that different loci can evolve under different trees due to biological processes such as lineage sorting, leading to discord between gene trees and species trees [32]. Given this, methods such as DEPP [6], which take discordance into consideration, may provide better accuracy. However, we have not tested any of these phylogenetic placement methods in this context, and methods like BSCAMPP(e) may provide good accuracy and speed even in the presence of discordance between gene trees and species trees.
2. The input to either SCAMPP or BSCAMPP must include a multiple sequence alignment that has all the query sequences as well as the sequences that are already in the placement tree. Here we offer some suggestions for which methods you should use, considering both runtime and accuracy. Alignments of datasets that contain both full-length and short sequences need to be done carefully, typically using two steps: first the full-length sequences are aligned and then the short sequences are added into the alignment of the full-length sequences. In the applications that involve phylogenetic placement, typically the leaf sequences in the reference tree are full-length but the query sequences may be short. We recommend using the most accurate methods available to align the full-length reference sequences, as this impacts accuracy. In particular, we recommend MAGUS [17, 18], which is an improvement on PASTA [19] and SATé [20, 21]; MAGUS can be used to align datasets with up to 1,000,000 sequences, especially if used in its recursive form. Once the multiple sequence alignment of the full-length sequences is obtained, adding the query sequences into the alignment should be performed using methods explicitly designed for this step. Some of the methods that have been developed to add sequences into alignments include MAFFT-add [22], UPP [23] and its faster version UPP2 [24], WITCH [25] and its faster version WITCH-ng [26], and EMMA [27]. Of these, we recommend WITCH or WITCH-ng when

the query sequences are short, and EMMA when the query sequences are full-length.

3. The input to SCAMPP or BSCAMPP must include a placement tree, which must also have numeric parameters computed using a maximum likelihood method. The tree itself can be computed using any preferred method, or can even be a taxonomy you obtain, as long as the numeric parameters are computed using an appropriate maximum likelihood method. If you want to estimate the tree topology and parameters at the same time, we recommend methods like RAxML-ng [28] and IQ-tree [29], although for very large trees you may wish to use FastTree 2 [30] (a much faster maximum likelihood software that is not as reliable as RAxML or IQ-tree 2 [31]). For numeric parameter estimation, the choice of method depends on whether you are using EPA-ng or pplacer. If you will use EPA-ng within these pipelines, then you should compute these parameters using RAxML-ng, but if you will use pplacer, then you can compute these parameters with either FastTree 2 or RAxML. We have found FastTree 2 to produce useful parameters for phylogenetic placement, and do not have any evidence that RAxML provides improved placements.
4. When the query sequence can be placed into the tree, there can be multiple positions that have high statistical support. The output will, therefore, contain more than a single best placement for a given query sequence, and will also provide the support for these placements. In particular, the jplace file [16] returned by SCAMPP and BSCAMPP will contain the information that would have been returned by pplacer or EPA-ng. That is: a number of possible placement edges per query sequence with (1) the query placement's corresponding likelihood weight ratio (confidence score), (2) the raw likelihood that was computed, (3) the placement edge, (4) the distal length along the placement edge where the new pendant edge will attach, and (5) the length of the new pendant edge. The number of possible placement edges returned by SCAMPP and BSCAMPP will be the same as the number returned by the underlying placement method run on the subtree.
5. SCAMPP (and BSCAMPP) both have several other algorithmic parameters beyond the ones discussed in this paper. While these parameters have default values based on fairly extensive experimental settings, it is possible that modifications to these parameters could provide advantages for accuracy or runtime, under some circumstances. The README on the GitHub page provides more information on these, and

the advanced user may wish to explore the impact of changing these defaults.

6. The subtree size parameter *B* has default settings that depend on both the divide-and-conquer strategy (i.e., SCAMPP and BSCAMPP) and the phylogenetic placement method. While the user can modify these, we do not recommend changes as they may result in worsened accuracy. For BSCAMPP used with EPA-ng, increasing *B* beyond its default setting of 2000 should only be considered if the query sequences are relatively long as we have seen placement accuracy degrade with short sequences when *B* exceeds 2000. For BSCAMPP used with pplacer, it is possible that increases to larger values may provide good accuracy, but we have not explored this question in sufficient depth to provide advice with high confidence.
7. When running BSCAMPP, the important algorithmic parameter to consider is the number of votes for each query sequence. This parameter helps determine the quality of the subtrees that are chosen for placement. In general, increasing the number of votes could lead to a better set of subtrees, but will increase the runtime. Our study has examined placement trees with up to 180,000 leaves, and this default has worked well; nevertheless, it is possible that increasing the number of votes beyond the default (5) may be beneficial if the reference tree is very large.
8. The main question is whether to use one of the ML-based phylogenetic placement methods as is (i.e., not within SCAMPP or BSCAMPP), or within one of these frameworks. And, if within one of the frameworks, which one? Here we present some basic advice, reflecting the tradeoff between accuracy and runtime among these four methods (see also Figs. 2 and 3). If running time is not too important an issue and the tree is not so large as to make pplacer infeasible (i.e., probably at most 80,000 leaves), we recommend using pplacer. However, when the tree is too large for pplacer and running time is not important compared to accuracy, then we would recommend SCAMPP(p). If runtime is the most significant factor, then using BSCAMPP(e) is recommended, as it is by far the fastest of the ML-based methods we have studied and not much less accurate than SCAMPP(p). The choice between the remaining two methods—SCAMPP(e) and BSCAMPP(p)—depend on the relative importance of accuracy and running time, but BSCAMPP(p) is faster and may have an edge for accuracy. All this said, the larger the number of query sequences the bigger the advantage in using BSCAMPP(e) compared to the other methods, making BSCAMPP(e) probably the most

appropriate choice for metagenomic analyses involving many sequencing reads.

9. In some cases, some query sequences will not be placed into the given reference tree. Typically, this would occur when there is so little similarity between the query sequence and any leaf in the tree, a case that might happen with short read sequencing technologies for reads that are distantly related to all the sequences in the tree so that even the alignment step fails to detect homologous sites. The user has two options at this point: either try to realign the query sequence to the reference sequences using a more sensitive alignment method, or discard the query sequence if realigning fails to recover sufficient homologies.
10. To take advantage of parallel placements of query sequences, BSCAMPP would be the preferred method since it will group sequences together for placement into subtrees. Furthermore, EPA-ng has been shown to have superior parallel efficiency than pplacer [14]. Therefore, we recommend using EPA-ng as the phylogenetic placement method to be used with BSCAMPP to optimize parallel efficiency.

Acknowledgments

This work was supported in part by NSF grants 2006069 and 2316233 (to TW). EW was supported in part by an Ira and Debra Cohen Fellowship from the Siebel School of Computing and Data Science at the University of Illinois Urbana-Champaign.

References

1. Nguyen ND, Mirarab S, Liu B, Pop M, Warnow T (2014) TIPP: taxonomic identification and phylogenetic profiling. Bioinformatics 30(24):3548–3555
2. Shah N, Molloy EK, Pop M, Warnow T (2021) TIPP2: metagenomic taxonomic profiling using phylogenetic markers. Bioinformatics 37(13):1839–1845. https://doi.org/10.1093/bioinformatics/btab023
3. Czech L, Stamatakis A, Dunthorn M, Barbera P (2022) Metagenomic analysis using phylogenetic placement—a review of the first decade. Frontiers Bioinf 2:871393
4. Zaharias P, Warnow T (2022) Recent progress on methods for estimating and updating large phylogenies. Phil TransRoyal Soc B 377(1861):20210244
5. Jiang P, Balaban M, Zhu Q, Mirarab S (2023) DEPP: deep learning enables extending species trees using single genes. Syst Biol 72(1):17–34
6. Jiang Y, McDonald D, Perry D, Knight R, Mirarab S (2024) Scaling DEPP phylogenetic placement to ultra-large reference trees: a tree-aware ensemble approach. Bioinformatics 40(6):btae361
7. Stamatakis A (2014) RAxML version 8: a tool for phylogenetic analysis and post-analysis of large phylogenies. Bioinformatics 30(9):1312–1313
8. Shen C, Wedell E, Pop M, Warnow T (2024) TIPP3 and TIPP3-fast: Improved abundance profiling in metagenomics. PLoS Comput Biol 21:e1012593. https://doi.org/10.1371/journal.pcbi.1012593

9. Blanke M, Morgenstern B (2021) App-SpaM: phylogenetic placement of short reads without sequence alignment. Bioinf Adv 1(1):vbab027
10. Wedell E, Shen C, Warnow T (2025) BSCAMPP: Batch-scaled phylogenetic placement on large trees. IEEE Trans Comput Biol Bioinform 22:1593-1605. https://doi.org/10.1109/TCBBIO.2025.3562281
11. Balaban M, Jiang Y, Roush D, Zhu Q, Mirarab S (2022) Fast and accurate distance-based phylogenetic placement using divide and conquer. Mol Ecol Res 22(3):1213–1227
12. Turakhia Y, Thornlow B, Hinrichs AS, De Maio N, Gozashti L, Lanfear R, Haussler D, Corbett-Detig R (2021) Ultrafast Sample placement on Existing tRees (UShER) enables real-time phylogenetics for the SARS-CoV-2 pandemic. Nat Genet 53(6):809–816
13. Matsen FA, Kodner RB, Armbrust EV (2010) pplacer: linear time maximum-likelihood and Bayesian phylogenetic placement of sequences onto a fixed reference tree. BMC Bioinf 11(1):538
14. Barbera P, Kozlov AM, Czech L, Morel B, Darriba D, Flouri T, Stamatakis A (2019) EPA-ng: massively parallel evolutionary placement of genetic sequences. Syst Biol 68(2):365–369
15. Wedell E, Cai Y, Warnow T (2022) SCAMPP: Scaling alignment-based phylogenetic placement to large trees. IEEE/ACM Trans Comput Biol Bioinf 20(2):1417–1430. https://doi.org/10.1109/TCBB.2022.3170386
16. Matsen FA, Hoffman NG, Gallagher A, Stamatakis A (2012) A format for phylogenetic placements. PLOS One 7(2):e31009
17. Smirnov V, Warnow T (2021) MAGUS: multiple sequence alignment using graph clustering. Bioinf 37(12):1666–1672
18. Smirnov V (2021) Recursive MAGUS: scalable and accurate multiple sequence alignment. PLoS Comp Biol 17(10):e1008950
19. Mirarab S, Nguyen ND, Guo S, Wang L-S, Kim J, Warnow T (2015) PASTA: ultra-large multiple sequence alignment for nucleotide and amino-acid sequences. Comp Biol 22(5):377–386
20. Liu K, Raghavan S, Nelesen S, Linder CR, Warnow T (2009) Rapid and accurate large-scale coestimation of sequence alignments and phylogenetic trees. Science 324(5934):1561–1564
21. Liu K, Warnow T, Holder MT, Nelesen SM, Yu J, Stamatakis AP, Linder CR (2012) SATé-II: very fast and accurate simultaneous estimation of multiple sequence alignments and phylogenetic trees. Syst Biol 61(1):90
22. Katoh K, Frith MC (2012) Adding unaligned sequences into an existing alignment using MAFFT and LAST. Bioinformatics 28(23):3144–3146
23. Nguyen ND, Mirarab S, Kumar K, Warnow T (2015) Ultra-large alignments using phylogeny-aware profiles. Genome Biol 16(1):124
24. Park M, Ivanovic S, Chu G, Shen C, Warnow T (2023) UPP2: fast and accurate alignment of datasets with fragmentary sequences. Bioinformatics 39(1):btad007
25. Shen C, Park M, Warnow T (2022) WITCH: improved multiple sequence alignment through Weighted Consensus Hidden Markov Model alignment. J Comp Biol 29(8):782–801
26. Liu B, Warnow T (2023) WITCH-NG: efficient and accurate alignment of datasets with sequence length heterogeneity. Bioinf Adv 3(1):vbad024
27. Shen C, Liu B, Williams KP, Warnow T (2023) EMMA: a new method for computing multiple sequence alignments given a constraint subset alignment. Alg Mol Biol 18(1):21
28. Kozlov AM, Darriba D, Flouri T, Morel B, Stamatakis A (2019) RAxML-NG: a fast, scalable and user-friendly tool for maximum likelihood phylogenetic inference. Bioinformatics 35(21):4453–4455
29. Minh BQ, Schmidt HQ, Chernomor O, Schrempf D, Woodhams MD (2020) A Von Haeseler, and R Lanfear. IQ-TREE 2: new models and efficient methods for phylogenetic inference in the genomic era. Mol Biol Evol 37(5):1530–1534
30. Price MN, Dehal PS, Arkin A (2010) FastTree 2—approximately maximum-likelihood trees for large alignments. PLOS One 5(3):e9490
31. Park M, Zaharias P, Warnow T (2021) Disjoint tree mergers for large-scale maximum likelihood tree estimation. Algorithms 14(5):148
32. Maddison WP (1997) Gene trees in species trees. Syst Biol 46(3):523–536

Check for updates

Chapter 4

Impact of Data Error on Phylogenetic Network Inference from Gene Trees Under the Multispecies Network Coalescent

Mehdrad Tamiji, Nicolae Sapoval and Luay Nakhleh

Abstract

Phylogenetic network inference has become an essential tool in evolutionary biology, offering a framework to model complex evolutionary events such as hybridization and horizontal gene transfer. However, a critical but often overlooked challenge is the presence of error in empirical datasets, including sequencing errors, misalignments, and inaccuracies in gene tree estimation. This issue is particularly pressing in the context of phylogenetic networks, which can contain an arbitrary number of parameters and are thus highly susceptible to overfitting. Errors in the input data can lead to artificially inflated network complexity, misrepresenting evolutionary history with non-biological reticulations.

In this study, we systematically examine how different sources of data error influence network inference and show that many widely used methods are vulnerable to these distortions. We find that inaccuracies in gene tree estimation and sequence alignment degrade the reliability of inferred networks. These issues are exacerbated when the number of reticulations that an algorithm can infer exceeds the true number of reticulations in the phylogenetic network. Our analysis underscores the importance of accounting for data error when applying network inference methods and provides practical recommendations for minimizing its impact. By highlighting the vulnerabilities of different approaches and demonstrating how errors propagate through the inference process, we offer practical recommendations for optimizing data processing pipelines. Our findings emphasize the necessity of integrating realistic error models into species network inference methods to enhance their reliability and applicability to real-world biological datasets.

Keywords Phylogenetic networks, Multispecies network coalescent, Maximum likelihood, Maximum pseudo-likelihood, Bayesian inference, Data error, Robustness

1 Introduction

Evolutionary histories are traditionally represented by phylogenetic trees, which model the divergence of species or genes through branching events. In a species tree, branches correspond to speciation events, while in a gene tree, they represent the ancestry of loci descended from a common ancestor. However, several evolutionary processes—including admixture, gene flow, hybridization, hybrid

Gustavo Caetano-Anollés (ed.), *Evolutionary Genomics*, Methods in Molecular Biology, vol. 2981,
https://doi.org/10.1007/978-1-0716-4836-0_4,

speciation, introgression, and horizontal gene transfer—can give rise to non-tree-like histories. As a result, species trees alone may provide incomplete or inaccurate representations of evolutionary history [1–5]. Forms of gene flow, such as introgression and hybrid speciation, have been shown to significantly shape evolutionary patterns. When such events occur, a phylogenetic network—a rooted, directed acyclic graph—provides a more accurate model of evolutionary relationships [6]. In this graph, the root represents the most recent common ancestor of all taxa, and internal nodes denote either divergence or reticulation events. Figure 1 illustrates a phylogenetic network over five taxa, where the blue-colored node inherits genetic material from two distinct ancestral lineages. It is important to note that this layout is purely a visual simplification; graph-theoretically, the underlying structure remains a directed acyclic graph, as used in formal phylogenetic network analysis.

The identification of introgression in the presence of incomplete lineage sorting (ILS) motivated the generalization of the multispecies coalescent (MSC) model [7], which defines a probability distribution over gene trees given a species tree and demographic parameters, to the multispecies network coalescent (MSNC) model [8, 9]. This generalization allowed for the development of a wide array of methods for inferring phylogenetic networks under the multispecies coalescent model while also accounting for phenomena such as ILS (e.g., [3, 9–16]). The MSNC allows for modeling evolutionary histories in the presence of both ILS and reticulation, and it enables the derivation of the probability distribution of gene trees on a species network.[1] Thus, contrasting gene tree distributions under the MSC and MSNC can provide a signal for detecting reticulation and also tease apart the separate roles that ILS and reticulation play.

Methods for inferring species networks under the multispecies network coalescent (MSNC) can be broadly categorized into three classes [17]: summary methods, site-based methods, and coestimation methods. Summary methods estimate networks using inferred gene trees; site-based methods rely on bi-allelic markers; and coestimation methods jointly infer networks and gene trees directly from sequence alignments [18]. Coestimation approaches, such as those described in [12, 14], extend Bayesian frameworks like StarBEAST [19] to phylogenetic networks by sampling from the posterior distribution over species networks and gene trees. However, these methods are computationally intensive and often face convergence and mixing challenges due to the transdimensional nature of network space. Site-based methods, which began with the analytic species tree approach of Bryant et al. [20], were later extended to

[1] The term 'network' is used to mean many things in biology. In this chapter, we use 'species network' (and, simply just 'network'), as the counterpart to 'species tree,' referring specifically to the explicit phylogenetic network of a set of species.

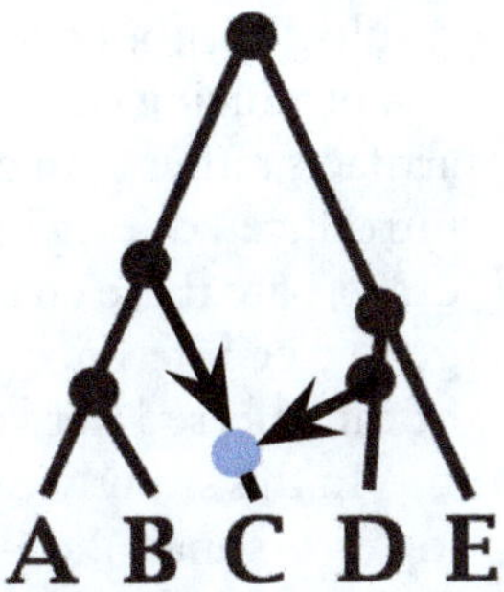

Fig. 1 Phylogenetic network of taxa A, B, C, D, and E. The blue node denotes a reticulation event, while the rest of the structure is interpreted in the same way as a phylogenetic tree. While directions are explicitly shown only for the two reticulation edges, all edges are directed away from the root of the network toward the leaves

networks by Zhu et al. [13], and subsequently improved through pseudolikelihood formulations [15] and more efficient inference algorithms [21]. Additional tests for detecting hybridization using bi-allelic markers include the D-statistic [22], its various extensions [23, 24], and invariant-based approaches such as HyDe [25, 26]. Summary methods for the MSNC, originally proposed by Yu et al. [9], employ maximum likelihood estimation from gene tree topologies, with or without branch lengths, and were later extended to a Bayesian framework [11]. To improve scalability, pseudolikelihood approaches were introduced [3, 27], enabling network inference from larger gene tree datasets without the computational burden of full likelihood estimation.

The space of possible network topologies is vast, and their complexity can grow rapidly with the number of taxa and reticulation events. While this complexity can reflect genuine biological processes, it can also arise spuriously due to methodological limitations, gene tree estimation error (GTEE), or model violations. For example, inaccurate or biased gene trees can mislead inference methods into inferring reticulations that do not correspond to actual evolutionary history [28]. Additionally, methods that assume a single gene tree per locus or fail to account for variation in substitution rates across loci may introduce artifacts into the inferred networks [4]. Model misspecification—such as neglecting incomplete lineage sorting (ILS) or overlooking rate heterogeneity—can lead to overfitting, resulting in unnecessary reticulations or incorrect inference of introgression directionality [29]. Notably, recent studies have shown that even fully Bayesian methods, which are theoretically robust, may be misled under rate heterogeneity unless substitution rate variation is explicitly modeled [28, 30]. These issues underscore the importance of statistical rigor in model formulation and biological interpretation of results. Without such diligence, species network inference methods risk producing overly complex networks that reflect methodological artifacts rather than genuine evolutionary signals. In particular, errors introduced during sequencing and multiple sequence alignment can exacerbate these challenges, as they may distort site patterns, mislead gene tree inference, and ultimately inflate the apparent complexity of the inferred networks [31].

In this study, we examined how data errors—specifically sequencing errors and indels—affect species network inference from gene tree estimates under the multispecies network coalescent (MSNC) model. We found that indels substantially increase alignment error and gene tree estimation error (GTEE), particularly in shorter sequences, leading to noticeable shifts in gene tree distributions. Despite these challenges, both maximum likelihood and maximum pseudo-likelihood methods performed well when the number of allowed reticulations matched the true value. However, allowing more reticulations led to overfitting, especially under error, while Bayesian inference via MCMC showed greater robustness due to prior regularization. Although none of the methods could automatically recover the true number of reticulations, tracking changes in (pseudo-)likelihood helped identify a reasonable model complexity, with pseudo-likelihood showing sharper declines in gain and a lower tendency to overfit. These results suggest that current methods—particularly pseudo-likelihood inference—are reasonably robust to moderate data error and can support informed model selection while remaining computationally efficient.

The rest of the chapter is organized as follows. Section 2 provides a brief overview of phylogenomic network inference, a summary of the three network inference methods we study, a detailed description of the simulation setup used to assess the impact of error, and finally information on data and code availability. In Sect. 3, we present our empirical findings, highlighting how different types of errors affect inference accuracy across methods. Section 4 offers a discussion of the implications of our results, emphasizing practical recommendations for data analysis and suggesting directions for improving model robustness in the presence of error. Afterwards, we propose a set of guidelines for biologists to aid in the design and interpretation of phylogenetic analyses and outline directions for future research to address open challenges and expand upon our findings.

2 Materials and Methods

This section outlines the methods used to study the impact of data error on species network inference, including a review of inference models, the simulation setup, and details on data and code availability.

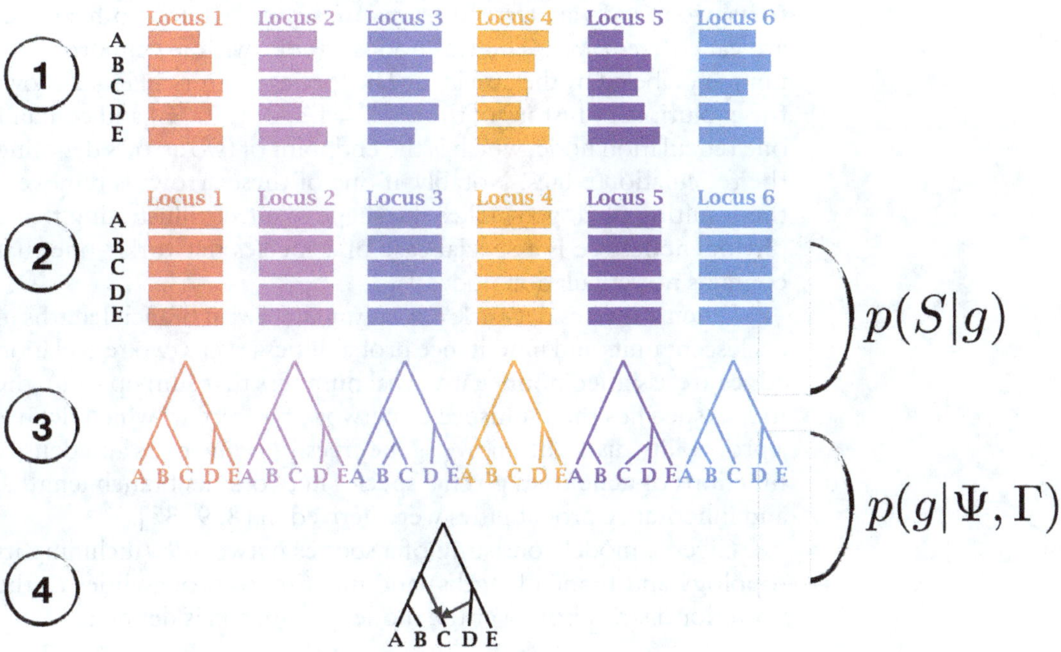

Fig. 2 Phylogenomic pipeline of species network inference. (1) Orthologs from five species—A, B, C, D, and E—are obtained across multiple loci (six in this case). (2) For each locus, the sequences are aligned to produce one multiple sequence alignment (MSA) per locus. (3) For many species phylogeny inference methods, the input consists of gene trees, typically one per locus. For these methods, a gene tree is constructed from the MSA of each locus. The methods examined in this chapter assume that the gene trees are also rooted. (4) A species network is inferred from the set of gene trees. In likelihood-based approaches, the MSA S and gene tree G of each locus are related by the likelihood function $p(S|g)$, and each gene tree g is related to the species network Ψ and its inheritance probabilities by the likelihood function $p(g|\Psi, \Gamma)$

2.1 Species Network Inference Under the Multispecies Network Coalescent: A Brief Overview

A typical phylogenomic inference pipeline for species networks is illustrated in Fig. 2. To infer a species network on a set $\mathcal{X}$ of species, sequence data are obtained from multiple loci. For the phylogenomic methods studied in this chapter, we assume the loci are independent and recombination-free. The sequences for each locus are then aligned, and from a modeling and inference perspective, the data are represented as a set $S = \{S_1, \ldots, S_m\}$ of multiple sequence alignments (MSAs) from M loci, where S_i denotes the MSA for locus i.

Phylogenetic networks have been defined in the literature to refer to various models [32]. In this work, we focus on species networks that are explicit phylogenetic networks, i.e., ones that explicitly model the evolutionary history of a set of taxa; Fig. 2. In this context, a *species network* Ψ on a set $\mathcal{X}$ of taxa is a rooted, directed, acyclic graph with a set V of nodes (or vertices) and a set E of edges (or branches). Except for the root node, which has no parents, each node in V has either one or two parents. Nodes with two parents are called *reticulation nodes*, and all other nodes—including the root—are referred to as *tree nodes*. The edges connecting a reticulation node to its two parents are termed *reticulation edges*, while all other edges are called *tree edges*. The leaf nodes (nodes with no children) are uniquely labeled by the taxa in $\mathcal{X}$. The species network in Fig. 2 shows the evolutionary history of the set $\mathcal{X} = \{A, B, C, D, E\}$ and contains one reticulation node, which is the endpoint of two arrows denoting the reticulation edges. Notably, if one of these arrows is removed, the resulting phylogeny takes the shape of a tree, illustrating that a phylogenetic tree is a special case of a species network—one that contains no reticulation nodes.

When a species network Ψ is augmented with branch lengths in coalescent units and inheritance probabilities (every two reticulation edges are assigned nonnegative real numbers that sum up to 1), the model specifies the multispecies network coalescent, which defines a probability distribution over gene trees [9]. The mass and density functions of gene trees given a species network, its branch lengths, and inheritance probabilities were derived in [8, 9, 33].

Given a model consisting of a species network Ψ (including its topology and branch lengths) and inheritance probabilities Γ, the posterior distribution over the model parameters is defined as

$$\begin{aligned} p(\Psi, \Gamma | S) &\propto p(S|\Psi, \Gamma) p(\Psi) p(\Gamma) \\ &= p(\Psi) p(\Gamma) \prod_{i=1}^{m} \int_G p(S_i|g) p(g|\Psi, \Gamma) dg, \end{aligned}$$

where $p(S_i|g)$ denotes the probability of the sequence alignment S_i given a particular gene tree g [34], and $p(g|\Psi, \Gamma)$ represents the density of the gene tree (including its topology and branch lengths) given the model parameters [9].

If the set G of gene trees has been estimated for the $m$loci, with gene tree g_i estimated from sequence alignment S_i, we can treat G as the data. The posterior distribution of the model is then given by

$$p(\Psi, \Gamma | G) \propto p(G|\Psi, \Gamma) p(\Psi) p(\Gamma) = p(\Psi) p(\Gamma) \prod_{i=1}^{m} p(g_i|\Psi, \Gamma).$$

Furthermore, under this formulation, the maximum likelihood estimate (MLE) of the species network and its parameters is given by

$$\text{argmax}_{\Psi, \Gamma} \prod_{i=1}^{m} p(g_i|\Psi, \Gamma).$$

Calculating the likelihood function $(\prod_{i=1}^{m} p(g_i|\Psi, \Gamma))$ is computationally very expensive [35], which has motivated the development of a maximum pseudo-likelihood approach [27] and a pseudo-likelihood-based Bayesian inference method [36].

2.2 Inference Methods

In this work, we focus on three methods for phylogenomic network inference from gene trees that are implemented in `PhyloNet` [37, 38]: maximum likelihood inference (ML) under the MSNC [9], maximum pseudo-likelihood inference (MPL) under the MSNC [27], and pseudo-likelihood-based Bayesian MCMC inference (PL MCMC) under the MSNC [11, 36]. All of these methods take as input a set of rooted gene trees. Details of these methods and their parameters are provided in the paragraphs below.

InferNetwork_ML is a likelihood-based method designed to infer species networks that model reticulate evolutionary histories while accounting for incomplete lineage sorting (ILS) [9]. The method operates by maximizing the likelihood (Eq. (3)) of observed gene trees under the MSNC. The process involves a heuristic search over the space of species networks with an upper bound on the maximum number of allowed reticulations. By default, the search begins with the species tree that minimizes the number of deep coalescences [39].

The script in Listing 1 provides a sample command for running the `InferNetwork_ML` method with *M* gene trees, a maximum of three reticulations, and utilizing a single CPU core (via the `-pl 1` flag). The output of the command provides an estimated species network and its likelihood.

Listing 1: Example NEXUS file for inferring species networks using `InferNetwork_ML`.

```
#NEXUS
BEGIN TREES;
Tree gt0 = (OUT:0.325872,(A:0.108396,(C:0.093204,(D
:0.093060,(E:0.047520,B
:0.047520):0.045540):0.000144):0.015192):0.217476);
...
Tree gtM = (OUT:0.312048,((B:0.076896,A:0.076896):
0.044424,
(C:0.102276,(E:0.060192,D:0.060192):0.042084):
0.019080):0.190692);
END;
BEGIN PHYLONET;
InferNetwork_ML (gt0-gtM) 3 -pl 1;
END;
```

InferNetwork_MPL is a pseudo-likelihood-based method designed for inferring phylogenetic networks while accounting for incomplete lineage sorting (ILS) [27]. Unlike `InferNetwork_ML`, which directly maximizes the full likelihood of the network, `InferNetwork_MPL` approximates the likelihood using a pseudo-likelihood function based on the distribution of the rooted triples of taxa.

The script in Listing 2 provides an example command for running the method in `PhyloNet`. Similar to InferNetwork_ML, this command specifies *M* gene trees, sets a maximum of three reticulations, and runs the computation using a single CPU core (via the `-pl 1` flag). The output of the command includes an estimated species network and its associated pseudo-likelihood.

Listing 2:Example NEXUS file for inferring species networks using `InferNetwork MPL`.

```
#NEXUS
BEGIN TREES;
Tree gt0 = (OUT:0.325872,(A:0.108396,(C:0.093204,
(D:0.093060, (E:0.047520,B
:0.047520):0.045540):0.000144):0.015192):0.217476);
...
Tree gtM = (OUT:0.312048,((B:0.076896,A:0.076896):
0.044424,
(C:0.102276,(E:0.060192,D:0.060192):0.042084):
0.019080):0.190692);
END;
BEGIN PHYLONET;
InferNetwork_MPL (gt0-gtM) 3 -pl 1;
END;
```

MCMC GT Pseudo `MCMC_GT_Pseudo` is a Bayesian inference method for species networks that leverages pseudo-likelihood [11, 36]. Unlike `InferNetwork_ML` and `InferNetwork_MPL`, which maximize the likelihood and pseudo-likelihood, respectively, `MCMC_GT_Pseudo` employs a Markov chain Monte Carlo (MCMC) framework to sample from the posterior distribution of species networks.

The script in Listing 3 provides an example command for running the method in `PhyloNet`. This command specifies *M* gene trees, sets a maximum of three reticulations, and utilizes a single CPU core. The `-pseudo` flag indicates the use of pseudo-likelihood during sampling. Additional parameters for MCMC sampling are also specified: `-cl` sets the total chain length, `-bl` defines the number of burn-in iterations (i.e., discarded samples to ensure convergence), and `-sf` determines the sampling frequency. The output of the command provides all sampled species networks, along with summary statistics for all topologies in the 95% credible set, including the MAP and average branch length estimates.

Listing 3:Example NEXUS file for inferring species networks using `MCMC GT Pseudo`.

```
#NEXUS
BEGIN TREES;
Tree gt0 = (OUT:0.325872,(A:0.108396,(C:0.093204,
(D:0.093060, (E:0.047520,B
   :0.047520):0.045540):0.000144):0.015192):0.21747
6);
...
Tree gtM = (OUT:0.312048,((B:0.076896,A:0.076896):
0.044424,
```

```
(C:0.102276,(E:0.060192,D:0.060192):0.042084):0
.019080):0.190692);
END;
BEGIN PHYLONET;
MCMC_GT (gt0-gtM) -cl 1000000 -bl 100000 -sf 1000 -mr 3
-pl 1 -pseudo;
END;
```

2.3 Simulation Setup

To comprehensively evaluate the impact of various modes of error on species network inference, we designed an extensive experimental setup (Fig. 3) that enables the assessment of individual contributions from each potential error source in a progressively more realistic manner under a variety of parameter settings.

We used three model networks in our simulation study, each with a different number of reticulations (Fig. 4).

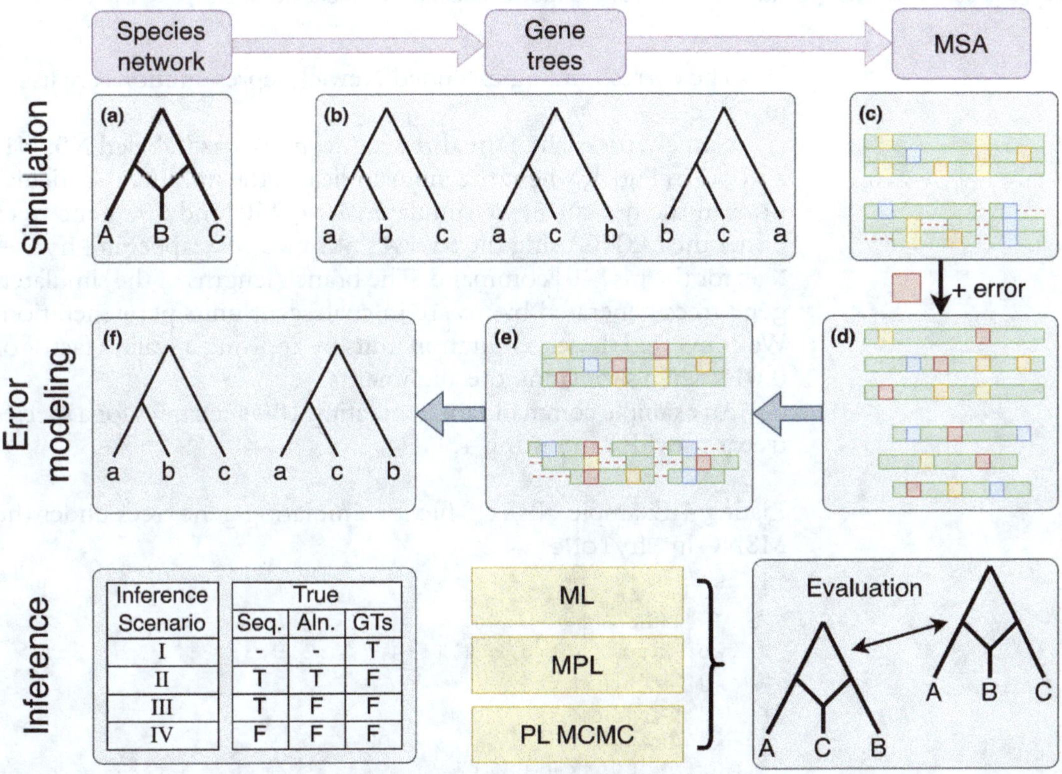

Fig. 3 Schematic representation of the experimental setup. **(a)** Model species networks are designed by hand. **(b)** Gene trees are simulated on the species networks under the MSNC model. **(c)** Alignments with and without indels are simulated for each gene tree using INDELible. **(d)** Sequence error is added to individual sequences in the alignment. **(e)** Sequences are re-aligned using MAFFT to obtain empirical MSAs. **(f)** Gene trees are inferred under ML with IQ-TREE. In total, four scenarios (I–IV) are analyzed: (I) inference from true gene trees, (II) inference from gene trees inferred from true alignments, (III) inference from gene trees inferred from inferred alignments of true sequences, and (IV) inference from gene trees inferred from alignments of sequences with error. For species network inference, we ran three methods: maximum likelihood (ML), maximum pseudo-likelihood (MPL), and Bayesian MCMC with pseudo-likelihood (PL MCMC)

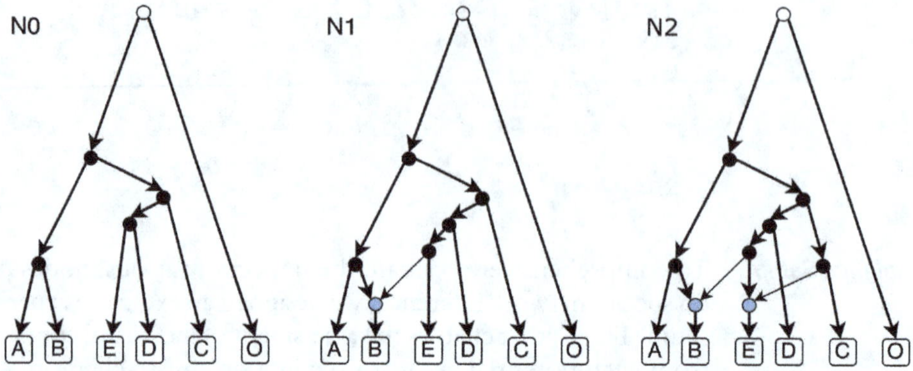

Fig. 4 Schematic representation of three species networks with varying numbers of reticulations. The three networks shown—N0, N1, and N2—on five ingroup taxa (A–E), one outgroup (O), and 0, 1, and 2 reticulation nodes, respectively. All networks have a total height of 16 coalescent units, with the branch from the root to the most recent common ancestor of the ingroup spanning 10 coalescent units. Reticulation nodes are shown in blue, and thinner arrows indicate reticulation edges with an inheritance probability of 0.3

The corresponding extended Newick representations are listed in Table 1.

Namely, for each of the three model networks (labeled N0, N1, and N2 in Fig. 4, where the index indicates the number of reticulations in the network), we simulated 100, 250, and 500 gene trees under the MSNC using the `SimGTInNetwork` wrapper in `PhyloNet` for the ms [40] command. The branch lengths of the simulated gene trees generated by ms are in coalescent units per generation. We converted these to mutation units by applying a scaling factor of 0.018 when simulating the alignments.

An example command for generating 100 gene trees for a species tree is provided in Listing 4.

Listing 4:Example NEXUS file for simulating gene trees under the MSNC in `PhyloNet`.

```
#NEXUS
BEGIN TREES;
Networknet=(((A:2,B:2)T1:4,((D:3,E:3)T2:2,C:5)T3:1)
T4:10,OUT:16);\\
END;
BEGIN PHYLONET;
SimGTInNetwork net 100 -ms /msdir/ms;
END;
```

For each simulated gene tree, we simulated an MSA under the GTR model using `INDELible` [41]. We considered three alignment lengths: 250, 500, and 1000. For all alignments, we used the following common GTR parameters: (a) substitution rates AC = 0.2173, AG = 0.9798, AT = 0.2575, CG = 0.1038, CT = 1.0, and GT = 0.207; (b) base frequencies A = 0.2112, C = 0.2888, G = 0.2896, and

Table 1
Extended Newick strings for the three model networks used in the simulations

Network	Extended Newick string
N0	(((A:2,B:2)T1:4,((D:3,E:3)T2:2,C:5)T3:1)T4:10,O:16);
N1	(((A:2,(B:0.9)X1#H1:1.1::0.7)T1:4,((D:3,E:2,X2#H1:1.1::0.3)T2:1)T3:2,C:5)T4:1)T5:10,O:16);
N2	(((A:2,(B:0.9)X1#H1:1.1::0.7)T1:4,((D:3, ((E:1)X3#H2:1::0.7,X2#H1:1.1::0.3)T2:1)T3:2,(C:3,X4#H2:2::0.3)T6:2)T4:1)T5:10,O:16);

T = 0.2104. For simulating indels, we used a power-law (POW) distribution with α=1.5and a maximum indel length of 5. We considered three indel rate values: 0, 0.05, and 0.1. Together, these parameters define our ground truth simulation data for gene trees and MSAs (Fig. 3b, c).

To use `INDELible`, one needs to create a `control.txt` file and then run `INDELible` using that file. Listing 5 shows an example `control.txt` configuration. The output is a single FASTA alignment of 1000 nucleotides, which may include gaps due to indels.

Listing 5:An example `control.txt` file for sequence generation in `INDELible`.

```
[TYPE] NUCLEOTIDE 1
[SETTINGS]
[randomseed] 2478
[output] FASTA
[insertaslowercase] FALSE
[MODEL] model1
[submodel] GTR 0.2173 0.9798 0.2575 0.1038 1.0 0.207
[statefreq] 0.2112 0.2888 0.2896 0.2104
[indelmodel] POW 1.5 5
[indelrate] 0.1
[TREE] tree1
(OUT:0.294516,((B:0.047880,A:0.047880):0.069264,(C:
0.090612,(E:0.072468,D
:0.072468):0.018144):0.026532):0.177372);
[PARTITIONS] part1
[tree1 model1 1000]
[EVOLVE] part1 1 address_of_saved_file
```

In order to mimic potential sequencing and/or assembly errors, we simulated sequence errors (Fig. 3d) in the simulated gene sequences by introducing random nucleotides with probabilities of 0 (no error), 0.01, and 0.1. Every erroneous nucleotide had the uniform probability of being any of the four nucleotides (i.e., once a nucleotide was selected for error introduction, it was replaced with a different nucleotide with probability 0.75).

We considered progressively more complex scenarios (Fig. 3I–IV), starting with inference from true gene trees and ending with inference from gene trees reconstructed from alignments of erroneous sequences. For MSA inference, we used MAFFT [42], specifically the iterative refinement algorithm for global multiple sequence alignment (`G-INS-i`, default). An example alignment command is shown in Listing 6.

Listing 6:MAFFT command used for sequence alignment.

```
mafft --globalpair --maxiterate 1000 --ep 1.2 not-
aligned.fasta > aligned.
fasta
```

To infer gene trees from true alignments (scenario II), inferred alignments (scenario III), and inferred alignments over erroneous sequences (scenario IV), we used `IQ-TREE` [43] with the ModelFinder Plus (MFP) option. Since IQ-TREEIQ-TREE generates unrooted trees, we subsequently rooted the inferred gene trees using the designated outgroup taxon. A sample command for running IQ-TREE with bootstrap is shown in Listing 7.

Listing 7:IQ-TREE command used for gene tree inference.

```
iqtree -s address_of_sequence -m MFP -st DNA -B 1000
```

2.4 Data and Code Availability

All simulated data produced for our analyses have been archived and are available on Zenodo [44]. All control files, extended Newick strings for the species networks, and scripts used to generate data, run analyses, and plot results are available on GitHub [45]. `PhyloNet` is open source and available on GitHub [46].

3 Results

This section presents our experimental findings, beginning with an analysis of the characteristics of the simulated data. We then examine the likelihood trajectories across different error scenarios. Subsequent subsections explore the distributions of MCMC samples, the topological accuracy of inferred networks, and the computational costs associated with the inference methods.

3.1 Characteristics of the Simulated Data

First, we summarized the characteristics of the inferred MSAs and gene trees in order to assess the levels of error introduced by standard inference procedures across varying indel rates in the true sequences and rates of erroneous nucleotides.

Multiple sequence alignment quality was quantified using the sum-of-pairs (SP) score, computed as the average (over pairwise alignments induced by the MSA) ratio between correctly aligned base pairs in the inferred alignment and the total number of aligned

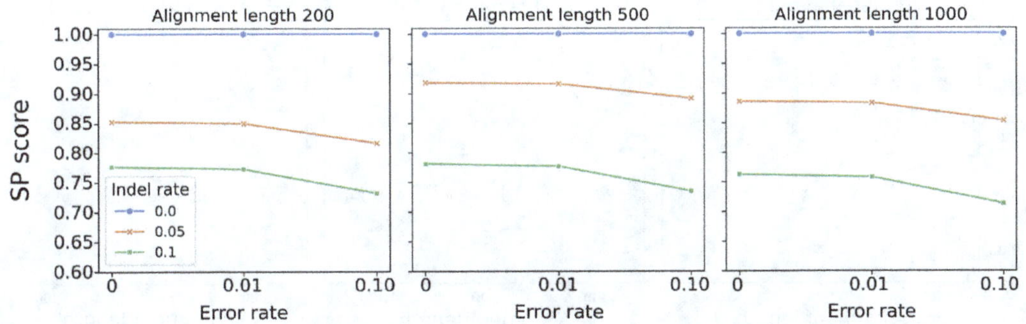

Fig. 5 Alignment quality for varying indel and error rates. Variability in the average SP scores (y-axis) of the alignments based on the indel rate (indicated by color), error rate (x-axis), and alignment length (subplots)

base pairs in the reference alignment. The results are shown in Fig. 5.

We observe that when the indel rate is set to 0, i.e., the simulated alignment has no indels, the average SP score of all alignments remains close to 1 regardless of the error rate. Furthermore, we note that the indel rate alone has a greater impact on alignment quality, resulting in a mean SP score drop of 0.114 from an indel rate of 0 to 0.05, and a drop of 0.227 from 0 to 0.1. In comparison, increasing the error rate from 0 to 0.1 has no impact on indel-free alignments, and results in an average SP score decrease of 0.032 for alignments with an indel rate of 0.05, and 0.047 for those with an indel rate of 0.1 (Fig. 5). Notably, there is little variation in alignment quality across independent replicates, with the mean standard deviation of SP scores between replicates being 0.0004, a maximum of 0.0012, and a minimum of 0.

Gene tree estimation error (GTEE) was quantified by measuring the average normalized Robinson–Foulds (nRF) distance [47] between ground-truth simulated gene trees and gene trees inferred from: error-free true alignments (scenario II), error-free MAFFT alignments (scenario III), and MAFFT alignments of sequences with erroneous nucleotides (scenario IV). The results are shown in Fig. 6.

We observe that in all cases, GTEE decreases as alignment length increases. In the case of true alignments (scenario II), indels have a mixed effect on gene tree inference accuracy, while in scenarios III and IV, the presence of indels consistently increases GTEE. Consistent with the alignment accuracy results, indels generally have a greater impact on GTEE than the error rate alone.

We calculated the *Kullback–Leibler (KL) divergence* between the empirical distributions of inferred and true gene tree topologies to assess the impact of data error on the inferred gene tree distribution. This measure is particularly relevant because species network

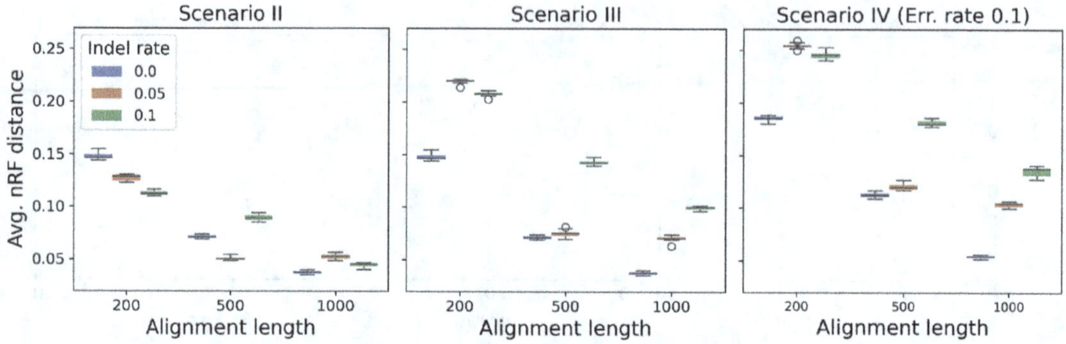

Fig. 6 Gene tree estimation error under different data error modes. Boxplots of normalized RF distances (y-axis) between ground-truth gene trees and inferred gene trees for the three scenarios: error-free true alignments (leftmost panel), MAFFT alignments of error-free sequences (middle panel), and MAFFT alignments of sequences with erroneous nucleotides at an error rate of 0.1 (rightmost panel). Alignment length is shown on the x-axis, and colors represent varying indel rates in the ground-truth alignment

inference under the MSNC utilizes the distribution of gene trees. Thus, distortions in this distribution—caused by alignment error, sequencing error, or indels in our study—can directly affect the accuracy of network inference. For each replicate, we enumerated the unique topologies across gene trees and computed their empirical frequencies. Let P denote the true distribution and Q the inferred one. The KL divergence is then computed as:

$$D_{\mathrm{KL}}(P \parallel Q) = \sum_{k} P(k) \log\left(\frac{P(k)}{Q(k)}\right),$$

where k ranges over all observed topologies. To handle cases where $Q(k)=0$, we applied Laplace smoothing by adding a small constant ($\varepsilon = 1e-5$) to all gene tree frequencies prior to calculating the divergence.

Figure 7 shows KL divergence values across increasing alignment lengths (x-axis) for three evolutionary models (rows) under three different scenarios (columns).

For model N0 (top row), KL divergence consistently decreases with longer alignments across all scenarios and indel rates. In scenario II, the KL divergence at indel rate 0.0 appears unexpectedly high, but this is due to averaging across different numbers of gene trees: shorter alignments with only 200 gene trees show higher divergence, skewing the average upward. Scenario III follows a similar downward trend with alignment length but exhibits slightly higher divergence values than scenario II, especially at higher indel rates. Scenario IV yields the highest KL divergence overall, with divergence increasing

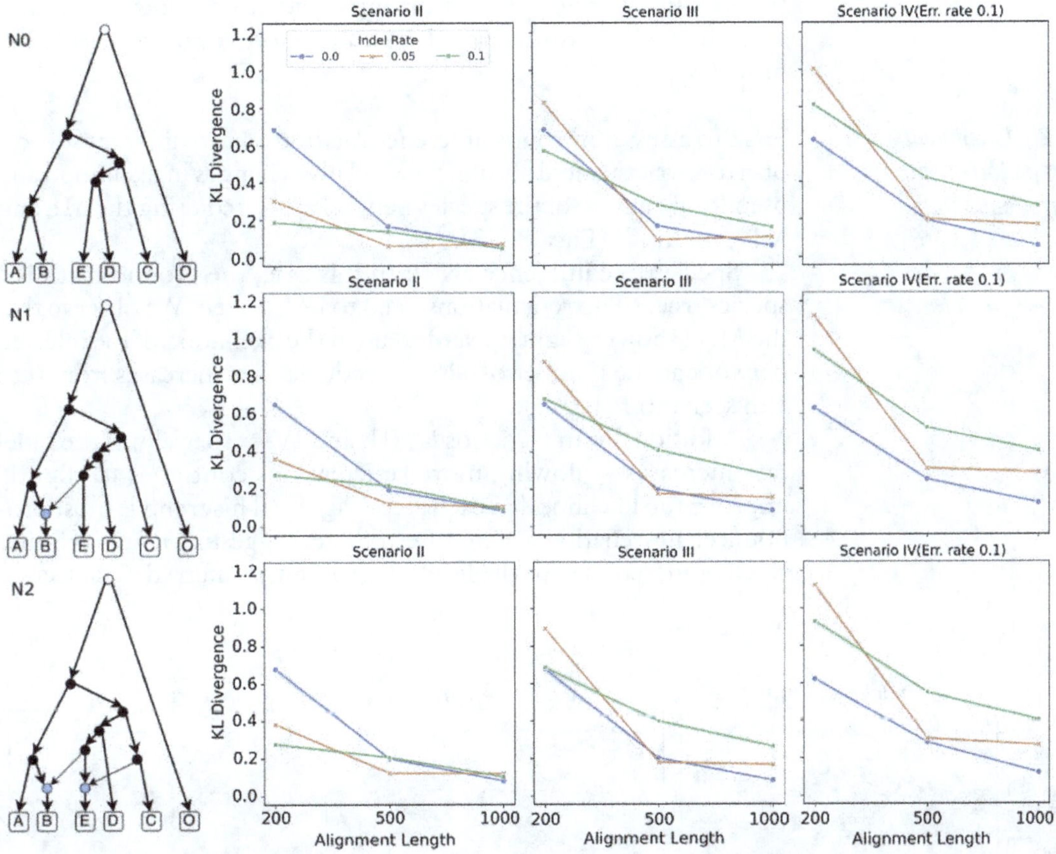

Fig. 7 Changes in the distributions of gene tree topologies under the data error scenarios. The KL divergence between inferred and true gene tree topology distributions as a function of alignment length (x-axis), across different models (rows) and scenarios (columns). Each line represents a different indel rate

clearly with indel rate—particularly for shorter alignments—highlighting the compounding effect of sequencing errors and indels.

Model N1 (middle row) shows patterns similar to N0, but with moderately higher KL divergence values overall, reflecting the greater complexity of the evolutionary history. Scenario IV again yields higher KL divergence than scenarios III and II, particularly at high indel rates (0.1). Notably, scenario III shows a more substantial improvement with increasing alignment length than scenario IV, indicating that sequencing errors reduce the benefits of longer alignments.

Model N2 (bottom row) exhibits the largest KL divergence values across all scenarios, with scenario IV showing the steepest divergence at high indel rates. While KL values decrease with alignment length in all cases, the curves corresponding to high indel rates (orange and green) remain clearly separated from those with no indels (blue), especially in scenario IV. These results underscore the sensitivity of gene tree distribution fidelity to both indel rates and sequencing errors, particularly under more complex reticulation histories.

3.2 Likelihood Trajectories with Erroneous Data

First, to assess how robust inference methods are to different sources of error, we evaluated absolute and relative changes in likelihood and pseudo-likelihood of the species networks inferred using the ML and MPL methods (Figs. 8–13).

Species tree inference (N0): In this case, the ground truth is a species tree, so no reticulations need to be inferred. We observe that the MLE shows a slight upward trend in the likelihood of the inferred network as the number of allowed reticulations increases from 0 to 1 in scenario I (Fig. 8).

Additionally, in scenarios II, III, and IV—especially as the indel rate increases—allowing more reticulations continues to slightly improve the likelihood at each step (Fig. 8). This trend is most pronounced in scenario IV with 0.1 error rate, suggesting that overfitting to error increases with the level of error in the inferred gene trees.

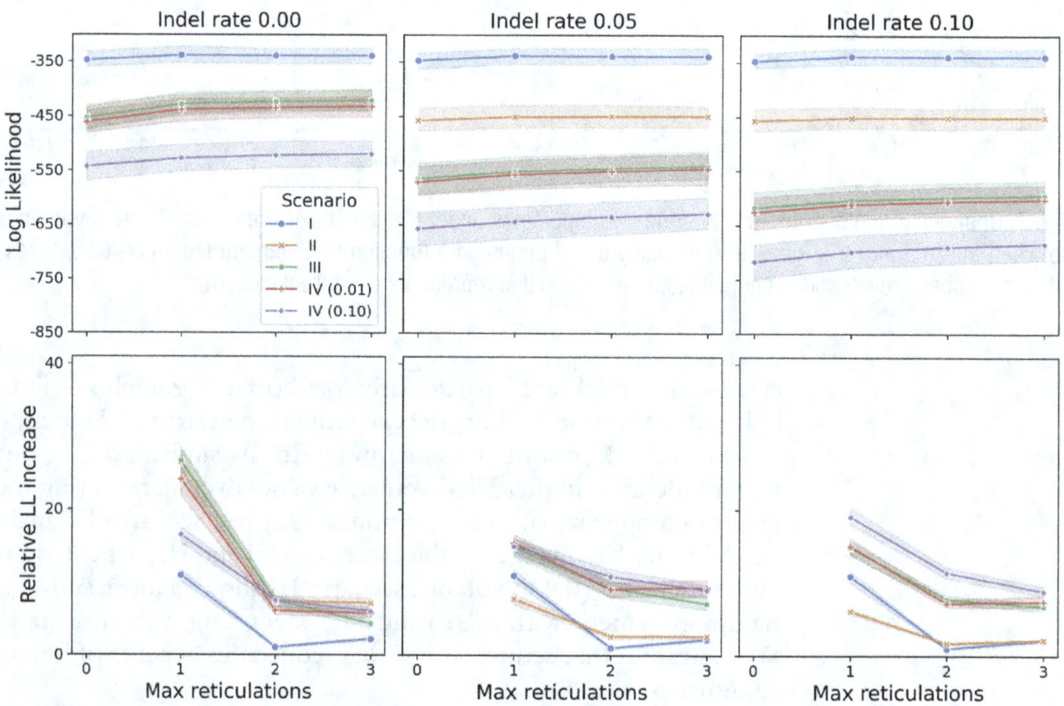

Fig. 8 Likelihood and likelihood change for N0 species network in the presence of data error. Top panels: Log-likelihood of the MLE species network (ground truth: N0) as the function of the number of maximum reticulations allowed for candidate networks (x-axis). Bottom panels: Relative change in log-likelihood when the number of allowed reticulations increases by one, up to the maximum shown on the x-axis. In all subplots, colors correspond to inference scenarios, and in the case of scenario IV (modeling sequence error), the error rate setting

Conversely, the MPLE shows no upward trend in scenario I (Fig. 9). Furthermore, in scenarios II, III, and IV, MPLE log pseudo-likelihood values remain remarkably stable.

Species network inference with 1 reticulation (N1): Here, the ground truth is a species network with one reticulation. The MLE exhibits a substantial increase in likelihood once the inference is allowed to include 1 reticulation (Fig. 10).

Further increases in the number of allowed reticulations yield only minor improvements in likelihood (Fig. 10). As with the N0 case, the noisiest setting (scenario IV with 0.1 error rate) leads to the highest relative increase in likelihood from 1 to 2 reticulations allowed per network (Fig. 10, bottom panels).

Similarly to N0, MPLE is more stable than MLE between 1 and 3 allowed maximum reticulations (Fig. 11). There is no notable upward trend in the pseudo-likelihood even in scenario IV with 0.1 error rate. In fact, the relative pseudo-likelihood gain beyond 1 reticulation remains negligible in cases with non-zero indel rate (Fig. 11, bottom panels).

Species network inference with 2 reticulations (N2): Here, the ground truth includes two reticulations. As with N1, we see a large

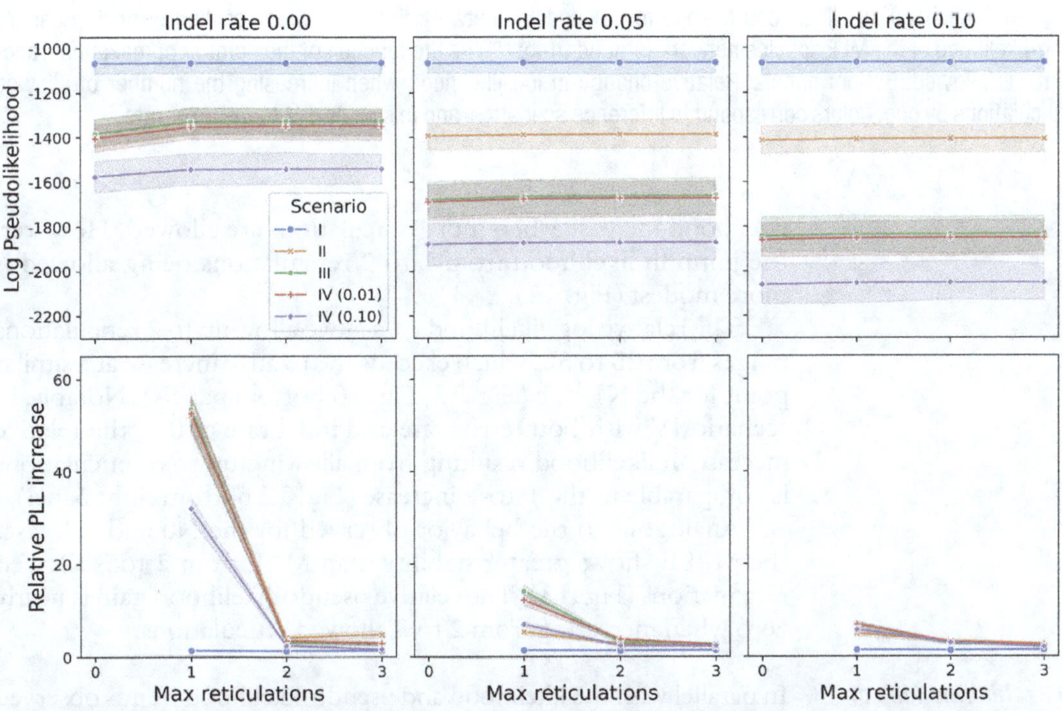

Fig. 9 Pseudo-likelihood and pseudo-likelihood change for N0 species network in the presence of data error. Top panels: Log pseudo-likelihood of the MPLE species network (ground truth: N0) as the function of the number of maximum reticulations allowed for candidate networks (x-axis). Bottom panels: Relative change in log pseudo-likelihood when the number of allowed reticulations is increased by one, up to the maximum shown on the x-axis. In all subplots, colors correspond to inference scenarios, and in scenario IV, the error rate setting

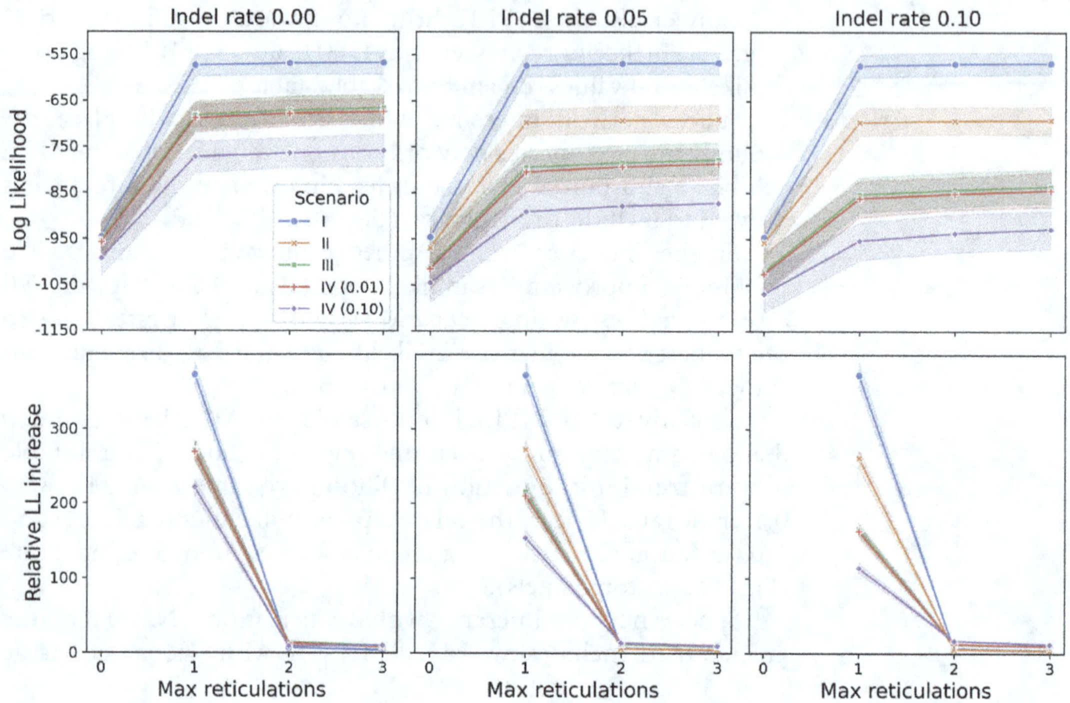

Fig. 10 Likelihood and likelihood change for N1 species network in the presence of data error. Top panels: Log-likelihood of the MLE species network (ground truth: N1) as the function of the number of maximum reticulations allowed. Bottom panels: Relative change in log-likelihood when increasing the number of allowed reticulations by one. Colors correspond to inference scenarios, and in scenario IV, to the error rate

likelihood increase when more reticulations are allowed. However, the jump in likelihood from 1 to 2 reticulations being allowed is more modest (Fig. 12).

The relative log-likelihood gain as we allow up to 2 reticulations ranges from 25 to 50, which exceeds the relative increase at a similar point for the N1 case (Fig. 12, Fig. 10 bottom panels). Notably, in scenario IV with both error rate and indel rate at 0.1, the relative increase in likelihood resulting from allowing up to 3 reticulations is comparable to the 1-to-2 increase (Fig. 12 bottom right panel).

Analogous to the behavior observed for the N0 and N1 case, the MPLE shows greater stability than MLE from 2 to 3 allowed reticulations (Fig. 13). The relative pseudo-likelihood gain is nearly zero when increasing from 2 to 3 allowed reticulations.

3.3 Distributions of MCMC Samples

In parallel with the likelihood and pseudo-likelihood trends observed in ML/MPL inference, we tracked the log pseudo-likelihood of the MAP estimates obtained from the MCMC runs (Figs. 14–16).

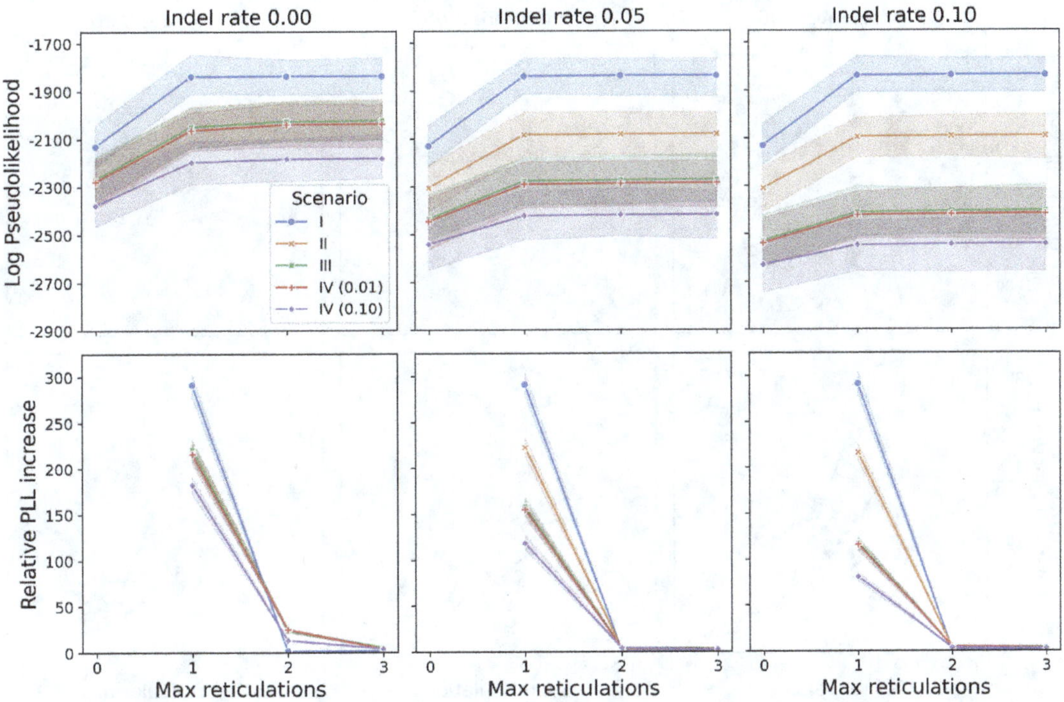

Fig. 11 Pseudo-likelihood and pseudo-likelihood change for N1 species network in the presence of data error. Top panels: Log pseudo-likelihood of the MPLE species network (ground truth: N1) versus maximum reticulations allowed (x-axis). Bottom panels: Relative change in log pseudo-likelihood when increasing the number of allowed reticulations by one. In all subplots, the colors correspond to inference scenarios; for scenario IV, the colors also indicate the associated error rate

For the N0 case, the pseudo-likelihood of the MAP estimates remained consistent as the number of allowed reticulations increased, with the only exception being a slight increase for data in scenarios II–IV without indels (Fig. 14).

In the N1 case, similar to the behavior observed with MPL inference, the pseudo-likelihood of the MAP estimate increased when one reticulation was allowed and remained relatively constant as the number of allowed reticulations increased further (Fig. 15).

For the N2 case, the pseudo-likelihood of the MAP estimates did not show a significant increase as the number of allowed reticulations went up from 1 to 2. This contrasts with the behavior observed in the ML and MPL inference results (Figs. 13 and 16).

3.4 Topological Accuracy of Inferred Networks

We also evaluated topological error (Fig. 17) and topological accuracy (Fig. 18) to evaluate method performance on erroneous data. Due to the high number of parameter combinations, we focused on inference scenarios (I–IV; Fig. 3) and inference method (ML, MPL,

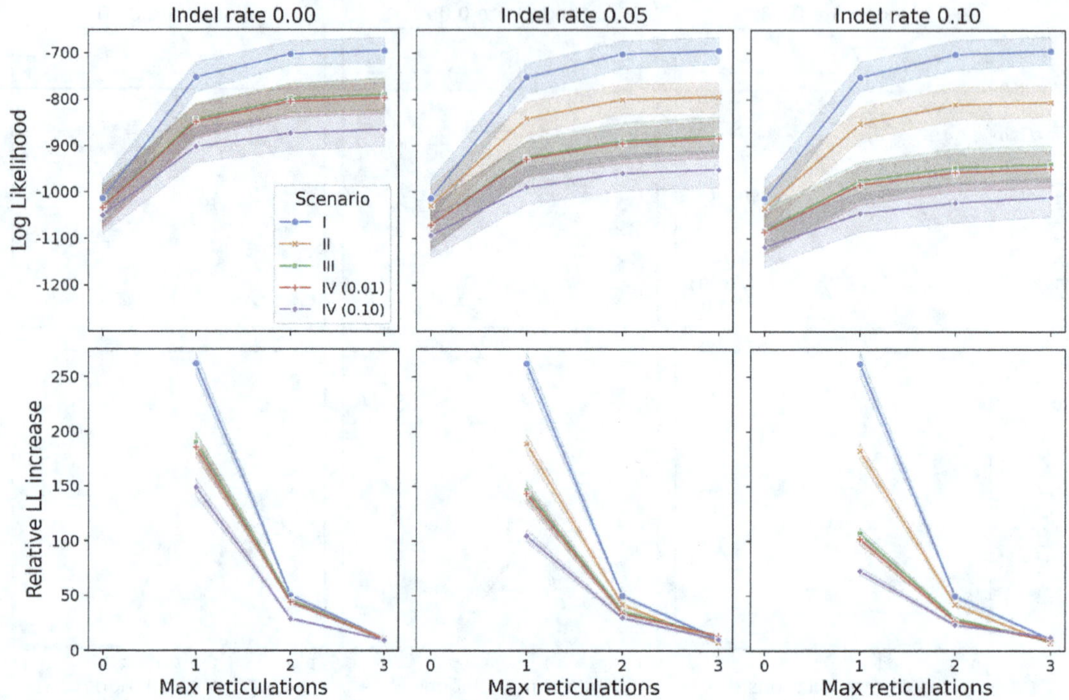

Fig. 12 Likelihood and likelihood change for N2 species network in the presence of data error. Top panels: Log-likelihood of the MLE species network (ground truth: N2) versus maximum reticulations allowed (x-axis). Bottom panels: Relative change in log-likelihood when increasing the number of allowed reticulations by one. In all subplots, the colors correspond to inference scenarios; for scenario IV, the colors also indicate the associated error rate

MCMC), while visualizing variation due to indel rate, alignment length, and number of gene trees as inter-quartile ranges. We used the Robinson–Foulds distance [47] when both ground-truth and inferred topologies were trees, and the metric from [48] when at least one of the topologies was a network.

We found that both ML and MPL methods achieved the lowest error when the maximum number of allowed reticulations matched the true number (Fig. 17). In contrast, MCMC MAP estimates showed similar error across 1–3 maximum reticulations zone. Notably, all methods struggled in accurately recovering the topology of the N2 network under data error (Fig. 17).

Topological accuracy results show that ML and MPL methods perform well for N0 and N1 when the maximum number of reticulations equals the ground truth value, even in the presence of data error (Fig. 18). On the other hand, in the N1 and N2, MCMC MAP estimates were less accurate than ML and MPL when the number of

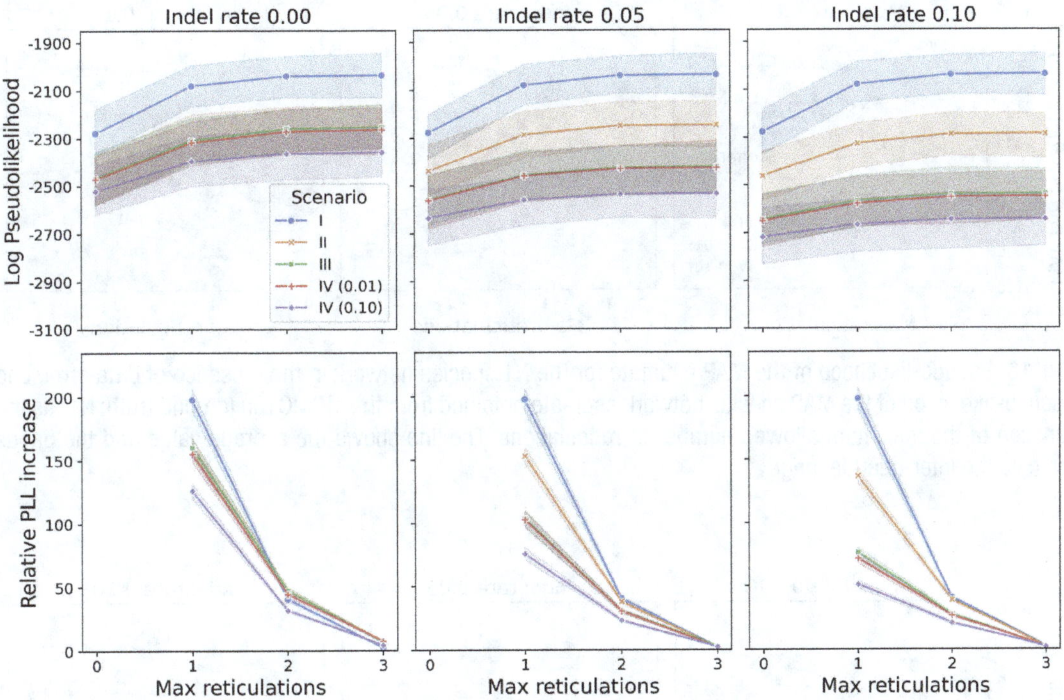

Fig. 13 Pseudo-likelihood and pseudo-likelihood change for N2 species network in the presence of data error. Top panels: Log pseudo-likelihood of the MPLE species network (ground truth: N2) as the function of maximum reticulations (x-axis). Bottom panels: Relative change in log pseudo-likelihood as allowed reticulations increase by one. Scenario and error rate encodings are consistent with earlier figures

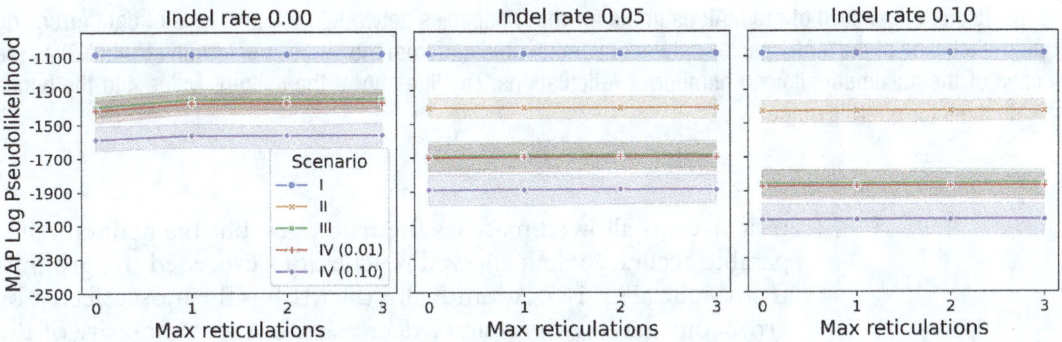

Fig. 14 Pseudo-likelihood of the MAP estimate for the N0 species network in the presence of data error. Log pseudo-likelihood of the MAP species network estimate obtained from the MCMC run (ground truth: N0) as the function of the maximum allowed number of reticulations. The line shows the average value and the bands indicate the inter-quartile range

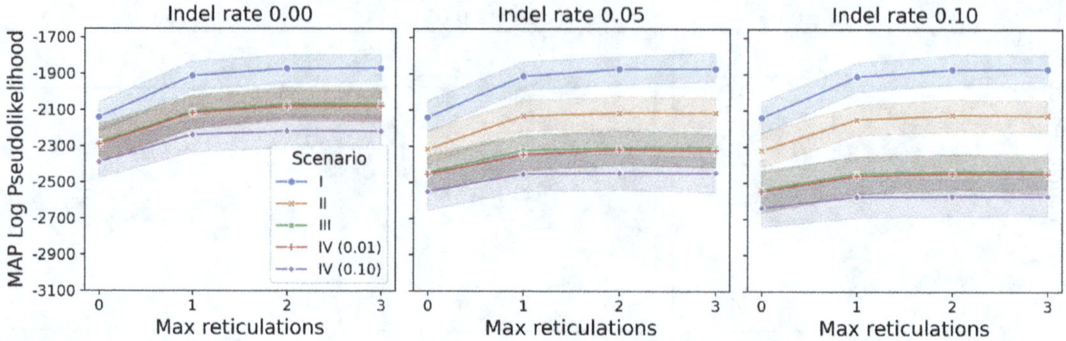

Fig. 15 Pseudo-likelihood of the MAP estimate for the N1 species network in the presence of data error. Log pseudo-likelihood of the MAP species network estimate obtained from the MCMC run (ground truth: N1) as the function of the maximum allowed number of reticulations. The line shows the average value and the bands indicate the inter-quartile range

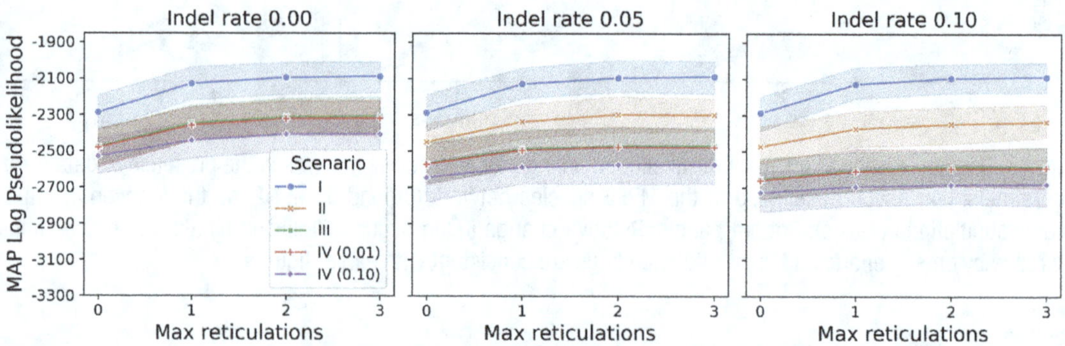

Fig. 16 Pseudo-likelihood of the MAP estimate for the N2 species network in the presence of data error. Log pseudo-likelihood of the MAP species network estimate obtained from the MCMC run (ground truth: N2) as the function of the maximum allowed number of reticulations. The line shows the average value and the bands indicate the inter-quartile range

reticulations allowed matches the true value but maintained comparable accuracy when allowed reticulations exceeded the ground-truth value (Fig. 18). Additionally, the MLE is the most affected by erroneous data, with accuracy decreasing as the complexity of the inference scenario increases (from I to IV (0.1)). Conversely, pseudo-likelihood-based methods (MPL and MCMC) are less sensitive to specific levels of error (Fig. 18), showing a notable change in accuracy primarily between scenario I (ground-truth gene trees) and the remaining scenarios (where gene trees are inferred).

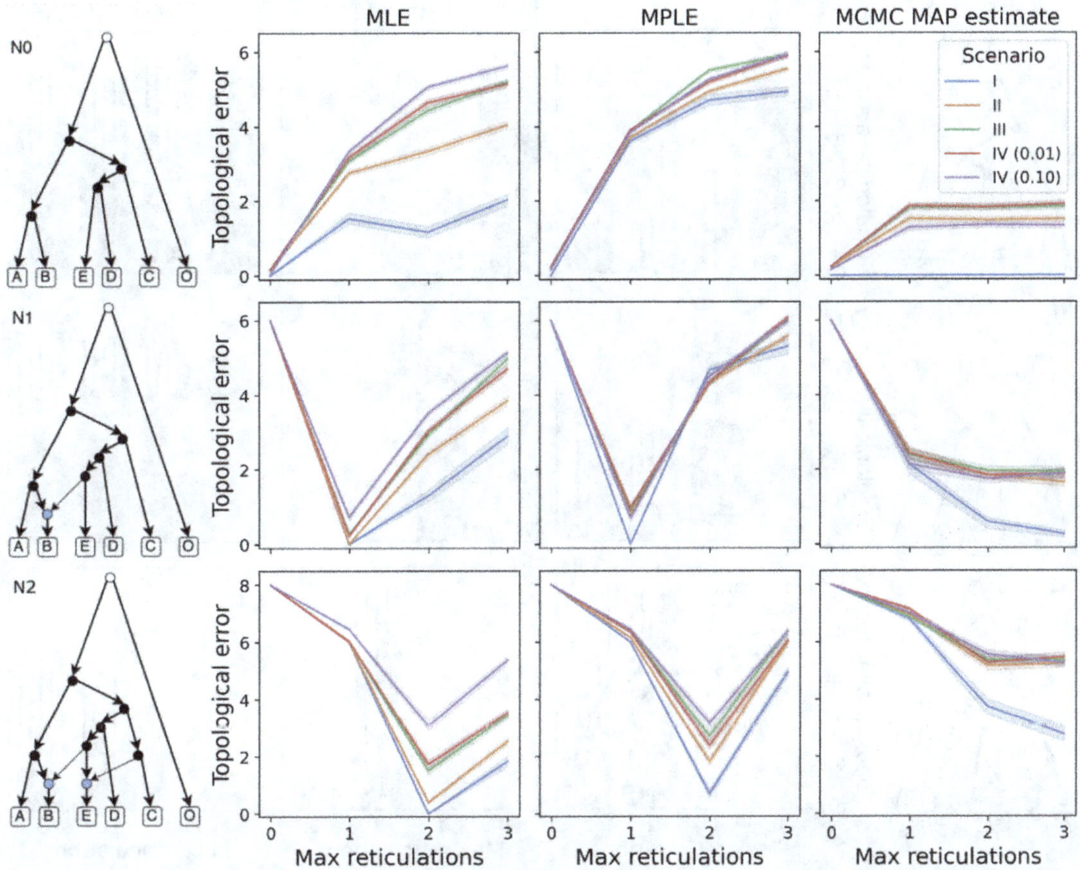

Fig. 17 Topological error across inference methods and ground-truth species networks in the presence of data error. Each row shows the mean topological error for each method (ML, MPL, MCMC) across varied indel rates, alignment lengths, number of input gene trees, and replicate runs. Colored bands indicate inter-quartile ranges. The first row corresponds to N0 (species tree), the second to N1 (network with 1 reticulation), and the third to N2 (network with 2 reticulations). The x-axis denotes the maximum allowed number of reticulations, and colors indicate inference scenarios I–IV, with scenario IV further subdivided by error rates (0.01 and 0.1)

3.5 Computational Costs of Inference

Error in the data has no impact on the computational cost of inference for methods based on heuristic search of the species network space. However, the three methods we evaluated differ substantially in runtime under their default settings, as shown in Table 2.

Among the methods, maximum-pseudo-likelihood inference is consistently the least resource-intensive method, with an average runtime of under 2 min. In contrast, Bayesian MCMC inference is the most computationally intensive, requiring nearly an order of magnitude more CPU time than MPL inference.

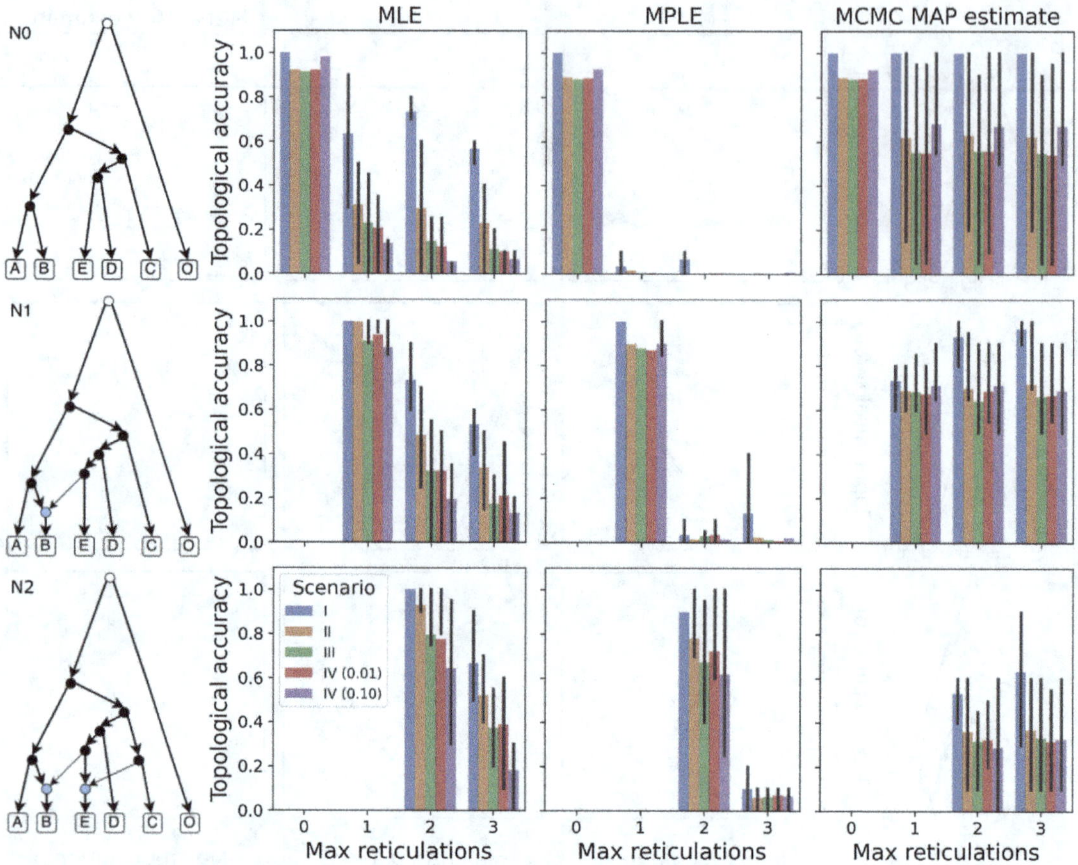

Fig. 18 Topological accuracy across inference methods and ground-truth species networks in the presence of data error. Each row shows the mean number of correctly inferred topologies among 10 replicate runs for ML, MPL, and MCMC methods. Error bars represent inter-quartile ranges, reflecting variance due to indel rates, alignment lengths, and number of input gene trees. Rows correspond to N0, N1, and N2 networks, respectively. The x-axis indicates the maximum number of reticulations allowed, and colors indicate inference scenarios I–IV, with scenario IV further subdivided by error rates (0.01 and 0.1). Results are omitted when the number of allowed reticulations is less than the true value

Table 2

CPU time required for running each of the evaluated inference methods on a single dataset

	CPU time (s)				
Method	**Mean**	**St. dev.**	**25%**	**50%**	**75%**
InferNetwork_ML	636.54	3169.78	77.15	145.69	371.96
InferNetwork_MPL	100.36	81.05	44.92	73.57	128.67
MCMC_GT_Pseudo	626.60	246.54	424.53	623.20	777.40

Columns represent mean runtime, standard deviation, and the 25th, 50th, and 75th percentiles. All times are in seconds.

4 Discussion

In this work we explore the impact of data error on species network inference from gene tree data under the MSNC. Specifically, we examined how indels and sequencing errors affect alignment and gene tree inference, and how these propagate to distort gene tree distributions.

We found that indels have a more substantial effect on alignment accuracy and, consequently, on gene tree estimation error (GTEE) and shifts in gene tree distributions. This effect is particularly pronounced for shorter sequences, where indels can increase the average normalized RF distance between ground-truth and inferred gene trees by up to 0.1. While longer sequences mitigated this effect slightly, the maximum indel length was not scaled with sequence length, and hence the impact of indels might be underrepresented for longer sequences which can harbor larger indel events.

We also note that both maximum likelihood and maximum pseudo-likelihood inference methods perform well, and are partially robust to errors when the maximum number of reticulations allowed for the method matches the true number of reticulations in the species network. Allowing more reticulations often led to drop in accuracy, especially in the presence of error, as the models overfit to noise by introducing additional reticulations during the heuristic search. Notably, Bayesian inference under pseudo-likelihood via MCMC shows stronger performance in the presence of error even when the number of allowed reticulations exceeds the true number of reticulations. This is likely due to the default value of the prior for number of reticulations which is a Poisson distribution with parameter 1.

Although none of the evaluated methods can automatically identify the correct number of reticulations, tracking the (pseudo-)likelihood increase as a function of allowed reticulations provided useful signal about reasonable cutoff point. While error leads to overfitting and continued likelihood improvement with additional reticulations, the relative increase consistently goes down. Pseudo-likelihood, in particular, showed steeper declines in gain compared to full likelihood, especially when the ground truth was a species tree, suggesting it is less prone to overfitting due to its simpler modeling assumptions.

Taken together, these findings suggest that current network inference methods are reasonably robust to moderate data error and offer sufficient information for appropriate model choice. Maximum pseudo-likelihood inference also offers the best computational efficiency, while showcasing some of the desired properties for distinguishing true reticulation events from overfitting to error, making it especially well-suited for large-scale genomic studies.

The most closely related prior study is that of Meijun et al. [31]. While their work focused on networks with a single reticulation and did not explore how data error interacts with inference complexity, our study systematically analyzed networks with up to two reticulations under a variety of error conditions—including alignment, sequencing, and gene tree estimation error. Unlike their approach, which didn't infer the number of reticulations, we evaluated the behavior of likelihood and pseudo-likelihood values as a function of increasing reticulation complexity. This allowed us to demonstrate how overfitting arises in the presence of error and to identify inflection points in model fit. We also incorporated KL divergence to quantify how error distorts gene tree topology distributions—an aspect not covered in their analysis. Finally, our use of `MCMC_GT_Pseudo`, a pseudo-likelihood-based method that yields posterior distributions over networks and benefits from prior regularization. Together, these contributions provide a more comprehensive perspective on species network inference under data error and offer practical guidelines for interpreting network inference in empirical settings.

These results offer actionable recommendations for biologists applying species network inference to empirical datasets. First, the pronounced impact of indels and sequencing errors—especially in shorter alignments—highlights the importance of high-quality sequence assembly and alignment in minimizing downstream inference errors. Longer alignments or filtering by alignment quality and gene tree support can reduce noise. Second, since current tools do not estimate the number of reticulations automatically, practitioners are advised to evaluate changes in the (pseudo-)likelihood trajectory, using diminishing gain as a heuristic to avoid overfitting. Our findings also reinforce the value of pseudo-likelihood-based methods—such as `InferNetwork_MPL` and `MCMC_GT_Pseudo`—for a good trade-off between computational efficiency and robustness to error, making them suitable for large-scale genomic analyses where data quality may vary.

Throughout our experiments, we used the single maximum likelihood estimate (MLE) gene tree per locus. To assess the reliability of these inferred gene trees and their agreement with the ground truth, bootstrap analysis is essential. For each replicate in the simulated dataset, we applied IQ-TREE ultrafast bootstrap method [49] with 1000 replicates to compute branch support values. Branches with bootstrap support below 0.7 were collapsed. From each collapsed gene tree, we extracted the set of bipartitions and compared them to those of the corresponding true gene tree. A collapsed gene tree was considered a refinement of the true gene tree if all of its bipartitions were present in the true tree. We then calculated the percentage of gene trees satisfying this refinement condition across replicates, reporting it as the *percentage of refinement*. The results are shown in Fig. 19.

Each column allows a direct comparison of the effects of alignment error and sequencing error on refinement accuracy. In scenario II, we observe the highest refinement percentages across all models, indicating that gene tree inference is most accurate when alignments

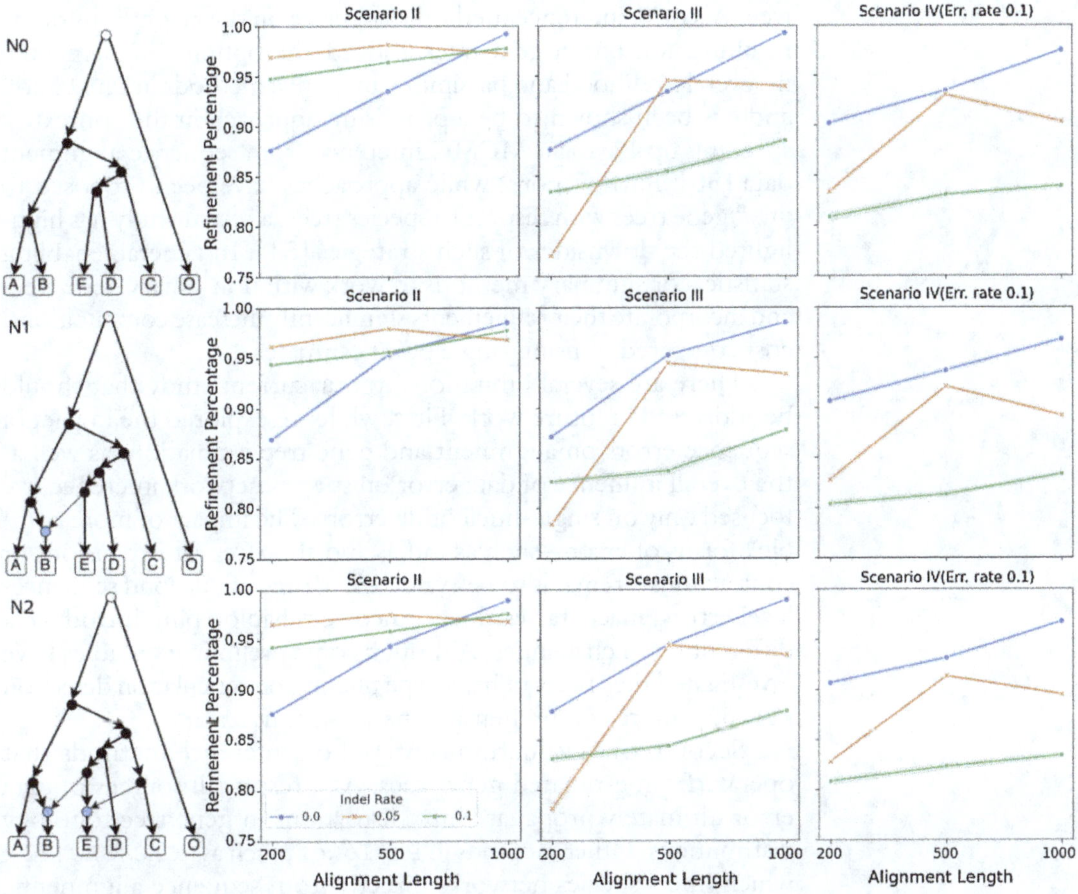

Fig. 19 Bootstrap refinement of inferred gene trees under data error. Bootstrap-based refinement accuracy across different scenarios (columns) and evolutionary models (rows). Refinement is defined as the proportion of collapsed inferred gene trees whose bipartitions are a subset of the corresponding true gene tree. Collapsing was performed using a 0.7 bootstrap support threshold

are error-free. In scenario III, refinement percentages decline slightly, reflecting the impact of alignment ambiguity. Scenario IV exhibits the lowest refinement percentages in all models. A closer comparison between scenario III and scenario IV further highlights the effect of sequencing error in the presence of varying indel rates. For moderate (0.05, orange) and high (0.1, green) indel rates, refinement percentages drop more substantially in scenario IV than in scenario III, particularly for shorter alignments. This pattern implies that sequencing error and indels interact synergistically to reduce inference accuracy, exacerbating the disruption caused by each individually. These effects are consistent across all model conditions (N0–N2).

These results indicate that one potential solution for dealing with gene tree error in practice is to first collapse poorly supported branches in the gene tree estimates, and then allow the species tree/network inference method to operate on the resolutions of the resulting non-binary gene trees. Indeed, this option is implemented in several likelihood and parsimony inference methods in PhyloNet and has been shown to be a promising approach in the context of speeding up Bayesian MCMC inference from sequence alignment data [50]. Furthermore, while approaches have been proposed to "fix" gene trees with respect to species trees, a recent study has highlighted the downsides of such strategies [51]. In general, enabling statistical or summary methods to work with non-binary gene trees and incorporate their refinements significantly increase computational cost compared to using single point estimates.

There are several limitations to our current study that should be addressed in future work. First, while we explored the impact of sequence errors on alignment and gene tree estimation, as well as the overall influence of data error on species network inference, we focused only on single-nucleotide errors. The impact of more complex forms of error—such as indels and short repeat expansions or contractions—remains to be evaluated. Moreover, beyond sequence-level errors, inaccuracies due to incorrect haplotyping introduce a different set of challenges. Although some preliminary studies have investigated the effects of haplotype phasing on reticulation detection [11, 52], more comprehensive analyses are necessary.

Second, our evaluation centered on inference methods that operate from estimated gene trees. As a result, all sources of data error ultimately propagate into distortions in gene tree topology distributions. Other methods in PhyloNet, such as MCMC_SEQ [12], which infers species networks directly from sequence alignments, and MLE_BiMarkers [15] and MCMC_BiMarkers [13], which infer from bi-allelic markers, work directly with sequence data, rather than gene tree estimates. We expect that these methods—especially those using bi-allelic markers—may exhibit increased sensitivity to data error, and further investigation is required to fully understand the implications for inference.

Third, our simulations followed the MSNC model and assumed a single sampled individual per species. A more extensive evaluation should examine how gene duplication and loss events, as well as sampling multiple individuals within species, influence the robustness and accuracy of network inference.

Acknowledgments

This work was in part supported by the NSF grants DMS/NIGMS-2153704 and DBI-2030604.

References

1. Roch S, Snir S (2013) Recovering the tree-like trend of evolution despite extensive lateral genetic transfer: a probabilistic analysis. J Comput Biol 20:93–112. https://doi.org/10.1089/cmb.2012.0234
2. Davidson R, Vachaspati P, Mirarab S, Warnow T (2015) Phylogenomic species tree estimation in the presence of incomplete lineage sorting and horizontal gene transfer. BMC Genom 16, S1. https://doi.org/10.1186/1471-2164-16-S10-S1
3. Solís-Lemus C, Ané C (2016) Inferring phylogenetic networks with maximum pseudolikelihood under incomplete lineage sorting. PLoS Genet 12:e1005896. https://doi.org/10.1371/journal.pgen.1005896
4. Zhu J, Yu Y, Nakhleh L (2016) In the light of deep coalescence: revisiting trees within networks. BMC Bioinform 17, 415. https://doi.org/10.1186/s12859-016-1269-1
5. Cao Z, Liu X, Ogilvie HA, Yan Z, Nakhleh L (2023) Practical aspects of phylogenetic network analysis using phyloNet. In: Kubatko LS, Knowles LL (eds), Species tree inference. Princeton University Press, 9 89–119, https://doi.org/10.1515/9780691245157.
6. Nakhleh L (2011) Evolutionary phylogenetic networks: models and issues. Springer US, Boston, MA, pp. 125–158
7. Degnan JH, Rosenberg NA (2009) Gene tree discordance, phylogenetic inference and the multispecies coalescent. Trends Ecol Evol 24:332–340. https://doi.org/10.1016/j.tree.2009.01.009
8. Yu Y, Degnan JH, Nakhleh L (2012) The probability of a gene tree topology within a phylogenetic network with applications to hybridization detection. PLoS Genet 8:e1002660
9. Yu Y, Dong J, Liu K, Nakhleh L (2014) Maximum likelihood inference of reticulate evolutionary histories. Proc Natl. Acad. Sci. USA 111(46):16448–16453. https://doi.org/10.1073/pnas.1407950111
10. Yu Y, Matthew Barnett R, Nakhleh L (2013) Parsimonious inference of hybridization in the presence of incomplete lineage sorting. Syst Biol 62:738–751. https://doi.org/10.1093/sysbio/syt037
11. Wen D, Yu Y, Nakhleh L (2016) Bayesian inference of reticulate phylogenies under the multispecies network coalescent. PLoS Genet 12:e1006006. https://doi.org/10.1371/journal.pgen.1006006
12. Wen D, Nakhleh L (2018) Co-estimating reticulate phylogenies and gene trees from multi-locus sequence data. Syst Biol 67:439–457. https://doi.org/10.1093/sysbio/syx085
13. Zhu J, Wen D, Yu Y, Meudt HM, Nakhleh L (2018) Bayesian inference of phylogenetic networks from bi-allelic genetic markers. PLoS Comput Biol 14:e1005932. https://doi.org/10.1371/journal.pcbi.1005932
14. Zhang C, Ogilvie HA, Drummond AJ, Stadler T (2018) Bayesian inference of species networks from multilocus sequence data. Mol Biol Evol 35:504–517. https://doi.org/10.1093/molbev/msx307
15. Zhu J, Nakhleh L (2018) Inference of species phylogenies from bi-allelic markers using pseudolikelihood. Bioinformatics 34:i376–i385. https://doi.org/10.1093/bioinformatics/bty295
16. Flouri T, Jiao X, Rannala B, Yang Z (2020) A Bayesian implementation of the multispecies coalescent model with introgression for phylogenomic analysis. Mol Biol Evol 37:1211–1223. https://doi.org/10.1093/molbev/msz296
17. Elworth RAL, Ogilvie, HA, Zhu J, Nakhleh L (2019) Advances in computational methods for phylogenetic networks in the presence of hybridization. In: Warnow T (ed) Bioinformatics and phylogenetics. Computational Biology, vol 29. Springer, Cham. https://doi.org/10.1007/978-3-030-10837-3
18. Mirarab S, Nakhleh L, Warnow T (2021) Multispecies coalescent: theory and applications in phylogenetics. Ann Rev Ecol Evol Syst 52:247–268. https://doi.org/10.1146/annurev-ecolsys-012121-095340
19. Ogilvie HA, Bouckaert RR, Drummond AJ (2017) Starbeast2 brings faster species tree inference and accurate estimates of substitution rates. Mol Biol Evol 34(8):2101–2114
20. Bryant D, Bouckaert R, Felsenstein J, Rosenberg NA, RoyChoudhury A (2012) Inferring species trees directly from biallelic genetic markers: bypassing gene trees in a full coalescent analysis. Mol Biol Evol 29(8):1917–1932
21. Rabier CE, Berry V, Glaszmann JC, Pardi F, Scornavacca C (2020) On the inference of complex phylogenetic networks by markov chain Monte-Carlo. bioRxiv, https://www.biorxiv.org/content/10.1101/2020.10.07.329425v1
22. Durand EY, Patterson N, Reich D, Slatkin M (2011) Testing for ancient admixture between closely related populations. Mol Biol Evol 28(8):2239–2252. https://doi.org/10.1093/molbev/msr048
23. Elworth RAL, Allen C, Benedict T, Dulworth P, Nakhleh L (2018) DGEN: A test statistic for detection of general introgression scenarios. In: 18th international workshop on algorithms in bioinformatics (WABI 2018) (Leibniz

International Proceedings in Informatics (LIPIcs), vol 113, Article 19, pp 19:1–19:13). Schloss Dagstuhl – Leibniz-Zentrum für Informatik. https://doi.org/10.4230/LIPIcs.WABI.2018.19
24. Pease JB, Hahn MW (2015) Detection and polarization of introgression in a five-taxon phylogeny. Syst Biol 64(4):651–662
25. Blischak PD, Chifman J, Wolfe AD, Kubatko LS (2018) HyDe: a Python package for genome-scale hybridization detection. Syst Biol 67:821–829. https://doi.org/10.1093/sysbio/syy023
26. Kubatko LS, Chifman J (2019) An invariants-based method for efficient identification of hybrid species from large-scale genomic data. BMC Evol Biol 19:112. https://doi.org/10.1186/s12862-019-1439-7
27. Yu Y, Nakhleh L (2015) A maximum pseudo-likelihood approach for phylogenetic networks. BMC Genomics 16(10):1–10
28. Cao Z, Li M, Ogilvie H, Nakhleh L (2024) The impact of model misspecification on phylogenetic network inference. Bull Soc Syst Biologists 3(1), https://doi.org/10.18061/bssb.v3i1.9553
29. Solís-Lemus C, Yang M, Ané C (2016) Inconsistency of species tree methods under gene flow. Syst Biol 65:843–851. https://doi.org/10.1093/sysbio/syw030
30. Huang J, Thawornwattana Y, Flouri T, Mallet J, Yang Z (2022) Inference of gene flow between species under misspecified models. Mol Biol Evol 39(12):msac237
31. Gao M, Wang W, Liu KJ (2022) The impact of gene sequence alignment and gene tree estimation error on summary-based species network estimation. In: Proceedings of the 13th ACM international conference on bioinformatics, computational biology and health informatics, pp 1–17. Association for Computing Machinery, New York, NY, USA. https://doi.org/10.1145/3535508.3545559
32. Huson DH, Bryant D (2006) Application of phylogenetic networks in evolutionary studies. Mol Biol Evol 23(2):254–267
33. Yu Y, Than C, Degnan JH, Nakhleh L (2011) Coalescent histories on phylogenetic networks and detection of hybridization despite incomplete lineage sorting. Syst Biol 60(2):138–149
34. Felsenstein J (1981) Evolutionary trees from DNA sequences: a maximum likelihood approach. J Mol Evol 17(6):368–376
35. Zhu J, Liu X, Ogilvie HA, Nakhleh LK (2019) A divide-and-conquer method for scalable phylogenetic network inference from multilocus data. Bioinformatics 35(14):i370–i378
36. Sapoval N, Liu Z, Tamiji M, Li M, Nakhleh L (2025) Theoretical and empirical performance of pseudo-likelihood-based bayesian inference of species trees under the multispecies coalescent. bioRxiv, https://www.biorxiv.org/content/10.1101/2025.01.28.635282v1
37. Than C, Ruths D, Nakhleh L (2008) Phylonet: a software package for analyzing and reconstructing reticulate evolutionary relationships. BMC Bioinf 9, 322. https://doi.org/10.1186/1471-2105-9-322
38. Wen D, Yu Y, Zhu J, Nakhleh L (2018) Inferring phylogenetic networks using PhyloNet. Syst Biol 67:735–740. https://doi.org/10.1093/sysbio/syy015
39. Than C, Nakhleh L (2009) Species tree inference by minimizing deep coalescences. PLoS Comput Biol 5(9):e1000501
40. Hudson RR (2002) Generating samples under a Wright-Fisher neutral model of genetic variation. Bioinformatics 18:337–338. https://doi.org/10.1093/bioinformatics/18.2.337
41. Fletcher W, Yang Z (2009) Indelible: a flexible simulator of biological sequence evolution. Mol Biol Evol 26(8):1879–1888
42. Katoh K, Standley DM (2013) Mafft multiple sequence alignment software version 7: improvements in performance and usability. Mol Biol Evol 30(4):772–780
43. Minh BQ, Schmidt HA, Chernomor O, Schrempf D, Woodhams MD, Von Haeseler A, Lanfear R (2020) IQ-TREE 2: new models and efficient methods for phylogenetic inference in the genomic era. Mol Biol Evol 37(5):1530–1534
44. Tamiji M, Sapoval N, Nakhleh L (2025) Data from "Impact of data error on phylogenetic network inference from gene trees under the multispecies network coalescent". https://doi.org/10.5281/zenodo.15345227
45. Tamiji M, Sapoval N, Nakhleh L (2025) Scripts for "Impact of data error on phylogenetic network inference from gene trees under the multispecies network coalescent". https://github.com/nakhlehlab/data-error-species-nets/
46. PhyloNet (2025) https://github.com/NakhlehLab/PhyloNet/
47. Robinson DF, Foulds LR (1981) Comparison of phylogenetic trees. Math Biosci 53:131–147. https://doi.org/10.1016/0025-5564(81)90043-2
48. Nakhleh Luay (2009) A metric on the space of reduced phylogenetic networks. IEEE/ACM Trans Comput Biol Bioinform 7:218–222. https://doi.org/10.1109/TCBB.2009.2
49. Hoang DT, Chernomor O, von Haeseler A, Minh BQ, Vinh LS (2017) UFBoot2: improving the ultrafast bootstrap approximation. Mol Biol Evol 35(2):518–522. https://doi.org/10.1093/molbev/msx281
50. Wang Y, Ogilvie HA, Nakhleh L (2020) Practical speedup of bayesian inference of species

phylogenies by restricting the space of gene trees. Mol Biol Evol 37(6):1809–1818
51. Yan Z, Ogilvie HA, Nakhleh L (2023) "correcting" gene trees to be more like species trees frequently increases topological error. Genome Biol Evol 15(6):evad094
52. Huang J, Bennett J, Flouri T, Leaché AD, Yang Z (2022) Phase resolution of heterozygous sites in diploid genomes is important to phylogenomic analysis under the multispecies coalescent model. Syst Biol 71(2): 334–352

Check for updates

Chapter 5

PhyloSystemX: Enhancing the Analysis of Interaction Networks

Hugo Bonnefous, Yu Zhang, Philippe Lopez, François-Joseph Lapointe and Eric Bapteste

Abstract

PhyloSystemX is a computational tool designed to reconstruct ancestral biological interaction networks by integrating phylogenetic species trees with modern interaction data. While systems biologists commonly use networks to represent complex biological interactions, contemporary network analysis alone cannot reveal the evolutionary history that shaped these systems. PhyloSystemX addresses this gap by implementing parsimony-based algorithms (Dollo and Sankoff adjusted algorithm) to infer both ancestral vertices (homology groups) and their interactions across evolutionary time. Unlike tools limited to specific interaction types, PhyloSystemX can analyze any network whose vertices can be labeled with homology identifiers, including protein–protein interactions, gene co-expression networks, and metabolic regulation systems. By inferring ancestral networks, researchers can uncover evolutionarily conserved functional modules, identify lineage-specific innovations, and explore how biological systems adapt over time. The current implementation assumes vertical inheritance of traits, with limitations regarding introgression, hybridization, and horizontal gene transfer that future versions may address.

Keywords Evolutionary biology, Phylogenetics, Systems biology, Interaction networks, Ancestral network reconstruction, Network evolution

1 Introduction

1.1 Summary

With the advancement of multi-omics technologies, an increasing amount of data is being leveraged for biological analyses. The large amount of mined data notably allows for biological interactions to be explored at an unprecedented scale [1]. Across all levels of biological organization, from molecules to ecosystems, interactions that shape biological structures and processes at a given moment can be investigated [2]. Importantly, some of these interactions drive evolutionary dynamics [3], and/or are impacted by evolutionary changes. Therefore, studying interactions and evolution together is a next essential step for a comprehensive understanding of biological systems.

Gustavo Caetano-Anollés (ed.), *Evolutionary Genomics*, Methods in Molecular Biology, vol. 2981,
https://doi.org/10.1007/978-1-0716-4836-0_5,

Systems biologists commonly use networks to represent complex biological interactions [4]. A network consists of vertices (or nodes), which typically represent biological entities such as genes, proteins, metabolites, or species, and of edges, which can define functional, physical, or regulatory interactions between vertices. While a contemporary network in extant species can give insights into the contemporary functional relationship associated with that species, the analysis of a contemporary network cannot show the whole evolutionary history that led to that result. To better understand how biological networks evolve, a tool that can reconstruct ancestral interaction networks with some level of confidence is necessary [5].

Precisely, PhyloSystemX is a computational tool designed to reconstruct ancestral interaction networks—such as protein–protein interaction, genetic interaction, metabolic regulation, gene co-expression, or other networks evolved within vertically diverging lineages—by integrating a phylogenetic species tree with modern interaction data. The approach used by PhyloSystemX is based on the hypothesis of vertical inheritance and modification of species, which assumes that species inherit their traits from their ancestors with gradual changes over time. Factors such as introgression, hybridization, and horizontal gene transfer [6] between different species populations are currently not explicitly modeled in PhyloSystemX.

Bearing this important limit in mind, this chapter provides a step-by-step guide to using PhyloSystemX, emphasizing its ability to trace evolutionary histories of interactions through different algorithms when a reference phylogeny appears to be available, representing the evolutionary relationships between the hosts of these interaction networks. Unlike tools limited to specific interaction types, PhyloSystemX is broadly applicable to any network whose vertices can be labeled with homology labels (e.g., gene family names in a gene co-expression network, a protein family names in a protein protein interaction network). By inferring ancestral networks, researchers can uncover evolutionary conserved functional network modules and how these modules got established in vivo, identify lineage-specific evolutionary innovations and explore how biological systems adapt over time.

1.2 Overview of the Different Steps to Use PhyloSystemX

PhyloSystemX is a Python program that operates through a streamlined workflow encompassing several key steps.

The process begins with input preparation, followed by ancestral vertices inference, which implements Dollo parsimony to infer the presence/absence of members of a given homologous family along a species tree. The next step, ancestral interaction inference, can be performed by applying three distinct algorithms to reconstruct ancestral interactions along this species tree. Finally, the workflow concludes by producing labeled ancestral networks for each internal node of the reference phylogenetic tree as output.

1.3 Theoretical Basis

1.3.1 Mechanisms of Network Evolution

To understand in broad, general terms, how biological networks evolve, we must characterize the fundamental principles of network evolution. Biological networks evolve through alterations in their topological structure, a process termed "network rewiring," primarily driven by changes in vertices and their interactions (edges) as well as by gain and loss of nodes and edges [7, 8] (see Fig. 1).

Vertex changes involve the gain or loss of vertices, directly impacting network structure. Node loss in a molecular network can typically arise from gene loss, pseudogenization, or other forms of gene inactivation. Conversely, the addition of new nodes in a molecular network can result from gene duplication, horizontal gene transfer for a novel host species or activation of a given gene in a new regulatory context. Likewise, edges can be lost, for instance in case of functional redundancy, leading to the removal of some of the redundant associated interactions, or due to changes in the regulation of gene co-expression. Edges can also be gained when novel functional associations emerge and are selected, introducing new interactions in a host lineage. These changes can significantly reshape the network, affecting its robustness and adaptability [3].

To summarize, biological networks can evolve through the gain and loss of connections between nodes due to mutations, shifts in regulatory mechanisms, or changes in environmental pressures. Such network rewiring plays a crucial role in refining functional pathways, optimizing biological efficiency, and enabling adaptation to new ecological niches [9].

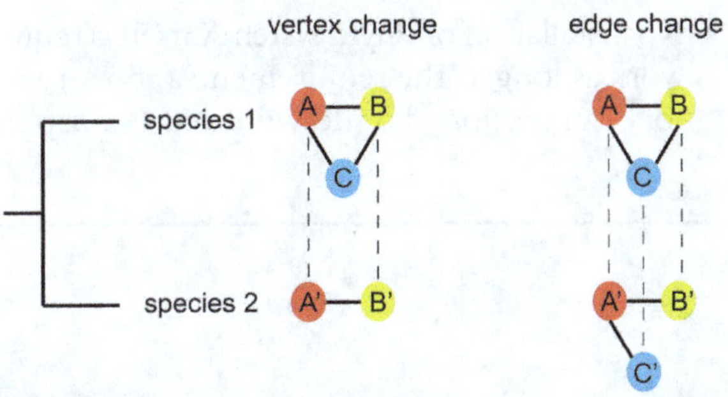

Fig. 1 Some mechanisms of network evolution. Continuous lines represent an interaction and dashed lines represent homology relationships. Vertex change is a gain or loss of at least one binding partner. Taking species 1 as the reference, the vertex change shown in species 2 is a vertex loss of C'. Swapping the reference species results in a gain of node C in species 1. Interaction rewiring is the gain or loss of an interaction but retention of both binding partners. Taking species 1 as the reference, the interaction reorganization event shown in species 2 is a loss of interaction between B' and C'. Swapping the reference species results in a gain of interaction between B and C in species 1

1.3.2 Methods for Inferring the Ancestral Networks

The reconstruction of ancestral networks is based on evolutionary inference concepts. In our approach, we mainly use parsimony-based methods to infer ancestral states. Parsimony is a well-established method in phylogenetics that assumes the simplest explanation—one that requires the fewest evolutionary changes—is the most likely. Maximum likelihood is also a powerful statistical framework for ancestral state inference and has been used in some specific network inference [10], but we do not adopt it in our current approach because well-defined evolutionary models for interaction networks evolution are still insufficiently developed. Maximum likelihood methods require a probability model that describes how different kinds of interactions evolve. While such models exist for sequence evolution (e.g., nucleotide or amino acid substitution models), there are unfortunately no universally accepted models for the evolution of protein–protein interaction, regulatory networks, or other complex interaction networks yet. Given these challenges, parsimony, although a strong assumption, offers a more practical and interpretable approach for inferring ancestral networks at least in first approximation and benchmark of the method confirms the efficiency of this approach.

2 Materials and Installation

2.1 System Requirement

PhyloSystemX requires Python 3.x and the following packages: os, math, argparse, dendropy, and numpy. These can be installed using pip install dendropy numpy argparse. Recommended hardware includes at least 4 GB of RAM (8 GB for larger datasets) and 10 GB of available disk space.

2.2 Installation

No installation of PhyloSystemX itself is required, the scripts should work as long as the requirements (above) are installed. The repository can be cloned or downloaded as a .zip file from GitHub [11].

3 Methods

3.1 Preparing Your Data for Input

3.1.1 Formatting the Dataset: Network Selection (Including Filtering)

A set of networks (e.g., protein–protein interaction network from difference Eukaryotes species), having evolved within vertically diverging lineages of interest, and coming from at least 2 species, are required infiles for a PhyloSystemX analysis. Details concerning the selection of input networks are provided in **Note 1**.

Networks to be fed to the tools are presented as edge files. Networks can be directed or non-directed. Edge weights are not taken into account by PhylosystemX (i.e., an edge is either present or absent: it is up to the user to impose minimum threshold values to consider an edge as present in a network).

The quality of networks is critical for the quality of the output (a network inference of homologous nodes/edges in ancestral species based on a list of edges in contemporary networks). Therefore, ideally, edges in extant networks should represent interactions that have been experimentally validated, e.g., a physical interaction between two proteins, rather than edges inferred to be present in a given species by homology. Typically, because experimentally demonstrated edges are present in species 1, it should not be a priori assumed that homologs of these edges are also by default present in its sister-species, species 2. If the networks upon which the inference of ancestral network is grounded were themselves to some extent obtained as a result of prior homology inference, using such a network would introduce a risk of circularity in the PhyloSystemX analysis. Typically, it might lead to overestimate the distribution of edges across extant and ancestral species.

For this reason, a first step of input preparation consists in cleaning extant networks from edges that are only based on homology-inference, when possible. For instance, for PPI network analyses, the STRING database [12] contains species-specific networks annotated with various evidence channels used to compute a confidence score for each PPI contains species networks in various formats, especially the format "proteins.links.full.v12.0.txt.gz" that provides various evidence channels associated with scores for each PPI edge, distinguishing between direct experimental evidence and interologs (i.e., presence/absence of edge inferred by homology). The file "proteins.links.full.v12.0.txt.gz" from STRING can be used to select edges only based on direct experimental evidence by filtering the edge list based on preferred channels of evidence (restricting networks to the "experimental data," "specialized databases," and "coexpression data" channels), or by recomputing new edge scores only based on relevant evidence channels as indicated in the FAQ 'How are the scores computed?" of the help of the string-db.org website [13]. In the particular case, relevant for PPI network analysis, based on the STRING database recommendations, a minimal edge score >0.7 corresponds to high confidence edges.

3.1.2 Mapping Homology Labels on Network Vertices

The second step to prepare a PhyloSystemX analysis requires to make networks from different species comparable with one another by introducing some common vertex labels on these networks, for instance homology labels. More precisely, vertices representing related entities, e.g., homologous proteins, must receive a common identifier: their homology group number, that will be the same across the networks presenting members of that homology family.

For instance, to analyze PPI networks, this mapping can be done as follows, using the OMA database [14] in addition to the STRING database. Specifically, the homology groups called ("rootHOGs") from the orthoXML file in the OMA database can provide useful homology labels, over large phylogenetic scales, that can be associated with individual protein sequences in various STRING PPI networks of interest. For this, a mapping between protein identifiers in OMA and in STRING databases must be performed. This mapping can be achieved by a simple pairwise sequence comparison. In practice, a protein stored in OMA and a protein stored in STRING can be considered as identical if they belong to the same species and their sequence comparison by a BLAST search [15] or by a Diamond search [16] returns an E-value <1e−5, a percentage identity >99% and a mutual coverage >99%. STRING proteins can then be renamed in the networks with the identifier of their "rootHOG" from OMA. When multiple proteins in STRING belong to the same "rootHOG," these proteins simply receive the same homology label, which in effect collapses these nodes into a single position in a homology-labeled PPI network to be used as infiles for a PhyloSystemX analyze. However, when one given protein associates with multiple "rootHOG" OMA group, a user decision is required to further proceed with the analysis of the nodes and edges involving this protein. A similar mapping can be conducted at a much finer-grained homology level, by mapping orthology labels (rather than homology labels) onto PPI networks, *see* **Note 2**. Again, the relevant information regarding orthology groups can be found in the orthoXML file from the OMA database. Other databases than OMA can likewise be use to extract homology or orthology labels, or homology groups can be computed by the user using various approaches based on sequence similarity networks [17].

This important step translates PPI networks into homology (or orthology) networks, described as an edge list between homology groups, returned as : "Homologygroup1;Homologygroup2" that can be used as direct infiles for the phylosystemX tool.

3.1.3 PhyloSystemX Input File Format Requirements

Before running PhyloSystemX, users must ensure that the following three input files are properly formatted and complete: the species tree file, the species-to-network mapping file, and individual species network files. All input data should be organized under a dedicated folder, recommended as PhyloSystemX on the desktop, with the following directory structure:

```
PhyloSystemX/
├── species_tree.newick   # species tree
├── table_leafname_filename.csv  # the species-to-network mapping file
├── PhyloSystemX.py      # Executable file
└── Network/       # Subdirectory for species network files
   ├── N1_network.txt
   └── N2_network.txt
```

1. **Species Tree File**

The species tree provides the program with the phylogenetic relationships of the species under study. A rooted tree in Newick format is required, as unrooted trees are not suitable for inferring ancestral groups. There is no limit on tree size, but it is recommended to include all species of interest, keeping in mind that very closely related species may introduce redundant information. The tree can contain polytomies (nodes with more than two descendants), and such structures are fully compatible with PhyloSystemX. Only the phylogenetic topology is needed; branch lengths, bootstrap values, are not required. Selecting an appropriate reference tree is critical, as it directly influences the accuracy of ancestral group inference. As shown, the Newick tree is presented in text format, along with a visual representation of its topology (see Fig. 2):

```
(N1,(N2,(N3,(N4,(N5,(N6,(N7,(N8,(N9,(N10,(N11,(N12,
(N13,(N14,(N15,N16)))))))))))))));
```

2. **Species-to-Network Mapping File**

This file links each species name in the phylogenetic tree to the corresponding network file location hosted by this species. It must be provided in CSV format with semicolon separators, with one species per line. Each line should follow the following structure:

```
species_name;absolute_path_to_network_file
```

Here, **species_name** must exactly match a leaf name in the phylogenetic tree to ensure dataset consistency. For example:

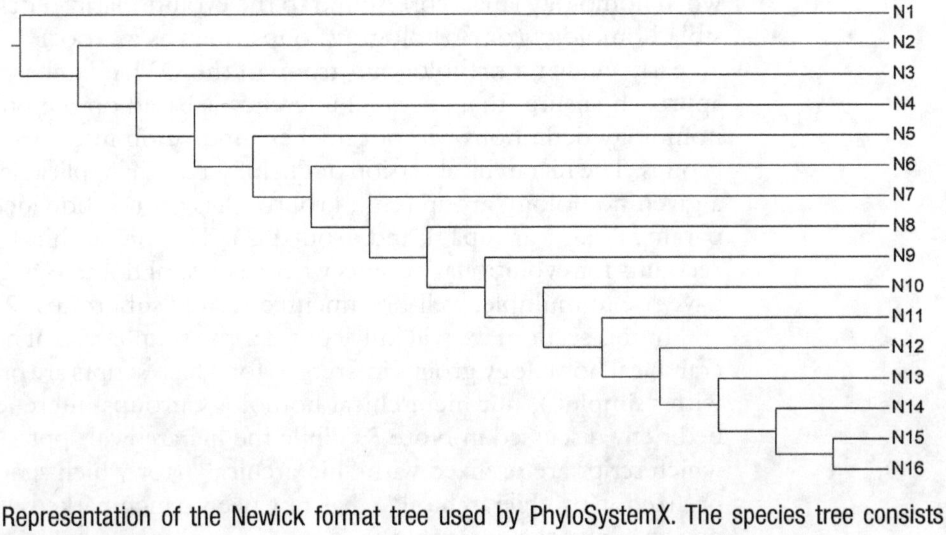

Fig. 2 Representation of the Newick format tree used by PhyloSystemX. The species tree consists of 16 species, labeled N1–N16

```
N1;/home/user/Desktop/PhyloSystemX/network/N1_net-
work.txt
N2;/home/user/Desktop/PhyloSystemX/network/N2_net-
work.txt
```

3. **Species Network Files**
Network files define homology-based interactions for each species and must be stored in the network/subdirectory within the main directory. Each file should contain interactions represented line by line in the format:

```
HomologygroupX;HomologygroupY
```

Additionally, every species included in the phylogenetic tree must have a corresponding network file named accordingly (e.g., N1_network.txt, N2_network.txt).

4. **Output Directory**
The software will generate output files by default in:

```
/home/user/Desktop/PhyloSystemX
```

Users can specify a custom absolute output path if desired, and the software will automatically create the directory if it does not exist.

3.2 Basic Usage of PhyloSystemX

The operation of the software consists of two main processes: inferring ancestral vertices (homology groups) and inferring ancestral interactions.

3.2.1 Inferring Ancestral Vertices

To infer ancestral vertices (homology groups), we have two presets. Both are based on the Dollo parsimony algorithm [18, 19], one in a non-hierarchical form and one in a hierarchical form [20, 21]. The non-hierarchical version of PhyloSystemX should be used when homology labels correspond to the evolutionarily oldest possible homology (or orthology) groups, such as a "rootHOG" or an early-evolving orthologous group in the OMA database. This approach ensures that ancestral inference is based on the broadest homology definitions without further subdivision into paralogous groups. The hierarchical version of PhyloSystemX is applicable when a given homology group (e.g., group1) has been subdivided into paralogs (e.g., group1.1 and group1.2). This hierarchical preset accounts for evolutionary events where ancestral homology groups gave rise to multiple, well-documented related subgroups [22].

In this section, we will only consider the simple case of non-hierarchical homology groups inference (for which scripts are prefixed with "simple"), but hierarchical homology groups inference will be briefly discussed in **Note 3**. While the hierarchical approach for which script are prefixed with "hierarchical" (for which scripts are prefixed with "hierarchical"), has not been benchmarked yet, the option allows the user to do so.

For each internal node of the species tree, the algorithm infers the existing vertex based on input data. By default, when only interaction network files are provided, vertices are inferred indirectly from the endpoints of edges. For instance, if an edge connects vertex A (representing homology group A) and vertex B (representing homology group B) in a given species, both homology groups are assumed to be present.

This computation for ancestral vertices is performed using the Dollo parsimony algorithm [18, 19]. This algorithm assumes that each vertex arises only once in evolution but can be lost multiple times across different lineages. This condition matches the definition of a homology (and of an orthology) group evolving by vertical descent, as these groups, have a single evolutionary origin and cannot be horizontally reacquired (see Fig. 3).

3.2.2 Inferring Ancestral Interactions

Two methods are suited for different scenarios. The first is the Dollo parsimony algorithm, which we refer to as "Dollo." The second approach is a modified version of the Sankoff parsimony algorithm (Sankoff classical algorithm that we will introduce in detail below) for networks [23, 24], referred to as the "Sankoff adjusted algorithm"-adjusted. We will explain in **Note 4** which of the two algorithms "Dollo" and "Sankoff adjusted" is more suitable for which goal.

1. **Dollo parsimony algorithm for inferring ancestral interactions**

Inferring ancestral interactions using Dollo parsimony is similar to inferring ancestral vertices. Under this approach, an interaction (an edge in a network) can be gained only once and cannot be reacquired in subsequent evolutionary processes.

If an interaction is observed in multiple extant species, Dollo parsimony assumes that it was present in their last common ancestor rather than gained independently in each lineage. Conversely, if an interaction is absent in some species but present in others, it

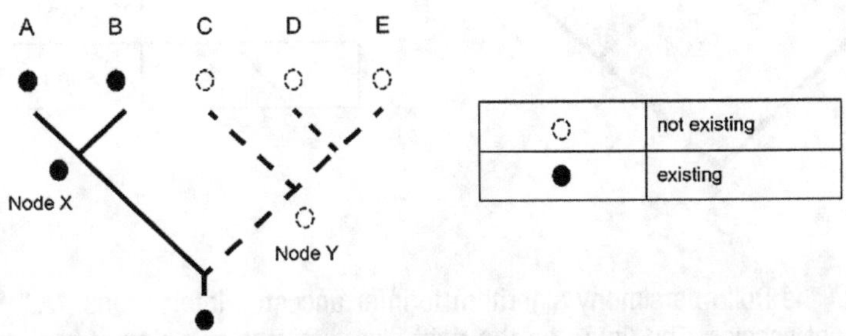

Fig. 3 Principles of the Dollo parsimony algorithm and its application for inferring ancestral vertices. "A," "B," "C," "D," and "E" represent different species. Each solid circle represents a vertex (homology group) that is present, while dashed hollow circles indicate the absence of that vertex. As shown in the left panel, once a vertex is lost at node Y, it will not reappear in descendant species (C–E), consistent with the Dollo principle that complex characters evolve only once and cannot be regained after loss

is inferred to have been lost in the lineages where it is missing (see Fig. 4).

2. **Description of the Sankoff algorithm (Sankoff classical algorithm) in inferring ancestral interactions**

The Sankoff algorithm is a parsimony-based method used to infer ancestral states in a phylogenetic tree. Unlike the Dollo parsimony method, which assumes that once a trait is lost, it cannot be regained, the Sankoff algorithm allows both multiple gains and losses of interactions within a lineage. This makes it particularly useful when state transitions (gains and losses) occur frequently. The following is the specific flow of the algorithm (see Box 1).

3. **Sankoff algorithms (Sankoff adjusted algorithm) for inferring ancestral interactions**

"Sankoff adjusted" is a modified version of Sankoff parsimony algorithm. The Sankoff adjusted algorithm addresses a critical limitation of Sankoff method by accounting for the dependency of edges (interactions) on vertices (nodes). Specifically, it distinguishes between edge loss due to vertex disappearance versus direct interaction loss. Therefore, an additional state is introduced into the state set to accurately express this distinction. The following is the specific flow of the algorithm (see Box 2).

3.2.3 Running PhyloSystemX

1. **Command-Line Parameters**

PhyloSystemX requires specific parameters to execute ancestral network inference. Below are a categorized list of required parameters (Table 1) with descriptions and default values and a categorized list of optional parameters (Table 2) with descriptions and default values:

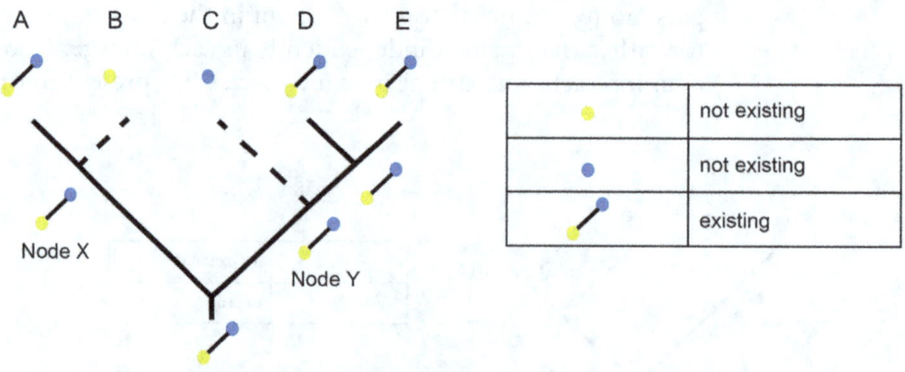

Fig. 4 Application of the Dollo parsimony algorithm to infer ancestral interactions. "A," "B," "C," "D," and "E" represent different species. The figure on the right illustrates the meaning of each symbol used in the figure. The absence of an interaction indicates that the interaction has been lost, and dashed lines represent evolutionary pathways where such losses occur — for example, from node X to species B and from node Y to species C. The Dollo parsimony algorithm assumes that interactions arise only once during evolution but may be lost multiple times in different lineages. Accordingly, the dashed lines highlight these inferred losses of interactions along evolutionary paths

Box 1: Sankoff Classical Algorithm Workflow

The algorithm minimizes the total cost of state transitions across a given species tree using dynamic programming. It consists of three main steps:

Initialization—Assign initial scores to leaf nodes based on observed interaction states.

Bottom-up pass (leaves to root)—Compute the minimum transition cost for each internal node.

Top-down pass (root to leaves)—Trace back the most likely ancestral states.

Define $s_t(v)$ to be minimum parsimony score of subtree rooted at v if vertex v has state t for the state of the interaction.

Step 1: Initialization

We define $s_v(k)$ as the minimum parsimony score for the subtree rooted at node v when v has state t.

For each leaf node v:

$$s_v(k) = 0, \text{ if } v \text{ has state } k$$

$$s_v(k) = \infty, \text{ otherwise}$$

This means that if a leaf node is observed in state k, its cost is zero; otherwise, it is infinite to reflect an impossible state (see Fig. 5A).

Step 2: Bottom-Up Pass (Leaves to Root)

Once leaf nodes are initialized, we propagate values up the tree to compute the minimum possible transition cost at each internal node.

Recurrence Relation

For each internal node v with children u and w, we calculate the minimum cost of assigning state t to v using:

$$s_t(v) = \min_i\{s_i(u) + \delta(i,t)\} + \min_i\{s_j(w) + \delta(j,t)\}$$

where:

$s_i(u)$ and $s_j(w)$ are the scores for child nodes u and w.

$\delta(i, t)$ and $\delta(j, t)$ are transition costs penalizing state changes from i to t and j to t.

The goal is to find the state t that minimizes the total transition cost.

Finding the Optimal Root State

After computing scores for all nodes, the optimal total parsimony score is:

$$s^* = min_i s_{root}(i)$$

This represents the minimum cost required to explain the observed data under the parsimony framework (see Fig. 5B).

Step 3: Top-Down Pass (Root to Leaves)

Once the total parsimony score is determined at the root, we trace back to assign the most likely ancestral states at each internal node.

Backtracking Procedure

1. The optimal root state is chosen as:

$$c^* = argmin_i s_{root}(i)$$

2. The most likely states for each internal node are assigned by following the minimum-cost paths down the tree. For each internal node v with children u and w, and having state t, we select the states i^* and j^* for its children as:

$$i^* = \underset{i}{argmin}\ [s_i(u) + \delta(i, t)]$$

$$j^* = \underset{j}{argmin}\ [s_j(u) + \delta(i, t)]$$

Here:

$s_i(u)$ and $s_j(w)$ are the parsimony scores of the child nodes with states i and j, respectively.

$\delta(i, t)$ and $\delta(j, t)$ are the transition costs from the state i or j to state t.

This ensures that the reconstructed history minimizes the total number of state transitions, leading to the most parsimonious scenario (see Fig. 5C).

Box 2: Sankoff Adjusted Algorithm Workflow

1. Dependence of the edges on vertices

In the network, the interaction (edge) can be lost in two ways:

(A) Loss of the edge itself, but the vertices of this edge still exist.

(B) Loss of at least one vertex, which removed all associated edges.

This distinction is crucial because when a vertex is lost, it becomes impossible to determine whether its associated interactions existed before the vertex loss. This introduces uncertainty into the inference process, which Sankoff adjusted algorithm specifically addresses (see Fig. 6A).

2. Handling the uncertain data

Since the disappearance of a vertex makes it impossible to determine whether its interactions using this vertex existed before the vertex loss, Sankoff adjusted algorithm annotates this information separately to avoid misinterpretation. Two scenarios must be distinguished:

1. "No edge, while both vertices remain"

This provides meaningful information about interaction loss.

2. "No edge because at least one vertex disappeared"

This does not reveal whether the interaction existed prior to the vertex and is thus treated as "unknown" (see Fig. 6B).

3. Computational Adjustments

Unlike Sankoff algorithm, which includes all observed data in computation, Sankoff adjusted ignores leaf nodes where edge presence is unknown due to vertex loss (see Fig. 6C).

2. **Example Command**

To run PhyloSystemX with two algorithms ("simple_both" mode), open a terminal (Command Prompt or PowerShell on Windows; Terminal on macOS or Linux). Make sure the Python script

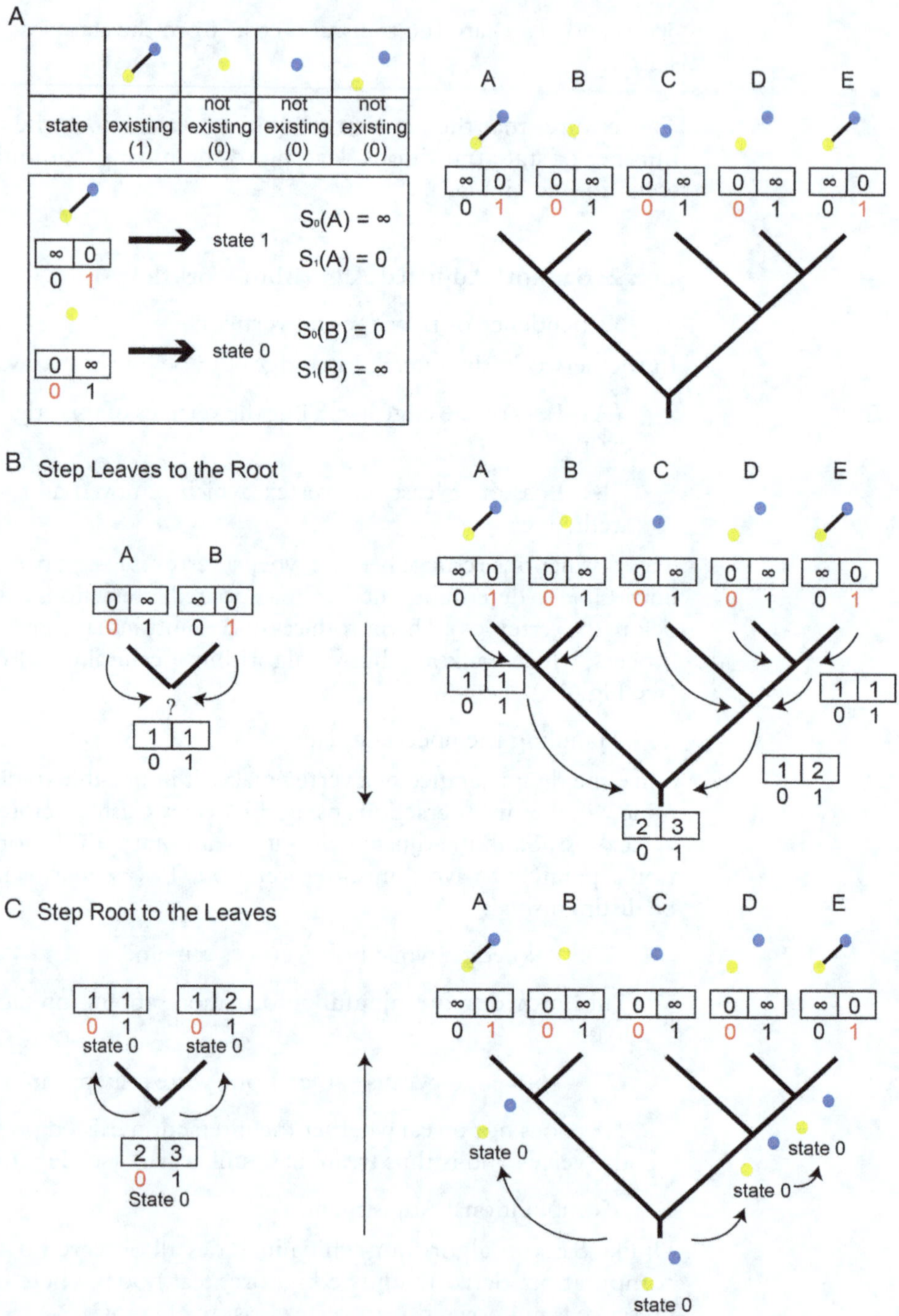

Fig. 5 Principles of the Sankoff parsimony algorithm to infer ancestral interactions. **(A)** illustrates our example dataset and shows the initial minimum parsimony costs assigned to each species (leaf node) for each possible interaction state. For each leaf node, the cost is set to 0 if the observed state matches the considered state, and ∞ (infinity) otherwise, representing an impossible assignment. **(B)** shows the process of propagating values from the leaves to the root. For each internal node, the algorithm calculates the minimum transition cost for assigning each possible state by considering all child node states and their associated costs. **(C)** illustrates the backtracking process from the root to the leaves. After determining the optimal root state, the states of internal nodes are inferred by calculating and following the minimum-cost paths. At each internal node, given the chosen parent state, the most likely states of its child nodes are selected by minimizing the sum of the child node's score and the transition cost

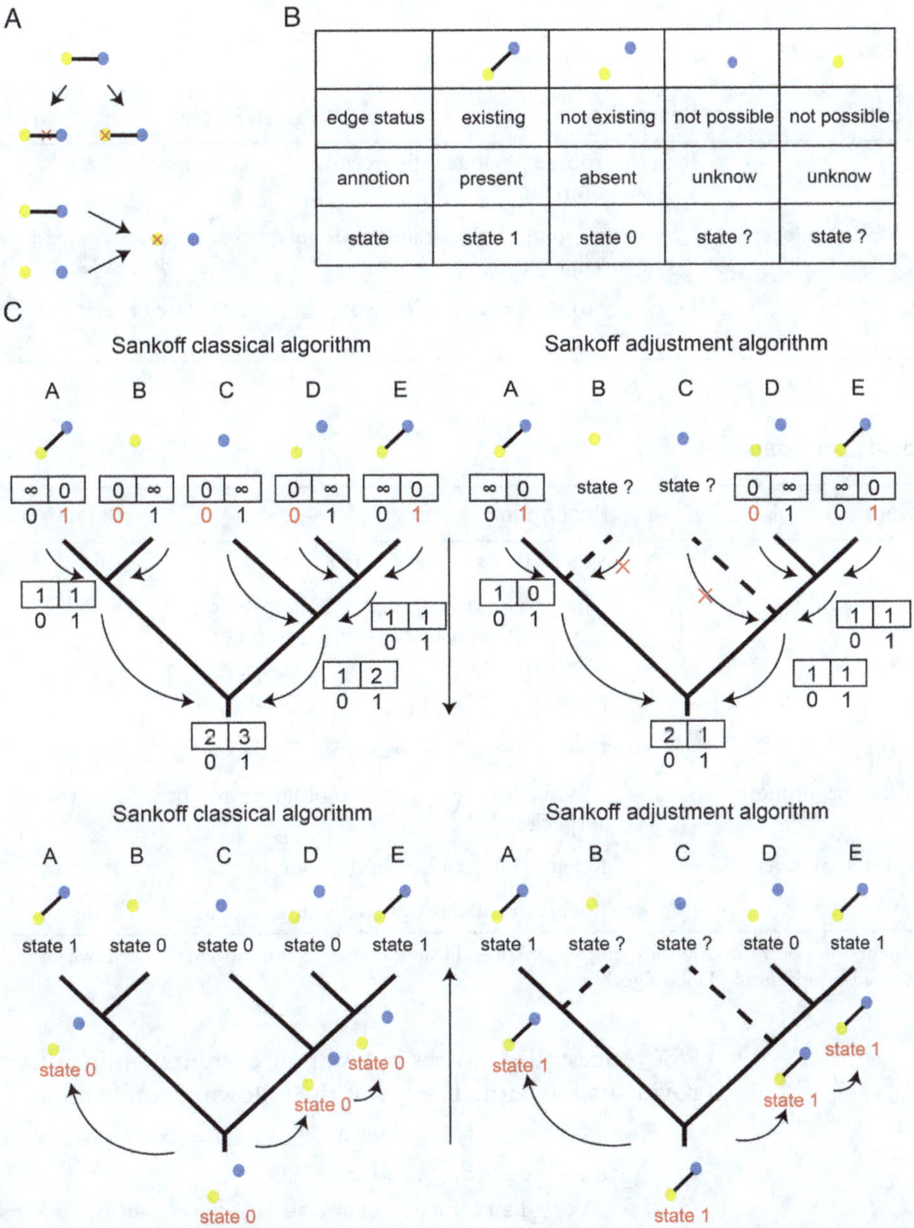

Fig. 6 Application of the Sankoff adjusted algorithm to infer ancestral interactions and comparison with Sankoff classical algorithms. **(A)** illustrates the two types of edge loss: one where the edge is lost but the involved vertices remain, and another where at least one vertex is lost, causing all associated edges to disappear. This distinction introduces uncertainty regarding the existence of interactions prior to the loss. **(B)** demonstrates how Sankoff adjusted algorithm differentiates between two uncertain edge loss scenarios: "No edge while both vertices remain" provides confirmed information about interaction loss, whereas "No edge due to vertex disappearance" leaves the interaction's existence uncertain and is treated as "unknown." **(C)** shows the difference between Sankoff adjusted algorithm and Sankoff classical algorithms. The dashed lines represent the edges ignored in the Sankoff adjusted algorithm during computation, those whose presence is uncertain due to vertex loss. The red mark shows the difference between the ancestral states inferred by the two algorithms

Table 1
Required parameters

Parameters	Description	Format
-tree	Path to a rooted phylogenetic tree in Newick format	.nwk file
-table_leafname_filename	CSV file mapping species names to their network file paths	.csv (e.g., species;path/to/network.csv)
-output_dirctory	Directory to store results. Created if missing	Absolute or relative path

Table 2
Key optional parameters

Parameters	Description	Default
-directed	Specify if edges are directed	False
-inference_method[a]	Choose algorithm:simple_Dollo,simple_Sankoff_adjusted or simple_both (run both)	Simple_Dollo
-cost_gain	Cost of gaining an edge (Sankoff_adjusted)	1
-cost_loss	Cost of losing an edge (Sankoff_adjusted)	1
-table_leafname_homolgygroups	CSV file listing known homology groups per species	Not required
-rename_internal_treenodes	Rename internal tree nodes for clarity	True
-verbose	Enable detailed progress logging	True

[a]This parameter offers only three options: simple_Sankoff_adjusted, simple_Dollo, and simple_both (which runs both Sankoff adjusted algorithm and Dollo algorithm).

(PSX_AncestralNetworks.py) is in the current working directory or provide its full path. Then, run the following command:

```
python3 /path/to/PSX_AncestralNetworks.py \
  -tree species_tree.nwk \
  -table_leafname_filename species_mapping.csv \
  -output_directory results/ \
  -inference_method simple_both \
  -cost_gain 1 \
  -cost_loss 1 \
  -table_leafname_orthogroups homology_groups.csv
```

Important

Replace /path/to/PSX_AncestralNetworks.py with the actual script path if it is stored elsewhere.

> Use absolute paths (e.g., /home/user/PhyloSystemX/network/) to avoid file-not-found errors.
>
> Here we use the optional parameter -table_leafname_orthogroups homology_groups.csv.

A list of potential runtime errors and suggested solutions is provided in **Note 5**.

3. **Accessing Help Documentation**

For a full list of parameters and additional usage details, you can display the help message by running:

```
Python3 /path/to/PSX_AncestralNetworks.py -h
```

3.2.4 PhyloSystemX Outputs

Upon completion, PhyloSystemX generates an output directory containing several key files.

1. **Modified Phylogenetic Tree (tree_internal.nwk)**

The first key output generated by PhyloSystemX is the modified phylogenetic tree file, tree_internal.nwk. This file presents the species tree in Newick format, with internal nodes uniquely renamed (e.g., 1_1, 2_2), each corresponding to an inferred ancestral network. The leaf names (such as N1, N2) are preserved for easy reference and cross-checking with the input data. This naming structure enables users to quickly identify which ancestral network files correspond to which nodes in the reference phylogeny. For example, the file structure could look like this:

```
[&R] (N1,(N2,(N3,(N4,(N5,(N6,(N7,(N8,(N9,(N10,(N11,
(N12,(N13,(N14,(N15,N16)'14_14')'13_13')'12_12')'11_
11')'10_10')'9_9')'8_8')'7_7')'6_6')'5_5')'4_4')'3_
3')'2_2')'1_1')'0_0';
```

This format shows the hierarchical structure of the phylogeny, with each internal node's name directly corresponding to its associated network result file, such as network_1_1.csv.

2. **Homology Group Annotations (homologygroups_ancestral.csv)**

The second output is the homology group annotation file (homologygroups_ancestral.csv), which contains a list of homology groups inferred for each node—both extant and ancestral. The file is provided in CSV format, where each line begins with the node name followed by a series of homology group identifiers separated by commas (e.g., 0_0;Homologygroup1, Homologygroup2). This output is essential for understanding which homology groups vertices are conserved, gained, or lost along evolutionary paths, and it helps verify the completeness and consistency of the input and output relationships within the analysis.

3. **Node-to-Network Mapping (node_network_map.csv)**
Another critical output is the node-to-network mapping file (node_network_map.csv), which links each internal node to its respective ancestral network result file. The format of each line follows a simple structure: node;path/to/network_file (e.g., `1_1;/output/ancestral_network/network_1_1.csv`). This mapping file is extremely helpful for automating further analysis and efficiently retrieving the corresponding ancestral network for any given node without confusion or manual searching.
4. **Ancestral Network Directory (ancestral_network/)**
Finally, the ancestral_network/ directory contains the reconstructed ancestral network files in CSV format, each named after the corresponding node in the reference phylogeny (e.g., network_1_1.csv). Each file lists inferred interactions in a simple format where each line represents an edge. The following is an example output to the Dollo algorithm:

```
Homologygroup1,Homologygroup2
```

For results generated from the simple_Sankoff_adjusted or simple_both modes, the network CSV files include additional information columns. *See* **Note 6** for details on results calibration and interpretation.

4 Notes

1. *Choosing relevant networks as input:* Network selection depends on a selection of reference species that is dictated by the evolutionary problem to be addressed. When phylogenetically equivalent species exist, but interactions within these species have been unevenly investigated by experimentalists, then inferences based on species with the largest and least sparse networks are likely to return more informative ancestral networks. Small networks (in terms of number of nodes or of numbers of edges) often convey less information than larger networks. In general, using larger networks with high-quality edges is highly recommended, unless only smaller networks are known for certain species that lie at critical position of the phylogenetic tree. For example, between *Drosophila melanogaster* (well-studied, dense protein network) and its close relative *Drosophila simulans* (sparsely documented network), *D. melanogaster* would be preferred despite their phylogenetic equivalence, as its comprehensive data reduces reconstruction uncertainty. However, the precision and the recall of both the Dollo algorithms and Sankoff adjusted algorithms are robust to missing data (up to 25% of missing edges and up to 25% of missing nodes), so networks of different sizes can be investigated together in the

same PhyloSystemX analyses. However, for 50% of missing edges, the recall of the Sankoff adjusted algorithm significantly drops [25].

2. *Choosing a relevant evolutionary scale:* Adjusting the granularity of the labels between homology vs orthology labels can be a good strategy to investigate the evolution of networks at different phylogenetic scales. Homology labels are potentially more widely shared than orthology labels. Using homology labels will however collapse multiple nodes of extant networks into fewer nodes than using orthology labels, therefore there should be an impact of the choice of homology vs orthology labels on the topologies of ancestral networks. Relying upon homology label can be a good way to investigate deep evolutionary history, i.e., when there are multiple duplications affecting a family and orthologs have recently evolved. However, inferring ancestral networks using homology labels rather than using orthology labels may support weaker functional inferences. When possible, we recommend to analyze the evolution of networks both using homology and orthology labels, separately.
3. *Hierarchical vs non-hierarchical version:* Non-hierarchical version of PhylosystemX should be used when homology labels correspond to the evolutionary oldest possible homology or orthology group, e.g., a rootHOG or an early evolving orthogroup in the OMA database. Hierarchical version of PhylosystemX can be used when a given homology group, e.g., group1, has been subdivided into paralogs, e.g., group1.1 and group1.2. Hierarchical analyses require networks annotated with nested group names, such as 1, 1.1, 1.1.1, and 1.1.1.1, and the option "inference_method" must be set to hierarchy_Dollo. Please note that while PhylosystemX efficacy has been benchmarked for analyses based on non-hierarchical groups, it has not been benchmarked yet against simulated data for analyses based on hierarchical groups.
4. *Choice of the inference algorithm:* For a given network input dataset, the Dollo algorithm tends to infer earlier edges than the Sankoff adjusted algorithm. Overall, the Sankoff adjusted algorithm has a higher precision than the Dollo algorithm, but the Dollo algorithm has a better recall than the Sankoff adjusted algorithm when more than 25% of existing edges are missing in input networks. The recall of the Sankoff adjusted algorithm is higher than the Sankoff classical algorithm when using optional information that can be added to the homology/orthology labeled network infiles (for an overall similar precision). Specifically, an infile indicated under the option table_listname_orthogroups of phylosystemX can be used with the Sankoff adjusted algorithm. The structure of this text file is indicated

in the help of PhylosystemX (PSX_AncestralNetworks.py -h). This file lists all homology/orthology groups present in all species of interest according to the OMA database, irrespective of their presence in their interaction networks. Typically, some proteins can be present in a species S but not connected to any other proteins of S, and therefore be absent in their S PPI network.

5. *Some possible running errors:* Some possible running errors include "File Not Found errors," which often result from incorrect file paths in the mapping file. To resolve this, double-check the specified paths and use absolute paths (e.g., C:\data\network.csv instead of network.csv) to ensure proper file referencing. Garbled text in files can occur due to incorrect text encoding; the solution is to ensure all files are saved in UTF-8 format to maintain compatibility and prevent character misinterpretation. Another common issue is the KeyError: "Not all labels matched to taxa," which occurs if species names in the phylogenetic tree do not match those in other input files. To fix this, verify that species names are consistent across all input files.

6. *Calibration of algorithm outputs through consensus:* In the Sankoff adjusted algorithm output, the third column represents the Sankoff state value, which can be either 0.5 or 1, where 1 indicates that the interaction is confidently inferred by the algorithm, and 0.5 indicates uncertainty in the inference. Based on this, users can construct two versions of the ancestor interaction network: a permissive version, where uncertain (0.5) interactions are treated as present, and a conservative version, where uncertain (0.5) interactions are treated as absent.

Example ("simple_Sankoff_adjusted" output):

```
Group1;Group2;Sankoff_state
1000;1153;1
1000;1167;1
1000;1176;0.5
```

In the simple_both output, in addition to the Sankoff state column, an extra column is provided for the Dollo algorithm state (values 0 or 1), enabling users to apply flexible filtering strategies, such as taking the intersection of the conservative Sankoff adjusted algorithm results and Dollo-inferred interactions or taking the intersection of the permissive Sankoff adjusted algorithm results and Dollo-inferred interactions.

Example ("simple_both"output):

```
Group1;Group2;Sankoff_state;Dollo_state
1000;1153;1;1
1000;1167;1;1
1000;1176;0.5;1
```

5 Conclusion

This chapter has introduced PhyloSystemX, a computational tool designed to reconstruct ancestral interaction networks by integrating species phylogenetic trees with interaction data of existing species. With the advancement of multi-omics technologies, the large-scale accumulation of biological interaction data has provided unprecedented opportunities for studying network evolution, and PhyloSystemX was developed to meet this need.

PhyloSystemX is based on the hypothesis of vertical inheritance and species modification, implementing two algorithms for ancestral network inference. The Dollo parsimony algorithm is suitable for conservative interaction inference, while the Sankoff adjusted algorithm specifically addresses vertex-edge dependency issues.

The primary strength of this tool lies in its versatility. Unlike tools limited to specific interaction types, PhyloSystemX is applicable to any network whose vertices can be labeled with homology labels, including protein–protein interaction, gene co-expression, metabolic regulation, and various other biological network types.

However, we must also acknowledge the limitations of PhyloSystemX. Currently, the tool does not explicitly model phenomena such as introgression, hybridization, and horizontal gene transfer, which may affect the accurate inference of interactions between certain species populations.

Looking ahead, as biological interaction data continue to accumulate and evolutionary models are refined, PhyloSystemX is expected to further improve its inference accuracy. The methodological framework and practical guidelines provided in this chapter equip researchers with tools to explore biological network evolution. By combining systems biology with evolutionary analysis, we believe that a more comprehensive understanding of the origins and evolution of biological complexity will emerge in the future.

References

1. Joyce AR, Palsson BØ (2006) The model organism as a system: integrating "omics" data sets. Nat Rev Mol Cell Biol 7:198–210. https://doi.org/10.1038/nrm1857
2. Yugi K, Kubota H, Hatano A, Kuroda S (2016) Trans-omics: how to reconstruct biochemical networks across multiple "omic" layers. Trends Biotechnol 34:276–290. https://doi.org/10.1016/j.tibtech.2015.12.013
3. Bapteste E, Huneman P (2018) Towards a dynamic interaction network of life to unify and expand the evolutionary theory. BMC Biol 16:56. https://doi.org/10.1186/s12915-018-0531-6
4. Alon U (2006) An introduction to systems biology: design principles of biological circuits. Chapman and Hall/CRC, New York
5. Navlakha S, Kingsford C (2011) Network archaeology: uncovering ancient networks from present-day interactions. PLOS Comput Biol 7:e1001119. https://doi.org/10.1371/journal.pcbi.1001119
6. Bapteste E, Lopez P, Bouchard F et al (2012) Evolutionary analyses of non-genealogical bonds produced by introgressive descent. Proc Natl Acad Sci USA 109:18266–18272. https://doi.org/10.1073/pnas.1206541109

7. Sun MG, Kim PM (2011) Evolution of biological interaction networks: from models to real data. Genome Biol 12:235. https://doi.org/10.1186/gb-2011-12-12-235
8. Dorogovtsev SN, Mendes JFF (2002) Evolution of networks. Adv Phys https://doi.org/10.1080/00018730110112519
9. Yan K-K, Wang D, Sethi A et al (2016) Cross-disciplinary network comparison: matchmaking between hairballs. Cell Syst 2:147–157. https://doi.org/10.1016/j.cels.2016.02.014
10. Rajan V, Zhang Z, Kingsford C, Zhang X (2021) Maximum likelihood reconstruction of ancestral networks by integer linear programming. Bioinformatics 37:1083–1092. https://doi.org/10.1093/bioinformatics/btaa931
11. Bonnefous H (2025) GitHub, https://github.com/Hugo-bonnefous/PhyloSystemX. Accessed 29 Apr 2025
12. Szklarczyk D, Kirsch R, Koutrouli M et al (2023) The STRING database in 2023: protein-protein association networks and functional enrichment analyses for any sequenced genome of interest. Nucleic Acids Res 51:D638–D646. https://doi.org/10.1093/nar/gkac1000
13. Help - STRING functional protein association networks. https://string-db.org/cgi/help. Accessed 29 Apr 2025
14. Altenhoff AM, Train C-M, Gilbert KJ et al (2020) OMA orthology in 2021: website overhaul, conserved isoforms, ancestral gene order and more. Nucleic Acids Res 49:D373–D379. https://doi.org/10.1093/nar/gkaa1007
15. Altschul SF, Gish W, Miller W et al (1990) Basic local alignment search tool. J Mol Biol 215:403–410. https://doi.org/10.1016/S0022-2836(05)80360-2
16. Buchfink B, Reuter K, Drost H-G (2021) Sensitive protein alignments at tree-of-life scale using DIAMOND. Nat Methods 18:366–368. https://doi.org/10.1038/s41592-021-01101-x
17. Watson AK, Lannes R, Pathmanathan JS et al (2019) The methodology behind network thinking: graphs to analyze microbial complexity and evolution. Methods Mol Biol 1910:271–308. https://doi.org/10.1007/978-1-4939-9074-0_9
18. Farris JS (1977) Phylogenetic analysis under Dollo's Law. Syst Biol 26:77–88. https://doi.org/10.1093/sysbio/26.1.77
19. Collin R, Miglietta MP (2008) Reversing opinions on Dollo's Law. Trends Ecol Evol 23:602–609. https://doi.org/10.1016/j.tree.2008.06.013
20. Altenhoff AM, Glover NM, Dessimoz C (2019) Inferring orthology and paralogy. In: Anisimova M (ed) Evolutionary genomics. Springer New York, New York, NY, pp 149–175
21. Altenhoff AM, Gil M, Gonnet GH, Dessimoz C (2013) Inferring hierarchical orthologous groups from orthologous gene pairs. PLoS One 8:e53786. https://doi.org/10.1371/journal.pone.0053786
22. Fernández R, Gabaldón T, Dessimoz C (2020). Orthology: definitions, inference, and impact on species phylogeny inference. In Scornavacca C, Delsuc F, and Galtier N, editors, Phylogenetics in the Genomic Era, chapter No.2.4, pp.2.4:1–2.4:14. No commercial publisher | Authors open access book. The book is freely available at https://hal.inria.fr/PGE
23. Sankoff D, Rousseau P (1975) Locating the vertices of a Steiner tree in an arbitrary metric space. Math Program 9:240–246. https://doi.org/10.1007/BF01681346
24. Sankoff D (1975) Minimal mutation trees of sequences. SIAM J Appl Math 28:35–42. https://doi.org/10.1137/0128004
25. Bonnefous H (2025) Etude de l'évolution des interactions moléculaires chez les eucaryotes par des méthodes de comparaison de réseaux dans le but d'identifier des processus ancestraux partagés associés au vieillissement. Thesis, Sorbonne université

Check for updates

Chapter 6

Discovering, Genotyping, and Annotating Structural Variants with the *Genome Analysis Toolkit* (GATK)

Derek Caetano-Anollés

Abstract

The *Genome Analysis Toolkit* (GATK) is a widely used software suite for identifying and annotating genomic variants from high-throughput sequencing data. Originally developed by the Broad Institute, GATK has become a cornerstone in genomic research thanks to its modular design, reproducible workflows, and integration with cloud platforms. Key tools such as *HaplotypeCaller*, *Mutect2*, and *Funcotator* enable researchers to detect germline and somatic variants, assess functional impact, and process large datasets with efficiency and accuracy. Its workflows support best practices for variant calling—both in human and non-model organisms—which makes it valuable across a broad range of disciplines ranging from clinical genomics to evolutionary biology. With the increasing scale and accessibility of sequencing technologies, tools like GATK play a critical role in transforming raw data into meaningful biological insights. This review outlines the toolkit's major components, common use cases, and practical considerations, offering a foundation for researchers looking to implement or better understand GATK in their own analyses.

Keywords Bioinformatics, Copy number variants (CNVs), Functional genomics, Genomic data analysis, Next-generation sequencing (NGS), Structural variants (SVs), Variant annotation, Variant calling

1 Introduction

Biological research has never been as easy to get into as it is now. The invention of PCR inspired an interdisciplinary revolution, whereby every dusty museum collection and specimen cabinet across the globe opened up into a genetic hunting ground for geneticists and medical researchers alike [1, 2]. The *Human Genome Project* unlocked brand new avenues of research that engendered entire careers for people hungry to expand and unpack the information it revealed [3]. Now, with the declining costs of genome sequencing [4, 5], what was once a lifetime of labor for previous generations of researchers [6] can be packaged into a quick and easy consumer kit [7, 8].

Diminishing costs and increased accessibility have opened the possibility for researchers to incorporate an increasingly varied toolkit

Gustavo Caetano-Anollés (ed.), *Evolutionary Genomics*, Methods in Molecular Biology, vol. 2981,
https://doi.org/10.1007/978-1-0716-4836-0_6,

into their own fields [7], but there are still hurdles to overcome. The final frontier for the true democratization of biology lies in the bioinformatics required to make sense of the massive quantities of data that we are now able to produce [9, 10]. Even the most Luddite wet lab specialists are finding themselves needing to spend more time weeding out bugs from analysis pipelines to stay relevant [11].

For most of these researchers, their main focus is on variant calling—the identification and detection of polymorphisms and variants in sequencing data [12]. Fortunately, a great number of variant calling resources have emerged over the years [13–15], and the most widely used among them is the *Genome Analysis Toolkit* (GATK) [16, 17].

2 The Genome Analysis Toolkit

The GATK—available at <*gatk.broadinstitute.org*>—is a comprehensive suite of tools designed for variant discovery and genotyping in high-throughput sequencing data. It has become an industry standard for identifying single nucleotide polymorphisms (SNPs) and insertions/deletions (indels) in both germline and somatic contexts. Beyond variant calling, GATK encompasses utilities for data preprocessing, quality control, and integrates the widely used Picard toolkit, facilitating tasks such as data formatting and duplicate marking [17, 18]. GATK is primarily written in Java, with some components utilizing Python and R for auxiliary tasks such as data processing and visualization. It relies on the Apache Spark™ framework for scalable parallel processing, particularly in its newer iterations [17].

GATK includes over 150 tools designed for variant discovery, genotyping, quality control, and sequence data analysis [16]. The main categories of tools in GATK include variant calling, copy number variant detection, coverage analysis, quality control and diagnostics, RNA-seq analysis, alignment processing, and structural variation detection. A whole book could be written about each of these tools—in fact, one has [17]—but several of GATK's most popular tools are by far *HaplotypeCaller* [19], *Funcotator* [20], and *Mutect2* [21]:

- *HaplotypeCaller* calls SNPs and indels through local *de novo* assembly of haplotypes in regions exhibiting variation.
- *Funcotator* is a functional annotation utility that enriches variant calls with information from diverse data sources, which aids in the interpretation of their potential biological impact.
- *Mutect2* detects somatic mutations in cancer genomics.

To streamline genomic analyses, GATK also provides "*Best Practice*" workflows, which are standardized pipelines that guide users through the recommended steps for various analysis types, including germline and somatic variant discovery, copy number variation, and structural variation detection [22]. These workflows are meticulously curated to ensure accuracy and reproducibility across different datasets and experimental designs.

It is important to note that GATK is currently in development, with new features being added all the time. As such, this chapter may contain outdated information depending on when you read it. For this reason, please seek out developer documentation and repositories for the individual tool(s) you may decide to use when running your own analyses.

2.1 Using GATK and GATK Workflows

GATK as a toolset is free to download and use. It can be run straight from a computer terminal or server console, making it a useful toolkit to include on any bioinformatician's tool belt. Some researchers may find themselves running large analyses on massive datasets, perhaps with confidential patient restrictions that may make it difficult to migrate from a local repository. For these cases, there are currently options for running GATK analyses on the cloud using cloud-stored data.

As of the writing of this chapter, GATK workflows are offered on *Terra* (available at <*terra.bio*>), a cloud-based bioinformatics platform designed to facilitate collaboration on large-scale genomic analyses, offered and maintained by the Broad Institute of MIT & Harvard. *Terra* enables researchers to access extensive datasets and execute GATK workflows (among others) without the need to manage computational infrastructure. *Terra* offers pre-configured workspaces that include GATK pipelines and Jupyter Notebooks, providing an interactive environment for data analysis and visualization, complete with sample data and step-by-step tutorials. While *Terra*'s integration with Jupyter Notebooks offers an educational platform where you may engage with tutorials that elucidate the logic and operation of each analysis step, it should be noted that *Terra* only serves as a GUI front-end to already-existent infrastructure provided by cloud service providers such as the *Google Cloud Platform* and *Microsoft Azure*. As such, use of *Terra* requires a paid subscription to these cloud service providers.

2.2 HaplotypeCaller

HaplotypeCaller is a tool within the GATK designed to identify germline SNPs and indels in DNA sequencing data. It achieves this by locally reassembling haplotypes in regions showing evidence of variation, allowing for accurate variant detection, especially in complex genomic areas where different types of variants are close to each other [23].

HaplotypeCaller works by first scanning the genome to identify regions with significant evidence of variation, termed "active regions." These are the areas where *HaplotypeCaller* will focus its analysis [24]. Next, for each active region, *HaplotypeCaller* constructs a De Bruijn-like graph to reassemble the sequence and determine possible haplotypes present in the data. It then realigns each haplotype against the reference sequence using the Smith-Waterman algorithm to pinpoint potential variant sites [24]. After this step is completed, *HaplotypeCaller* will perform pairwise alignments of each read against each identified haplotype using the Pair Hidden Markov Model (PairHMM) algorithm. This results in a matrix of likelihoods for haplotypes given the read data, which is then used to derive allele likelihoods for each potential variant site [24]. Finally, *HaplotypeCaller* applies Bayes' theorem to calculate the probability of each genotype at a site based on the allele likelihoods and assigns the most probable genotype to the sample [25].

The end result of using *HaplotypeCaller* is an output file in VCF format that contains all the SNPs and indels it was able to discover using the assigned parameters and inputs. *HaplotypeCaller* is ideal for the following use cases, although its specific parameters may be modified for others:

- Single-sample variant calling: *HaplotypeCaller* can be run on individual samples to produce variant calls in VCF format. This is suitable for analyses focusing on a single sample [23].
- Joint genotyping in cohorts: For studies involving multiple samples, it's better to run *HaplotypeCaller* in GVCF mode for each sample, generating GVCF files that contain information about all sites (not just those with detected variants). These GVCFs can then be combined and jointly genotyped using GATK's *GenotypeGVCFs* tool, improving the accuracy of variant calls across the cohort [26].
- Non-diploid organisms and pooled data: *HaplotypeCaller* also supports analyses of organisms with ploidies other than diploid and can handle pooled experimental data. You'll need to specify the sample's ploidy using the -ploidy argument. However, for cases with extreme allele frequencies, such as somatic (cancer) variant discovery, it's advisable to use specialized tools like *Mutect2* [27].

By leveraging local de novo assembly, *HaplotypeCaller* offers enhanced accuracy in variant detection, particularly in challenging genomic regions, making it a valuable tool for comprehensive genomic analyses.

2.3 Funcotator

Funcotator (shortened from "Functional Annotator") is a tool within the GATK designed to annotate genomic variants with functional information from various data sources [28]. It enhances the interpretability of variants by providing context about their potential biological impact. In addition, it is highly customizable and can incorporate both pre-packaged and user-defined data sources to tailor annotations to specific research needs.

Funcotator accepts variant data (in formats such as VCF), and it can produce annotated outputs in various formats (including VCF and MAF, the Mutation Annotation Format). The choice of output format depends on the downstream analysis requirements and the tools used in subsequent steps.

Annotations provided by *Funcotator* include information from sources like the GENCODE Project, which offers comprehensive gene annotations [28]. These annotations encompass details such as gene symbols, variant classifications, and protein changes, providing a robust framework for interpreting the functional consequences of variants.

Funcotator relies on organized data sources to provide accurate annotations. Each data source resides in a specific directory structure, with subdirectories corresponding to different genome references (e.g., hg19, hg38). Within each reference-specific directory, a configuration file defines the data source and outlines the criteria for matching it to variants. This setup enables *Funcotator* to efficiently manage and apply multiple annotations during variant analysis.

Pre-packaged data sources are available for both somatic and germline use cases. These can be downloaded using the *FuncotatorDataSourceDownloader* tool, which ensures data integrity and simplifies the setup process. For instance, to download somatic data sources, one would use the command:

```
./gatk FuncotatorDataSourceDownloader --somatic
--validate-integrity --extract-after-download
```

For germline data, one would use the "--germline" option instead of the "--somatic" option in the above command.

Funcotator also supports user-defined data sources, allowing researchers to incorporate custom annotations. This flexibility is facilitated through configuration files that specify the format and matching criteria for each custom data source. For example, a simple tab-separated values (TSV) file can be configured with a corresponding .config file detailing how *Funcotator* should interpret and apply the data.

2.4 Mutect2

Mutect2 is a tool within the GATK designed to identify somatic short mutations, including single nucleotide variants (SNVs) and small insertions and deletions (indels), through local assembly of haplotypes. It employs a Bayesian somatic genotyping model, distinct from the original *MuTect*, and utilizes the assembly-based machinery of *HaplotypeCaller* [28]. *Mutect2* is ideal for running the following types of analyses:

- Tumor-normal paired analysis: For studies that include matched tumor and normal samples, this mode produces accurate identification of somatic mutations by comparing the two [28].
- Tumor-only analysis: *Mutect2* includes a "tumor-only mode" for use when normal samples are unavailable, although this

approach may yield higher false-positive rates due to the absence of a matched normal for comparison [28]. It is always advisable to have matched tumor and normal samples, but it is not always feasible for all sample sets.

- Mitochondrial variant calling: *Mutect2* can be employed for detecting variants in mitochondrial DNA, benefiting from its sensitivity in calling variants at high sequencing depths [28].

Mutect2 identifies somatic mutations by first assembling local haplotypes in regions of interest, using a De Bruijn-like graph to reconstruct possible sequences [28]. This approach—similar to that used by *HaplotypeCaller*—enables accurate variant detection even in complex or ambiguous genomic regions.

Once candidate haplotypes are formed, *Mutect2* applies a Bayesian genotyping model to evaluate the probability of each mutation being somatic, contrasting tumor and normal samples to filter out germline variation. To further refine results, it cross-checks against a Panel of Normals (PoN) [29], removing recurrent technical artifacts, and integrates population-level germline variant databases to distinguish somatic calls from common inherited polymorphisms [28].

By integrating this external data, *Mutect2* enhances its ability to discriminate between rare somatic events and common germline polymorphisms, ensuring a more accurate call set for downstream analysis.

3 Typical GATK Pipelines

3.1 Using the Germline Short Variant Discovery Pipeline on GATK

The *Germline Short Variant Discovery* pipeline is designed to identify germline single nucleotide polymorphisms (SNPs) and insertions/ deletions (indels) across one or more individuals, resulting in a joint callset in Variant Call Format (VCF). In versions of GATK later than version 4.2, this pipeline was co-developed in collaboration with Illumina, integrating features from Illumina's *Dynamic Read Analysis for GENomics* (*DRAGEN*) platform to enhance performance and accuracy [30, 31].

The process begins with per-sample variant calling using GATK's *HaplotypeCaller* in GVCF mode [30]. This tool performs local *de novo* assembly of haplotypes in regions that show evidence of variation, enabling highly accurate detection of SNPs and indels—even in complex or repetitive genomic regions. The output is an intermediate GVCF file for each sample, which captures genotype likelihoods and variant information without yet committing to a final call.

Next, the GVCF files from multiple samples are consolidated using the *GenomicsDBImport* tool, which organizes the data into a scalable database format that can be queried [30]. This structure allows for more efficient processing in the joint genotyping step, where *GenotypeGVCFs* evaluates all samples together. Joint genotyping increases sensitivity by leveraging population-level data to confirm variant sites and build a complete genotype matrix across the cohort.

Following the joint genotyping step, the pipeline applies *Variant Quality Score Recalibration* (*VQSR*) to filter the variant calls [30]. Using machine learning models trained on known high-quality variants, *VQSR* assigns a probability to each call, effectively distinguishing true variants from likely false positives. This probabilistic filtering ensures a final VCF that balances sensitivity and precision.

The integration of *DRAGEN* features brings several enhancements to the GATK pipeline. *DRAGMAP*, an alt-aware aligner, replaces traditional mapping tools to improve read alignment accuracy, particularly in regions with structural complexity [31].

Additionally, the *DragSTR* model incorporates short tandem repeat (STR) variation into the calling process, refining genotype accuracy in repetitive areas. The *CalibrateDragstrModel* tool estimates STR-specific parameters used by *HaplotypeCaller* in "*DRAGEN* mode." Notably, *DRAGEN's* improved algorithms eliminate the need for traditional Base Quality Score Recalibration (BQSR), simplifying the pipeline without sacrificing quality. Together, these innovations significantly enhance the pipeline's performance, making it well-suited for comprehensive germline variant discovery across diverse genomic datasets.

By integrating these *DRAGEN* features, the GATK Germline Short Variant Discovery pipeline achieves enhanced performance, accuracy, and efficiency in identifying germline variants across diverse genomic datasets [30].

3.2 Detecting Germline Copy Number Variants with GermlineCNVCaller

Germline copy number variants (gCNVs), which involve deletions or duplications of genomic regions, are a significant source of genetic diversity and are often implicated in disease. Accurate identification of these variants is critical for both research and clinical applications. GATK's *GermlineCNVCaller* offers a robust framework for detecting gCNVs from sequencing data, such as whole exome (WES) and whole genome sequencing (WGS), using a Bayesian model that accounts for technical variation and improves the reliability of calls [32]. *GermlineCNVCaller* supports two primary modes of operation: cohort mode, and case mode [33].

In cohort mode, *GermlineCNVCaller* models read-depth biases and detects CNVs across multiple samples simultaneously, enhancing sensitivity and specificity by leveraging shared information [32, 33]. This mode is most effective with at least 30 samples to ensure a well-trained model. Case mode, by contrast, allows for individual sample analysis using a pre-trained model from a similar cohort. To maintain accuracy in this mode, it's important that the case sample closely matches the cohort in tissue type, library prep, and sequencing protocol [33].

At the core of *GermlineCNVCaller's* methodology is a probabilistic model that integrates both biological and technical influences on read depth [34]. The read-depth likelihood model evaluates the probability of observed coverage patterns given specific copy number states, factoring in sequencing depth, GC content, and mapping artifacts. This is paired with a hierarchical hidden Markov model (HHMM), which captures genomic structure and models the correlation of copy number states across regions [33, 34]. Together, these components enable *GermlineCNVCaller* to distinguish true CNVs from noise and detect both common and rare variants with high confidence.

Running *GermlineCNVCaller* involves several preparatory and analytical steps.

- First, preprocess the data by defining target intervals and collecting read counts, typically with *PreprocessIntervals* and *CollectReadCounts* [32, 34]. Optional annotations, such as GC content, can be followed as an additional step to refine bias modeling [34].
- Next, the *DetermineGermlineContigPloidy* tool performs ploidy determination and establishes the baseline copy number states for each contig, setting the stage for CNV detection [32].
- To actually call CNVs in *GermlineCNVCaller*, the decision has to be made to run the tool in either cohort or case mode to analyze the read-depth data, identify copy number changes, and segment the genome accordingly [33].
- Finally, for a post-processing step, these results are consolidated and exported in VCF format for downstream interpretation; this step involves refining the detected CNVs, merging adjacent variants, and formatting the data for interpretation and reporting.

GermlineCNVCaller provides several key advantages for gCNV detection. Its joint modeling of technical biases and biological variation improves both the sensitivity and specificity of CNV detection [34]. It is highly scalable, suitable for both small-scale analyses and large population studies, and supports scattered processing for efficiency [33]. *GermlineCNVCaller* is also flexible across sequencing platforms and protocols, with full compatibility for both WES and WGS data [32]. Altogether, it offers a powerful and adaptable solution for the accurate discovery of germline copy number variants.

In summary, GATK's *GermlineCNVCaller* provides a robust and flexible framework for the detection of germline copy number variants, leveraging advanced probabilistic modeling to account for technical variability and improve the accuracy of CNV calls [32].

4 Genome Aggregation Database (*gnomAD*)

The *Genome Aggregation Database* (*gnomAD*) [35] is one of the largest and most widely used public resources for human genetic variation. It aggregates and harmonizes petabytes of raw sequencing data from hundreds of thousands of individuals, drawn from a diverse range of studies. This immense dataset provides a critical reference for identifying genetic variants, estimating allele frequencies, and deepening our understanding of both rare and common variation across global populations.

To ensure consistency and accuracy, all sequencing data in *gnomAD* is uniformly processed through a standardized pipeline. Joint variant calling is performed using the *BWA-Picard-GATK* workflow [16], which includes read alignment to the human reference genome, marking of duplicate reads, base quality score recalibration, and robust variant calling and filtering. This unified approach reduces technical variability and enables high-confidence variant detection across both exome and genome datasets.

The *gnomAD* has become an indispensable tool for genetic research, clinical variant interpretation, and population genetics. By providing large-scale allele frequency data, it helps distinguish pathogenic mutations from benign polymorphisms and informs studies of genetic architecture in disease. As the database continues to grow—with the inclusion of new datasets and ongoing improvements in variant annotation—it remains a cornerstone of modern human genomics.

5 Conclusion

As genome sequencing continues to become more affordable and widespread, tools like the *Genome Analysis Toolkit* have become essential for making sense of the massive data volumes modern biology produces. With its expansive set of modular tools, well-supported workflows, and robust documentation, GATK has helped standardize variant discovery and annotation across disciplines, from evolutionary biology to precision oncology.

Core utilities like the ones we have explored here—*HaplotypeCaller*, *Funcotator*, and *Mutect2*—exemplify GATK's flexibility. Each is tailored for different stages and types of variant analysis, whether germline, functional, or somatic. The ability to run GATK locally or scale analyses on platforms like Galaxy or Terra further empowers researchers with varying resources and expertise to harness its capabilities.

As bioinformatics becomes a necessary skill across biological sciences, tools like GATK will likely continue to shape how discoveries are made. Its balance of accessibility, power, and reproducibility makes it a foundational resource in the age of genomic data.

Conflict of Interest

DC is a former employee of the Broad Institute of MIT and Harvard. He heavily contributed to documentation and user education for the Genome Analysis Toolkit (GATK). DC has no ongoing affiliation and no financial interest in the continued use of GATK or related tools and platforms.

References

1. Saiki RK, Scharf S, Faloona F, Mullis KB, Horn GT, Erlich HA, Arnheim N (1985) Enzymatic amplification of beta-globin genomic sequences and restriction site analysis for diagnosis of sickle cell anemia. Science 230(4732):1350–1354. https://doi.org/10.1126/science.2999980
2. Appenzeller T (1990) Democratizing the DNA sequence. Science 247(4946):1030–1032. https://doi.org/10.1126/science.2309117
3. Lander ES (1996) The new genomics: global views of biology. Science 274(5287):536–539. https://doi.org/10.1126/science.274.5287.536
4. Mardis ER (2006) Anticipating the 1,000 dollar genome. Genome Biol 7(7):112. https://doi.org/10.1186/gb-2006-7-7-112
5. Service RF (2006) The Race for the $1000 Genome. Science 311(5767):1544–1546. https://doi.org/10.1126/science.311.5767.1544
6. International Human Genome Sequencing Consortium (2001) Initial sequencing and analysis of the human genome. Nature 409:860–921. https://doi.org/10.1038/35057062
7. Check Hayden E (2014) Technology: the $1,000 genome. Nature 507:294–295. https://doi.org/10.1038/507294a
8. Greninger AL, Naccache SN, Federman S, Yu G, Mbala P, Bres V, Stryke D, Bouquet J, Somasekar S, Linnen JM, Dodd R, Mulembakani P, Schneider BS, Muyembe-Tamfum J, Stramer SL, Chiu CY (2015) Rapid metagenomic identification of viral pathogens in clinical samples by real-time nanopore sequencing analysis. Genome Med 7(1):99
9. Langmead B, Nellore A (2018) Cloud computing for genomic data analysis and collaboration. Nat Rev Genet 19:208–219. https://doi.org/10.1038/nrg.2017.113
10. Mboowa G, Sserwadda I, Aruhomukama D (2021) Genomics and bioinformatics capacity in Africa: no continent is left behind. Genome 64(5):503–513. https://doi.org/10.1139/gen-2020-0013
11. Gauthier J, Vincent AT, Charette SJ, Derome N (2018) A brief history of bioinformatics. Brief Bioinf 20(6):1981–1996. https://doi.org/10.1093/bib/bby063
12. Muzzey D, Evans EA, Lieber C (2015) Understanding the basics of NGS: from mechanism to variant calling. Curr Genet Med Rep 3:158–165. https://doi.org/10.1007/s40142-015-0076-8
13. Pirooznia M, Kramer M, Parla J, Goes FS, Potash JB, McCombie WR, Zandi PP (2014) Validation and assessment of variant calling pipelines for next-generation sequencing. Hum Genom 8: 14. https://doi.org/10.1186/1479-7364-8-14
14. Mahmoud M, Gobet N, Cruz-Dávalos DI, Mounier N, Dessimoz C, Sedlazecket FJ (2019) Structural variant calling: the long and the short of it. Genome Biol 20: 246. https://doi.org/10.1186/s13059-019-1828-7
15. Koboldt DC (2020) Best practices for variant calling in clinical sequencing. Genome Med 12: 91. https://doi.org/10.1186/s13073-020-00791-w
16. McKenna A, Hanna M, Banks E, Sivachenko A, Cibulskis K, Kernytsky A, Garimella K, Altshuler D, Gabriel S, Daly M, DePristo MA (2010) The Genome Analysis Toolkit: a MapReduce framework for analyzing next-generation DNA sequencing data. Genome Res 20:1297–1303. https://doi.org/10.1101/gr.107524.110
17. Van der Auwera GA, O'Connor BD (2020) Genomics in the cloud: using docker, GATK, and WDL in Terra, 1st ed. O'Reilly Media, https://www.oreilly.com/library/view/genomics-in-the/9781491975183
18. GATK Team (2025A) GATK - Genome Analysis Toolkit. GATK Support Documentation, https://gatk.broadinstitute.org
19. Poplin R, Ruano-Rubio V, DePristo MA, Fennell TJ, Carneiro MO, Van der Auwera GA, Kling DE, Gauthier LD, Levy-Moonshine A, Roazen D, Shakir K, Thibault J, Chandran S, Whelan C, Lek M, Gabriel S, Daly MJ, Neale B, MacArthur DG, Banks E (2017) Scaling accurate genetic variant discovery to tens of thousands of samples. *bioRxiv*, https://www.biorxiv.org/content/10.1101/201178v3
20. GATK Team (2025B) Funcotator. GATK support documentation, https://gatk.

broadinstitute.org/hc/en-us/articles/4418054208155-Funcotator
21. GATK Team (2025C) How to call somatic mutations using GATK4 Mutect2. GATK support documentation, https://gatk.broadinstitute.org/hc/en-us/articles/360035531132--How-to-Call-somatic-mutations-using-GATK4-Mutect2
22. GATK Team (2025D) Best practices workflows. GATK support documentation, https://gatk.broadinstitute.org/hc/en-us/sections/360007226651-Best-Practices-Workflows
23. GATK Team (2025E) HaplotypeCaller. GATK support documentation, https://gatk.broadinstitute.org/hc/en-us/articles/360037225632-HaplotypeCaller
24. GATK Team (2025F) HaplotypeCaller in a nutshell. GATK support documentation, https://gatk.broadinstitute.org/hc/en-us/articles/360035531412-HaplotypeCaller-in-a-nutshell
25. GATK Team (2025G) Assigning per-sample genotypes: HaplotypeCaller. GATK support documentation, https://gatk.broadinstitute.org/hc/en-us/articles/360035890511-Assigning-per-sample-genotypes-HaplotypeCaller
26. GATK Team (2025H) Calling variants on cohorts of samples using the HaplotypeCaller in GVCF mode. Genome Analysis Toolkit (GATK). GATK support documentation, https://gatk.broadinstitute.org/hc/en-us/articles/360035890411-Calling-variants-on-cohorts-of-samples-using-the-HaplotypeCaller-in-GVCF-mode
27. GATK Team (2025I) Does GATK work on non-diploid organisms? Genome Analysis Toolkit (GATK). GATK support documentation, https://gatk.broadinstitute.org/hc/en-us/articles/360035889691-Does-GATK-work-on-non-diploid-organisms
28. GATK Team (2025J) Mutect2. Genome Analysis Toolkit (GATK). GATK support documentation, https://gatk.broadinstitute.org/hc/en-us/articles/360037593851-Mutect2
29. GATK Team (2025K) FAQ for Mutect2. GATK support documentation, https://gatk.broadinstitute.org/hc/en-us/articles/360050722212-FAQ-for-Mutect2
30. GATK Team (2025L) Germline short variant discovery (SNPs + Indels). GATK support documentation, https://gatk.broadinstitute.org/hc/en-us/articles/360035535932
31. GATK Team (2025M) (How to) Run germline single sample short variant discovery in DRAGEN mode. GATK support documentation, https://gatk.broadinstitute.org/hc/en-us/articles/4407897446939--How-to-Run-germline-single-sample-short-variant-discovery-in-DRAGEN-mode
32. GATK Team (2025N) GermlineCNVCaller – GATK. GATK support documentation, https://gatk.broadinstitute.org/hc/en-us/articles/360037226352-GermlineCNVCaller
33. GATK Team (2025O) (How to) Call rare germline copy number variants – GATK. GATK support documentation, https://gatk.broadinstitute.org/hc/en-us/articles/360035531152--How-to-Call-rare-germline-copy-number-variants
34. Babadi M, Fu JM, Lee SK, Smirnov AN, Gauthier LD, Walker M, Benjamin DI, Zhao X, Karczewski KJ, Wong I, Collins RL, Bizzotto S, Satterstrom FK, Neale BM, Daly MJ, Talkowski ME (2023) GATK-gCNV enables discovery of rare copy number variants from exome sequencing data. Nat Genet 55(9):1589–1597. https://doi.org/10.1038/s41588-023-01449-0
35. Karczewski KJ, Francioli LC, Tiao G, Cummings BB, Alföldi J, Wang Q, Collins RL, Laricchia KM, Ganna A, Birnbaum DP, Gauthier LD, Brand H, Solomonson M, Watts NA, Rhodes D, Singer-Berk M, England EM, Seaby EG, Kosmicki JA, Walters RK, Tashman K, Farjoun Y, Banks E, Poterba T, Wang A, Seed C, Whiffin N, Chong JX, Samocha KE, Pierce-Hoffman E, Zappala Z, O'Donnell-Luria AH, Minikel EV, Weisburd B, Lek M, Ware JS, Vittal C, Armean IM, Bergelson L, Cibulskis K, Connolly KM, Covarrubias M, Donnelly S, Ferriera S, Gabriel S, Gentry J, Gupta N, Jeandet T, Kaplan D, Llanwarne C, Munshi R, Novod S, Petrillo N, Roazen D, Ruano-Rubio V, Saltzman A, Schleicher M, Soto J, Tibbetts K, Tolonen C, Wade G, Talkowski ME, Genome Aggregation Database Consortium, Neale BM, Daly MJ, MacArthur DG (2020) The mutational constraint spectrum quantified from variation in 141,456 humans. Nature 581(7809):434–443. https://doi.org/10.1038/s41586-020-2308-7

Chapter 7

Physics Determines and Evolution Shapes Basic Units of Protein Function and Allosteric Regulation

Igor N. Berezovsky

Abstract

We discuss here physics and evolution of globular proteins with a focus on the basic units of their structure and function, closed loops and elementary functional loops. A starting point of the journey here is a prebiotic evolution, in which short linear peptides and corresponding RNA duplexes with traits determined by demands to survive and to move on in the harshness of the Origin of Life emerged first. Next, we follow the fate of ring-like peptides, which apparently were the "Dayhoff fragments" that passed through abiogenic transition, formed first functional domains followed by their inclusion in multidomain structures, formation of complexes, assemblies, and molecular machines in later stages. I argue that physics, specifically polymer nature of protein chains, not only served as a determinant of the basic units of proteins—closed loops but also could play a decisive role in determining the size of another important structural unit—domain—based on the polymer nature of nucleic acids that governed the optimal ring closure size of DNA molecules. The protein closed loops served as scaffold for carrying elementary functions—descendants of ancient ring-like peptides, combinations of which provided the modern protein function universe. Another alliance between physics and evolution discussed here is the allosteric regulation of protein function, which is based on the structural dynamics underlying the allosteric signal transduction and its regulatory role. While the physics drive the structure-based conserved patterns of allosteric signalling determined by the folds, the evolution brings in a sequence diversity, allowing to alternate the allosteric communication and to make it archetypal for distinct functional (super)families using the same fold as the structural platform. Considering only few above aspects, I show that important lessons from the hand-in-hand walk of physics and evolution already helped to achieve the current state-of-the-art in understanding of protein structure and function. I conclude, however, that there is still even more to learn from Nature about long and remaining mainly mysterious 3.5 billion years endeavor.

Keywords Protein function, Protein evolution, Linear peptides, RNA duplexes, Dayhoff motifs, Polymer physics, Persistence length of the polymer, Closed loops, Elementary functional loops, Protein folds, Protein domains, Protein dynamics and allosteric regulation of protein function

1 Introduction

It has been almost 70 years since John Kendrew determined the first 3D structure of protein, myoglobin, using the X-ray analysis [1], followed by Max Perutz who obtained the structure of hemoglobin in

Gustavo Caetano-Anollés (ed.), *Evolutionary Genomics*, Methods in Molecular Biology, vol. 2981,
https://doi.org/10.1007/978-1-0716-4836-0_7,

1959 [2]. Less than a dozen of protein structures was obtained within the next 10 years, reflecting rather slow and difficult start of what one could call now *the protein research era*. Then, it picked a pace on both experimental and theoretical fronts. The former evolved from improving the resolution of X-ray structure of individual proteins to turning crystallography into the proteomic technique, adding the NMR and Cryo-EM as approaches for structure determination, achieving in them resolution comparable to that in the X-ray analysis, and turning them into high-throughput components of structural proteomics. On the theoretical side, it was a journey that actually started much earlier from the lock-an-key model of the protein function proposed by Emil Fisher in 1894 [3]. Fischer argued that a chemical reaction can be performed in the protein only if molecules with similar structures approach each other closely and interact; hence, there should be like lock-and-key fit between the enzyme and the substrate. Coming to the edge of 1960–1970, several important concepts, such as the Anfinsen's dogma, or thermodynamics hypothesis, that the native structure is determined only by the protein amino acid sequence [4] and the Levinthal's discussion on the existence of protein folding pathways driving the protein to the native structure [5] were proposed, forming a basis for future theoretical works. For example, numerous efforts, many of which remained on the level of toy models [6], were made to resolve so-called Levinthal paradox. The latter is actually only an interpretation of the original Levinthal's proposition [5] that "a pathway of folding means that there exists a well-defined sequence of events, which follow one another so as to carry the protein from the unfolded random coil to a uniquely folded metastable state." This statement rather agrees with a modern understanding of the protein folding as a kinetic co-translational process [7–11] supported by corresponding observations [12–15] and underlying current models [10, 16–19]. Many other models, on the other hand, became instrumental in studies of different aspect of protein structure and function, such as, to name a few, physical mechanisms of protein stability [20–27] and adaptation [20, 27–32], protein–protein interaction [20, 32], and functional dynamics and its regulation [33–36]. The latter has gained momentum after understanding of the universality of a dynamics nature of proteins [37–39], which was observed in NMR [33] and other experiments and, then, formally described in the energy landscape models [34]. Remarkably, one of the most intricate aspects of structure dynamics-based functional regulation, the allostery [40], was first hypothesized based on the biochemical data for only handful number of proteins [41], then formalized in two phenomenological models Monod–Wyman–Changeux [42] and Koshland–Nemethy–Filmer [43] already in 1960. The energy landscape view, in turn, provided a foundation for the development of more advanced models of allostery considering the ensemble of functional conformations instead of the unique native state [35]. Starting from useful, yet phenomenological models [44–48], we developed a rigorous physical models based on statistical mechanical consideration [48–54], which we used in the analysis of allosteric regulation [49, 50, 55–57] in prediction of allosteric sites [52, 54], in diagnostics [58–60], and in the development of new class of medicines—allosteric drugs [58, 61–63] and biologics [54, 57, 64], and implemented in web-resources [55, 56, 65, 66].

The brink of the twenty-first century brought a wealth of sequence/structure data obtained in massive genomic/proteomic efforts and further complemented by the leap in computational power that resulted in a recent breakthrough in the development of AI/ML-based methods. The natural result of this progress is perfectly represented by the RosettaAI-based protein design implemented in web-resources [55, 56, 65, 66] developed by David Baker and colleagues and in the AlphaFold model for the protein structure prediction [67] recognized by the 2023 Nobel Prize in Chemistry. Considering these already outstanding achievements and great promises of AI-based approaches, do we still need to talk about basic physics and evolution of the protein structure? The answer is yes, because we want to develop an efficient rational design of desired functions [68, 69] with necessary specifics and details [70, 71], making a shortcut in design efforts and using what is already discovered and optimized by Nature after long evolutionary struggle. To this end, it will be necessary to understand basic building blocks of the structure and function [21, 27, 72–76], how they formed modern folds, then how these folds evolved adapting to different environmental extremes [28–32] and comprising new functions [21, 70, 71, 74, 77–80]. All this knowledge will be instrumental in future efforts, providing natural functional and evolutionary optimized units for the AlphaFold-type predictions [81] and Rosetta-based design [67].

In this review based on my works and on relevant studies by others we will focus on the very basic, but fundamental aspects determining the universal basic structural and functional units of globular proteins and their evolutionary fate. First, we will briefly discuss how basic demands on molecular stability determined the codon and amino acid chronology [82–84], then formation of short peptides and RNA duplexes followed by their fusion in ring-like peptides with elementary functions [84, 85]. Next, the role of polymer physics in determining the size of ring-like peptides, descendants of which became the basic universal units of protein structure and function [21, 74, 86]—closed loops [17, 87–89] and eventually served as scaffolds for the basic units of the protein function—Elementary Functional Loops (EFLs, [70, 77–79]) will be discussed. The focus of this work will be on showing the relevance of *physics-determined and evolutionary-approved* closed loops to the major aspects of structure and function of contemporary soluble/globular proteins, from co-translational folding to formation of domain structure, emergence and evolution of current diversity of protein functions, and protein dynamics-based functional allosteric regulation.

2 From Prebiotic Chemical Evolution to Abiogenesis and Contemporary Protein Function

2.1 Prebiotic Linear Peptides as a Result of the Amino Acid and Codon Chronologies

To answer the question about minimal, quasi-independent, and stable common element of globular proteins, its major structural and functional characteristics, and rules and events that determined it, one should go 3.5 billion years back, asking what happened then and how it all started and trying to imagine how abiogenic transformation from primitive prebiotic functions and structures to sophisticated molecular catalysis and machines of modern life forms had occurred. Seminal Miller's experiments unraveling possible compositions of the "primordial soup" [90, 91] with a minimal set of first simple amino acids [71, 82, 83, 85], prompted researchers to think about first steps of prebiotic chemical evolution. Only basic physics and chemistry, operating in harsh environment on the scarce repertoire of small molecules [82, 83] and their components, determined strong limitations on characteristics of the first functional prebiotic compounds [71, 84, 85, 92, 93]. For example, an elegant scenario of a simultaneous emergence of first amino acids and their codons that gave rise to the RNA-protein world was proposed by Edward Trifonov [83]. He obtained the amino acid chronology, which is based on the chemical simplicity and reactivity of amino acids present in early proteins, showing how Miller's amino acids could become the first to emerge with respective codons being engaged as complementary pairs. The codon chronology started from the complementary codon pair GGC and GCC with increased stability that may form GC-rich RNA duplexes [83], and the complete chronology emerged as a result of single nucleotide changes from the previously engaged codons. Both GGC and GCC are codons for amino acids alanine (Ala) and glycine (Gly), which originated two corresponding sets of codons via the wobble single-nucleotide mutations (with only few exceptions when transitions took place [83]) in codons and by the complementary coping. Noteworthy, the Eigen and Schuster's assumption on the priority of thermostability and complementarity [94] is clearly manifested in the obtained chronology as the most stable codons were utilized first. Two resulting alphabets of amino acids originated from the Gly and Ala are encoded by the codons with the central purines or pyrimidines, respectively. The Gly alphabet contains predominantly polar/charged amino acids (G, D, S, E, N, R, K, Q, C, H, Y, and W), while the Ala alphabet contains mostly hydrophobic residues (A, V, P, S, L, T, I, F, M). Interestingly, the existence of polar/charged and hydrophobic alphabets of amino acids can explain an observation that simple binary code of polar and nonpolar amino acids was useful in design of simple α-helical folds via arrangement of polar/nonpolar residues in the appropriate order [95].

Two amino acid alphabets hypothesized above can originate poly-Gly and poly-Ala peptides of seemingly unknown structure determined by the short RNA duplexes of corresponding size. It appeared, however, that simple arguments on characteristics of $(Ala)_n/(Gly)_n$ homopeptides and respective RNA $(GCC:GGC)_n$ duplexes can help to estimate their sizes [84]. The shortest length can be inferred from the thermostability of GC-rich RNA duplexes, which are stable at 90–100°C starting from the ten base pairs size [96] that corresponds to three to four residues of the peptide size. On the other hand, the size of soluble peptide cannot exceed seven to eight residues [97], setting the upper limit for the peptide length. We performed the autocorrelation analysis of residues belonging to Ala and Gly alphabets in sequences of 23 complete proteomes, which revealed traces of mosaic structure in modern protein sequences that was formed by the peptides of these alphabets. The autocorrelation also showed that size of peptides was about five to six residues, which is in a good agreement with three to eight residues estimate for linear peptides obtained in the study of the evolutionary transition between prebiotic amino acids and early proteins [98]. What happened to these short peptides and nucleic acid duplexes and how their traces ended up being present in sequences of modern proteins? It would be hard to reconstruct the whole sequence of events, but one can imagine that physics could provide something better defined and stable enough to preserve important traits (e.g., one-two functional residues)—to become ready for abiogenic transformation and for the start of biological evolution. Our analysis, indeed, prompted us to believe that physics did that, and next two stages in the protein evolution were apparently at least affected if not determined by the polymer nature of proteins and nucleic acids [84].

2.2 Closed Loop of Nearly Standard Size Is the Universal Basic Element of Globular Proteins

It started in our studies of the energy hierarchy of protein domain structure [23–26, 99, 100] where I arrived to a question about basic common element of protein structure and function and its potential determinant(s) [21]. Recognizing that the protein polypeptide chain defines the overall architecture and topology of the protein fold, I assumed that tracing the protein backbones and quantifying differences between the numbers of their returns in different parts of structures reveal densely packed stable structural units (subdomains) of globular proteins. The distribution of backbone sub-trajectories with short distances between their ends—closed loops or returns of the polypeptide chain—revealed a preferential contour length of 25–30 residues (Fig. 1A, [87–89]). It appeared that globular proteins in both eukaryotes and prokaryotes have closed loops of nearly standard size, regardless of the protein size, architecture, and topology of domains/folds, secondary structure composition, and their functional involvement [21]. We have complemented studies of closed loops in structures of proteins with the analysis of autocorrelation between all amino acid residues in sequences of 23 complete prokaryotic proteomes (Fig. 1B), which showed characteristic distance between them matching to the size of closed loops [84]. Additionally, Fig. 1C presents a distribution of lengths of gapless segments in the Conserved Domain Database

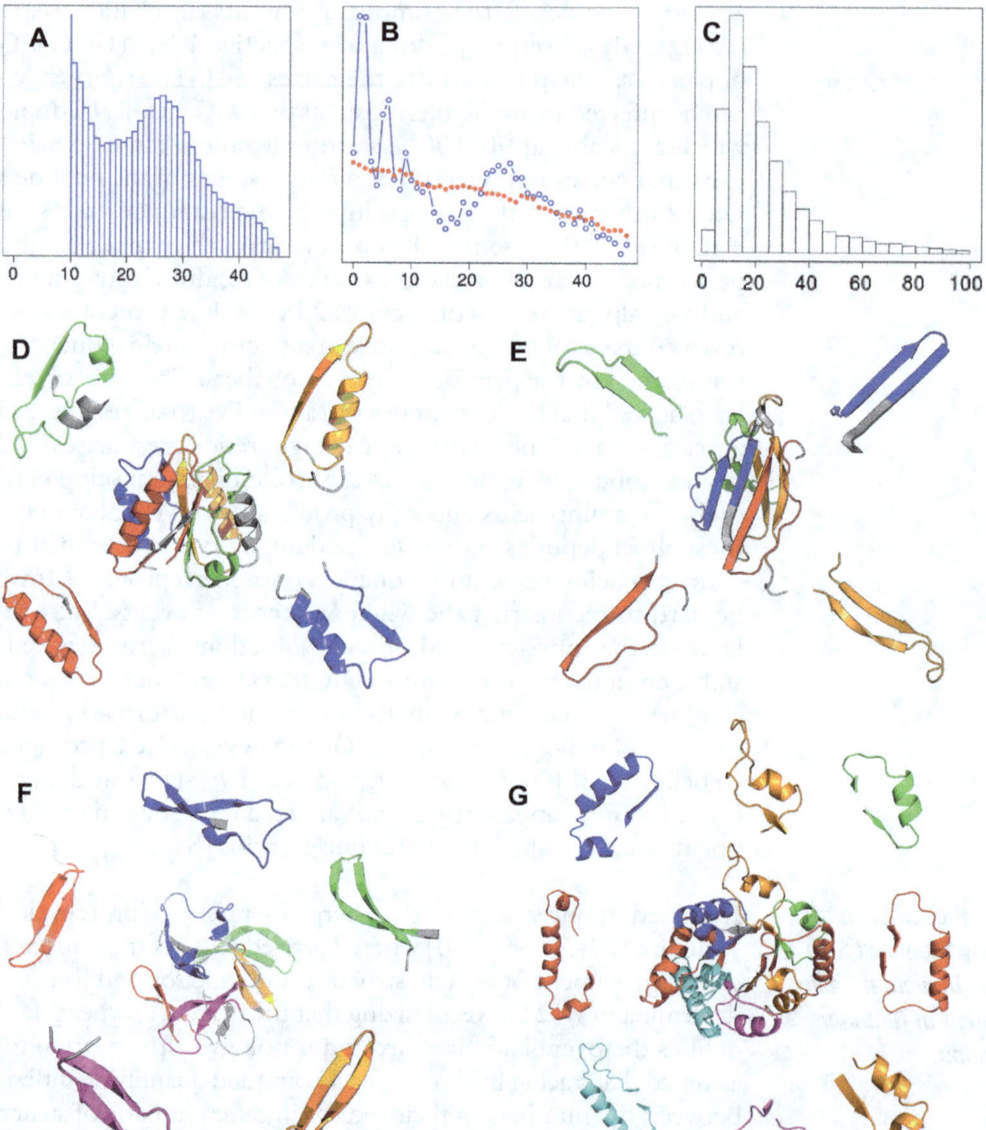

Fig. 1 Closed loops of nearly standard size of 25–30 residues is universal basic element of globular proteins. **A** Distribution of number of closed loops along the polypeptide chain with the contacting distance ≤ 8Å, calculated for the PDBSELECT database [87]. **B** Autocorrelation of all amino acid residues along the protein sequence in 23 complete proteomes (blue points), control—autocorrelation obtained on reshuffled proteomes (for details see [84]). **C** Distribution of lengths of gapless segments in the CDD alignments [101] of protein domain families. **D** Doubly wound fold, flavodoxin, *C. beijerinckii* (PDB ID: 2fox). **E** Immunoglobulin fold, Bence-Jones protein, *H. sapiens* (PDB ID: 2rhe). **F** Trefoil fold, Interleikin-1beta, *H. sapiens* (PDB ID: 1i1b). **G** α/β-barrel fold, triosephosphate isomerase, *T. brucei brucei* (PDB ID: 4tim)

(CDD, [101]) alignments of protein domain families, demonstrating that continuous functional signatures can be well incorporated into closed loops that serve as their structural carriers. Analysis of functional signatures in Pfam, Blocks, and Prosite databases [21] corroborates above conclusion. There is a number of works performed by others providing different direct and indirect evidences of the existence, evolutionary selection for, and functioning of closed loops on the basis of the following arguments: (i) sequence-based—estimated size of ancestral exons [102], recombinatorial swapping or protein sequences [103], sequence conservation [104], and size distribution of the insertions and deletions [105]; (ii) structure-based—centripetal structural modules of proteins [106], contact statistics and probability distribution [107, 108], protein fractal characteristics [109], Delaunay [110] and Voronoi [111] tessellations, comparison of different folds [112], decomposition into supersecondary structure modules [113]; (iii) experimental [12–14, 114], and (iv) other arguments [18, 115–118]. Figure 1D–G exemplifies four major folds, Doubly wound, Immunoglobulin, Trefoil, and α/β-barrel, represented through their closed loops, clearly illustrating persistence of the backbone returns and their lengths regardless of the type of fold and its secondary structure composition (details are in [21]). It clearly shows that different folds have few major types of loops: (i) with β-turn-α(-turn) motif, which actually dominate most designable and evolutionary advantageous α /β-folds, such as α/β-barrels/sandwiches/repeats, or (ii) β-turn-β structural motifs typical for many different types of all-β structures, such as β-trefoil/roll/prism/propeller/solenoid. This picture points to evolutionary connections between folds, which are built from similar closed loops. Indeed, the common ancestry was shown elsewhere for α/β-barrel and flavodoxin-like folds having conserved $(\alpha/\beta)_2$ motif [119], as well as for Rossman-fold enzymes [71, 79, 120], and for other α/β-folds [71, 79, 121, 122]. There are also many indications that the loop closure into β-hairpin-like structure may started the evolution of all-β folds. Remarkably, three commonly occurring closed loop-like structural patterns α-α, β-β, and β-α-β were found already in 1976 by Levitt and Chothia based on the analysis of less than a dozen of available at that time solved structures [123].

The closed loops or returns of the polypeptide chains, descendants of ancient ring-like peptides, in modern proteins are involved into the loop-and-lock structure of globular proteins [124]. The loop closure is facilitated by the van der Waals locks formed by the short 3–5 residue long segments of hydrophobic residues [124]. The closed loops also provide a foundation for discrete structure of protein domains and their alteration on different levels of hierarchy [125], which can be explored using the DHcL (Domain Hierarchy and closed Loops) web-server [126]. The interactions between van der Waals locks, in turn, contribute to the formation-stabilization of the protein globule's hydrophobic core complemented by additional stabilizing contacts in the core and on the globule's surface [16]. The hydrophobic core, chiefly formed by the van der Waals locks fixing the loop ends, underlies a simple model of co-translational folding, in which consecutive loop closure of the nascent polypeptide chain is followed by the arrangement of formed loops into the final folds/domains. This picture agrees with the Levinthal's idea on the "existence of a sequence of events...." [5] that drive the protein to the folded state. Nowadays, phenomenology of co-translational folding is supported by experimental data on the importance of non-local contact closing the loops for initiation of protein folding observed in the folding kinetics studies [12–14, 114]. Remarkably, the closed loop formation in the start of co-translational folding [127] on the ribosome is supported by the size of the exit tunnel [128, 129] of ribosomal 50S subunit that accommodates 30–35 amino acids and controls the initial formation of the "low entropy structures" [130, 131]. It is also corroborated by the sheer data on a diversity of events in the folding pathways, including interactions with auxiliary cellular machinery, such as multiamynoacyl-tRNA synthetase (MARS) complex [132] and other objects in the crowded cellular environment [133–135]. All these participants and interactions between them, should be considered in future models of co-translational protein folding [10, 11, 15, 16, 19].

2.3 Polymer Nature of Protein Backbones is the Physical Determinant of Closed Loops

Considering the loop closure as an important stage in the protein evolution [84, 85], one should also explain physical foundation for the formation of ring-like peptides. We found that polymer nature of protein backbones was a key determinant in establishing universal units of globular proteins—closed loops of nearly standard size of 25–30 amino acid residues [87]. Briefly, flexibility of any polymer chain can be considered in generalized form by referring to a smooth worm-like chain described by its persistence length p [136]—a distance, at which the chain loses memory about its direction or, in numbers, the deviation of the chain's direction exceeds 60° from the original one. Since, the chain has two directions, the statistical segment, so-called Kuhn segment [137], with $2p$ length is also used for characterizing flexibility of the polymer. According to Shimada and Yamakawa theory [138], polymer chain has a highest probability to close into the ring at 3–4 persistence length (1.5–2 Kuhn segments), which is determined by the trade-off between the high energy of deformation for shorter chains and excessive entropy term for longer ones. Since circularization experiments for polypeptide chains with natural amino acid sequences/compositions had never been performed, one can only estimate the contour length of such closed loops. The data for persistence length of homopeptides obtained in classical experiments and theoretical calculations of the Flory school [139–141] can be used for these estimations. In particular, it was shown that the length of the persistence length (P_l) for homopolymers with various amino acid compositions is between 8 and 12 residues [140–142], with only two notable exception for Gly ($P_l = 2$, [140]) and Pro (P_l = 116, [141]). Depending on the amino acid content in the copolymer Gly-Ala, the persistence length's value ranges between 9 and 2 amino acid residues [139]. We have concluded, therefore, that 8–12 amino acid residues as a characteristics length of the statistical Kuhn segment would be a fair estimate for further considerations [87]. To this end, the ring closure for random mixed polypeptide chains will be 12–24 residues, which should be roughly doubled, making it about 20–50 residues to take into account higher rigidity of the secondary structure elements present (and involving roughly half of the structure) in modern natural proteins [87]. An interesting, though not directly connected elegant model of the loop closure emerged in the work on peptide nucleic acids (PNAs)—a tool in molecular targeting with the protein-like chain and nucleotide bases as side residues [143]. Specifically, molecular fluorescent beacons, PNA fragments of 10–20 residues with fluorophore and quencher at their ends, are used to detect binding via emission when bound to the DNA target. At the same time the beacon is optically silent in the unbound state, forming the ring-like structure with, thus, quenched fluorophore [144]. This experimental model could be used in further studies of the loop closure, aiming to explore enthalpy and entropy balance in the closed loop stability [145].

Since both structure, sequence, and the data on functional signatures manifest the omnipresence of closed loops of nearly standard size of 25–30 residues in all types of proteins, one may conclude that the "loop closure" was an important event and a stage in "prebiotic protein evolution" that provided an advantage of stability and conformational specificity necessary for the start of biological evolution and selection. The ring-like prebiotic peptides may be considered, thus, as candidates for "Dayhoff fragments" that became the first proteins in the process of abiogenesis [68], then, according to seminal Margaret Dayhoff idea [146], fused in larger structures apparently forming first domains that eventually originated the most ancient LUCA set [68]. Descendants of the Dayhoff's ring-like peptide were also independently hypothesized in 1969 by the great biologist, Nobel Laureate Jacques Monod (citation is in [147]), already mentioned here as one of the inventors of Monod–Wyman–Changeux (MWC) models of allostery [42]. Monod proposed that "Since amino acids in the active side are linked together by relatively long loops (Fig. 2A), it is of some interest to estimate the minimum size required for each of these loops, when the conformation of the active site itself is prescribed. This should lead in particular to one lower bound for the molecular mass of a globular protein carrying one active size" [147]. And another great scientist, Nobel Laureate in physics, Pierre-Gilles de Gennes motivated by Monod's idea performed this estimate using simple statistical consideration. Figure 2A is a replica of the original figure in the chapter "Minimum number of amino acids required to build up a specific receptor with a folded polypeptide chain" of de Gennes book [147], showing how he imagined the "receptor site"

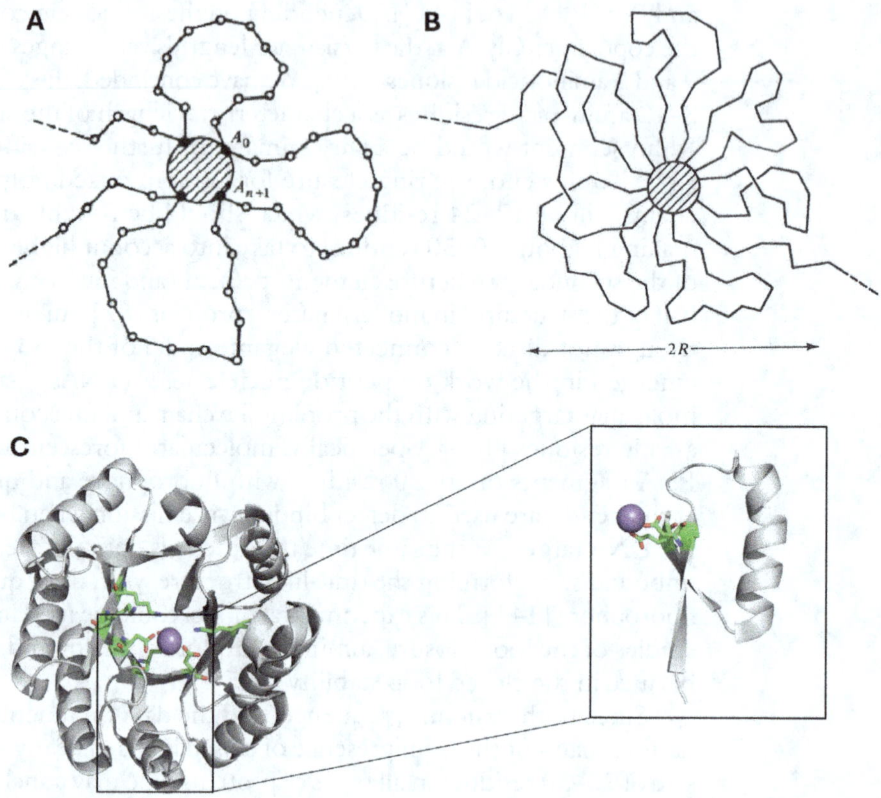

Fig. 2 Closed loops and elementary functional loops (EFLs). **A** de Gennes view on the Monod idea (Figure 2.1 in [147]). **B** Illustration of the de Gennes estimate of domain size (Figure 2.6 in [147]). **C** Elementary Functional Loop (EFL)—closed loops with one of few functional residues (functional in wide meaning—catalytic, interaction in signaling, or structural role [21, 70, 79])

with four amino acids formed by the loops. He found that the lower limit for the loop size will be 13 amino acid residues, which is in a perfect agreement with the estimate based on the ring closure of the polypeptide chain with mixed amino acid composition and free from the secondary structure elements.

2.4 Size of the Protein Domain and Its Potential Determinants

Next, de Gennes estimated a size of the whole molecule (Fig. 2B shows a replica of the original figure in [147]), to which he found applicable a first approximation of many-loop problem: the lower limit for total number of residues is, then, estimated as $(p-1)(n+1)$, where p is the number of catalytic residues and n is the size of loop in number of amino acids. Rough estimate of the functional protein with 3–5 catalytic residues located in the loops of size 13 gives a minimal size of 39–65 residues for most basic proteins. Indeed, since first structures were most probably built from the small number of most ancient amino acids, their sizes could not reach the one of smallest modern folds around 50–100 amino acid residues

[21, 79, 148]. The functions of these first domains should also be very basic, representing rather simple combination of few elementary steps of the modern enzymatic transformations [70, 77, 78]. Additionally, the 39–65 estimate is obtained for polypeptide chain free of the secondary structure elements. The same consideration applied to their potential rigidity as we used to estimate the contour length of closed loops will result in about 80–130 amino acids in a simple protein or what we nowadays call domain/fold. Interestingly, already in 1929 Svedberg addressed the question of the protein size in his seminal work "Mass and size of protein molecules" [149]. From the observation of a constant size increment of about 160 amino acid residues in sedimentation fractions, he concluded that this is a typical size of an elementary individual protein. In another remarkable work dated back to 1944, Bresler and Talmud purely theoretically formulated the "minimal condition" for the stable globular protein [150, 151], proposing that van der Waals interactions are the major forces for globular protein formation and the hydrophobic nucleus should be covered by the hydrophobic envelope. Accounting for the proportion between the hydrophobic and hydrophilic residues under these conditions, they found that "sharply limited size" of about 130 residues is the archetype for a stable globular protein [150]. Noteworthy, the size of 130–160 amino acid residues is well within the range of typical protein domain sizes, from 100 to 200 residues, observed, regardless of the domain/fold type, in the analysis of crystalized proteins [152–154].

To large extent unexpectedly, but not completely excluded this protein size can also be a consequence of the polymer nature of DNA. The DNA circularization experiments [155–157] conducted in 1980 provided an estimate of about 500 base pairs for the optimal ring-closure size, corresponding to about 130 amino acids in the encoded protein chain. We hypothesized in [84], therefore, that there was a "domain stage" in protein evolution where the size of circular DNA played a role of the determinant for the domain/protein size. While many important works have been devoted to the emergence [148] and evolution of domain structure [158, 159] and function [159–161], our focus in the next part is to show what we found in our research about formation of functions from the units on the basis of the most basic universal element of globular proteins—the closed loop.

2.5 Elementary Functional Loops (EFLs): Basic Units of the Protein Function

Based on our understanding of closed loops as universal structural units of globular proteins [87, 89, 125], we started from the efforts on resolving the proteomic code [72] and spelling the protein structure [73, 86] using this code. Eventually, we came to the idea of the Elementary Functional Loops (EFL)—basic units of protein function [70, 77]. We defined the EFL as the closed loop or return of the polypeptide chain with a certain function [21, 70, 79]. In a wider meaning, the function can be a catalytic one that provides an elementary step in enzymatic reaction, or formation of the interaction in signaling protein, or purely structural facilitating the overall protein scaffold. Figure 2C illustrates the EFL of the catalytic function, in which the closed loop delivers catalytic residues to the functional site. Our definition of the Elementary Function (EF) was later supported by the hypothesis on uniform binding that provided the flexibility for evolutionary design and resulted in diversity of contemporary protein function [162]. In a number of works we showed that it is possible to reconstruct ancestral prototypes of modern elementary functions [70] and developed a computational framework for such reconstruction [77].

2.6 Computational Framework for the EFL-Prototype Derivation

The evolutionary prototypes of EFLs unravel deep evolutionary connections between different enzymatic functions [70] not detectable by the direct comparative analysis of modern proteins sequences. Therefore, a rigorous statistical approach for derivation of the EFLs prototypes, which are simplified ancient signatures not existing in modern proteins [77], was required. On the formal procedure side, the fundamental distinction of prototypes from consensuses obtained in ancestral reconstructions is determined by their derivation from sequences of proteins from distant (super)families and even different protein folds [77]. As a result, prototypes can find their representatives in proteins with no apparent phylogenetic, structural, or functional connections, contrary to ancestral sequences. The ancestral sequences, however, reflect only origins of functional signatures in certain (super)families for which they were obtained. The ancestral reconstructions are typically performed via phylogenetic analysis of related functional (super)families, and they based on the evolutionary model with mutation and amino acid substitution rates [163–165]. We developed rigorous statistical approach for obtaining prototypes/profiles [70, 77] of elementary functions (EFs)— an iterative procedure that produces a profile in form of a position specific scoring matrix (PSSM). Kullback–Leibler divergence [166] is used for weighting profile positions according to their informational content in the scoring function, allowing thus to discriminate between the matches with specific signatures from the non-specific ones. The calculation of empirical distributions for the estimation of the scores' statistical significance is performed on reshuffled profiles. The elimination of redundancy and generalization of profiles is completed by the iterative hierarchical clustering of profiles [70].

Below is a brief description of the framework for the prototype derivation (Fig. 3). The procedure is typically applied to complete proteomes of several organisms. Since proteomic sequences contain many sources for biases and redundancies, such as homologous proteins, domain swapping, recombination, and multiplication, these redundancies have to be removed, as they reflect recent events in the evolution of proteins [160]. To remove redundancy originated from domain swapping, procedure compares 80-residue long sequence segments for identity, because the average domain size is 80–150 residues [149, 152–154]. To gradually remove redundancy between domains down to 40% identity, sequences are clustered with CD-HIT [167] several times (NR filter, Fig. 3). Low complexity regions in sequences which contain repeats or have highly biased amino acid composition are masked with SEG [168]. Non-redundant domains are cut with 10-residue steps into overlapping 50-residue segments. Two 10-residue flanks can be used for adjusting final 30-residue prototype. It was assumed that the cost of insertion or deletion in a functional signature is higher compared to at an arbitrary position in the whole protein sequence, based on the observation that gaps are not distributed uniformly and multiple sequence alignments of remote homologs (below 25% sequence identity) contain well-aligned blocks without gaps [169]. Therefore, gaps in profiles of elementary functional loops are not allowed.

The procedure starts from the search for sequences in complete proteomes that are most closely related to the initial sequence segments (origins, Fig. 3A) with the goal to obtain a seed alignment with a frequency matrix constituting a profile. To find additional sequence matches and to update the profile, obtained profiles are then matched to the complete proteomes iteratively. This profile-sequence search is repeated until the profile no longer changes, hence, considered converged (Fig. 3B–D). The final profiles represent families of EFLs with specific signatures, which can be gradually expanded in the iterative procedure to more distantly related, but statistically significant matches. The procedure is unique, having some notable differences originating from the specific requirements of the prototype derivation task, although it may resemble the PSI-BLAST pipeline ([170], see for details [70]).

2.6.1 Weighting of Profile Positions by Information

Position-specific scoring matrices (PSSM) are calculated to score the profiles (for details of PSSM calculation see Supplementary Information in [70]). The profile scoring function ranks matches based on the similarity between sequence of the matched segment and signature of the profile. An uneven contribution of positions in the profile is also considered. Degeneration of the profile towards the random compositional background or rare amino acids because of overestimation of pseudocounts [171] is prevented, as well as decrease of the profile sensitivity because of underestimation of pseudocounts. To discriminate between matches with specific and non-specific signatures, positions are weighted proportionally to Kullback–Leibler divergence (D_{KL}) [166], which reflects the information content on position i relative to the random background:

$$D^i{}_{KL} = \sum_{j=1}^{20} \left[f_{i,j} \log_2 \left(\frac{f_{i,j}}{c_j} \right) \right],$$ where f is observed amino acid frequencies on position i and c is proteomic amino acid composition.

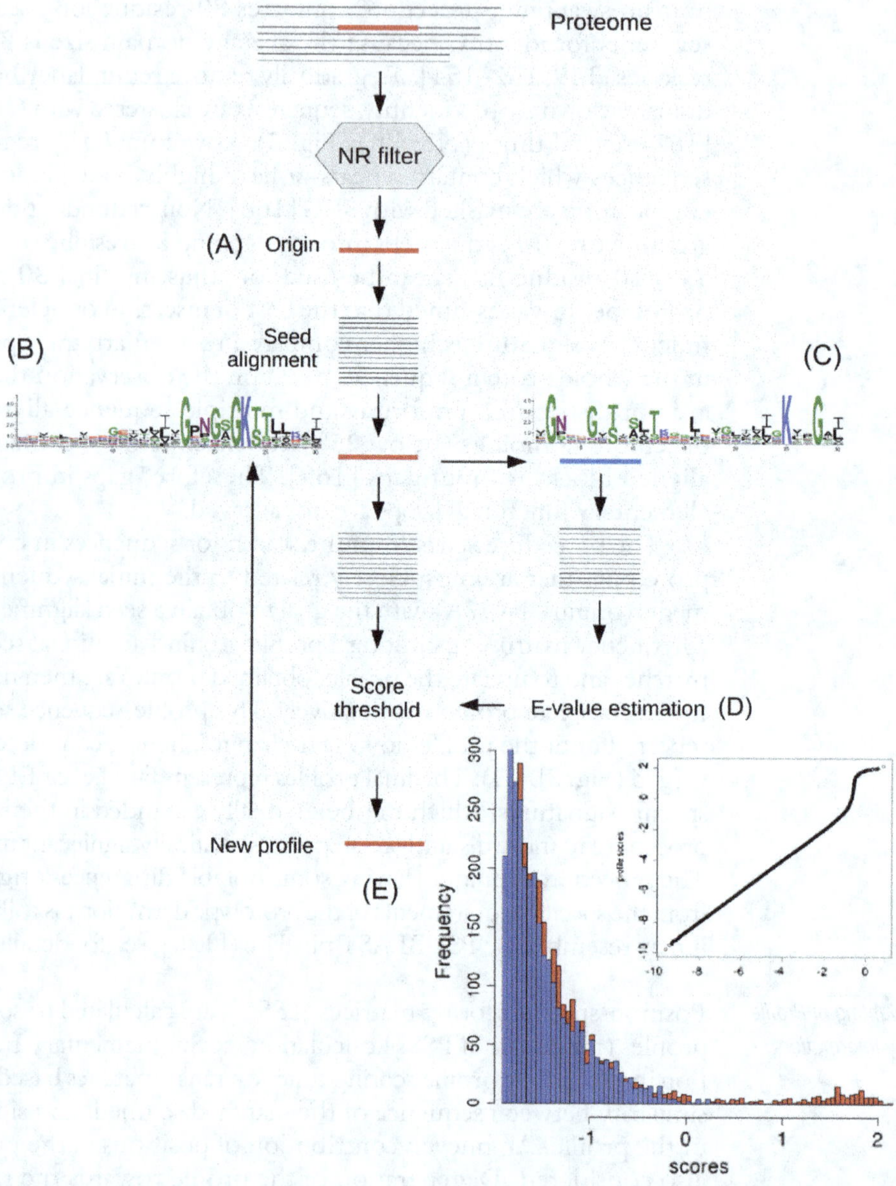

Fig. 3 The flowchart of procedure for the computational reconstruction of the prototype of elementary functional loop (EFL) . This is a schematic representation of the procedure. Non-redundant proteomes are matched against origins (**A**). As a result, seed alignments are obtained and profiles constructed (**B**). Profiles are iteratively updated with the significant matches collected on proteomes (**E**). Significance of the matches is estimated by comparing **D** the exact distributions of scores of the prototype with the score distribution of the prototype with reshuffled (**C**) positions (**D**). Procedure is repeated until prototypes converge (**C-E**).

The score of a sequence segment q to profile $P^{(n)}$ reads: $Score(q, P^{(n)}) = \frac{1}{n}\sum_{j=1}^{n} D_{\mathrm{KL}}^{i} m_{i,q_i}$, where $[m_{i,j}]_{i=1..n;j=1..20}$ is the corresponding PSSM with n positions. Profile positions with low information content ($D_{KL} < 1$ bit) do not contribute to the overall score, hence, they are omitted.

2.6.2 Empirical Calculation of the Background

The significance of a score is characterized by the -value, which is the number of false positive or unrelated matches above the particular score. The E-value is evaluated by comparing distributions of scores of the profile (s) with scores of the reshuffled profile (s_R): $E(s) = Np = N(1 - ecdf(s_R))$, where p is the p-value, N is the size of the combined proteome and $ecdf(s_R)$ is an empirical cumulative distribution function of the reshuffled profile scores. When randomized proteome is used as control instead of reshuffled profile, all the biases except amino acid composition are lost, resulting in an undesirable increase of the variance of scores for non-specific signals present in the sequence. Complete positional permutations that destroy relative distances between profile positions, similar to the combinatorial problem of pattern-avoiding permutations [172] are to be used for robust estimation of the -value below a certain significance threshold (e.g., $E \leq 1$) are used to construct the updated profile. Noteworthy, some of the profile positions have much higher information content than others, and these positions constitute the signature. The sequence identity is low on average and has a large variance, according to the comparison of all significant sequence matches to the consensus. Therefore, for proper -value estimation, i.e., for efficient separation of related matches from the unrelated ones, it is necessary to discriminate between informative and non-informative positions.

2.6.3 Hierarchical Clustering of Converged Profiles

The iterative procedure described above results in a set of converged profiles that should be further analyzed. First, there is a redundancy between these profiles, caused by the way in which their origins are obtained: they overlap with a step of 10 residues. Redundancy can also originate from the fact that different origins can converge to the same or very similar profiles, meaning that the origins are evolutionarily connected EFLs, but their similarity can only be detected with the help of the profile. The distance measure that considers all possible profile-profile alignments to hierarchically cluster the profiles introduced to resolve above problem. Since profile-profile comparison is more sensitive than profile-sequence comparison, more distant relations could be detected during profile clustering. This procedure results in the removal of redundancy and further generalization of the profiles. The profiles with the most generic signatures represent the functional characteristics of presumably original prototypes. All possible profile-profile alignments without gaps are performed by sliding one 50-residue long profile against the other 50-residue long profile and calculating pair-wise positional distances between all possible 30-residue windows $[a]_{30}$ and $[b]_{30}$, respectively. Distances between the pairs of corresponding positions are weighted proportionally to the information at each position (D_{KL}):

$$d([a]_{30}, [b]_{30}) = \sum_{i=1}^{30} \sqrt{(D_{\mathrm{KL}}^{a_i} + D_{\mathrm{KL}}^{b_i}) \sum_{j}^{20} (a_{i,j} - b_{i,j})^2}$$

The distance between two 50-residue profiles $[A]_{50}$ and $[B]_{50}$ is equal to the minimal distance between all possible sliding windows of size 30: $D([A]_{50}, [B]_{50}) = \arg\min[d([a]_{30}, [b]_{30})]_{a\in A; b\in B}$. Hierarchical clustering of profiles is an iterative procedure, where the most similar profiles ($\min[D(A, B)]$) are consecutively merged, resulting in a new and more generic profile.

2.6.4 Characterization of Prototypes

Prototypes are characterized and their elementary functions are assigned by obtaining their sequence matches in crystallized enzymes from ASTRAL/SCOP database [173, 174], which describe descendant of EFLs that diverged from the prototype. The elementary functions of derived prototypes and characteristic positions in their signatures are determined based on the known enzymatic function and mechanisms [175] identified in crystallized proteins (for details and further explanations see [70])

This framework was used for deriving collections of prototypes of EFs [70, 77, 80], implemented in the studies of the evolution of protein function in Archaea [78]. We illustrate here the catalog of EFLs collected in the above works with the involvement of one of the most common and ancient elementary functions—the P-loop EF of the phosphate binding in (di)nucleotide-containing ligands [70, 71, 79, 80]—in the ATP-binding domains. Figure 4 shows two ATP-binding domains: the ATPase domain of the ABC transporter [176] and the ATP-binding domain the Mre11-binding Rad50 protein [177]. ABC transporters are very common proteins present in all kingdoms of life, which transport molecules across membranes, and they are powered by the ATP provided by the cytoplasmic ATPase domain (A) shown here. In the Rad50 protein, the ATPase (B) works as a molecular switch, which controls the interaction with DNA through conformational changes upon the ATP binding and hydrolyzation. The value and the beauty of this example is in the fact that if the ATPase domain of the ABC transporter is a continuous cytoplasmic domain, the same domain in Rad40 is assembled in the "head" of the protein where two Rad50 subunits come together to make a dimer and form *discontinuous ATP-binding domain*. It is easy to see that three basic EFLs are present in both ATPases: Walker A (P-loop, red), Walker B (D-loop, green), and Q-loop (blue). However, in discontinuous ATPase of Rad50 the D-loop (green) is provided by one α-helical hand of the protein, while P-/Q-loops—by the other hand, making an excellent illustration of how independent EFLs can be fused to form required function and proving an idea of building a function from the descendants of ancient “Dayhoff peptides” [68, 71, 79, 92, 146].

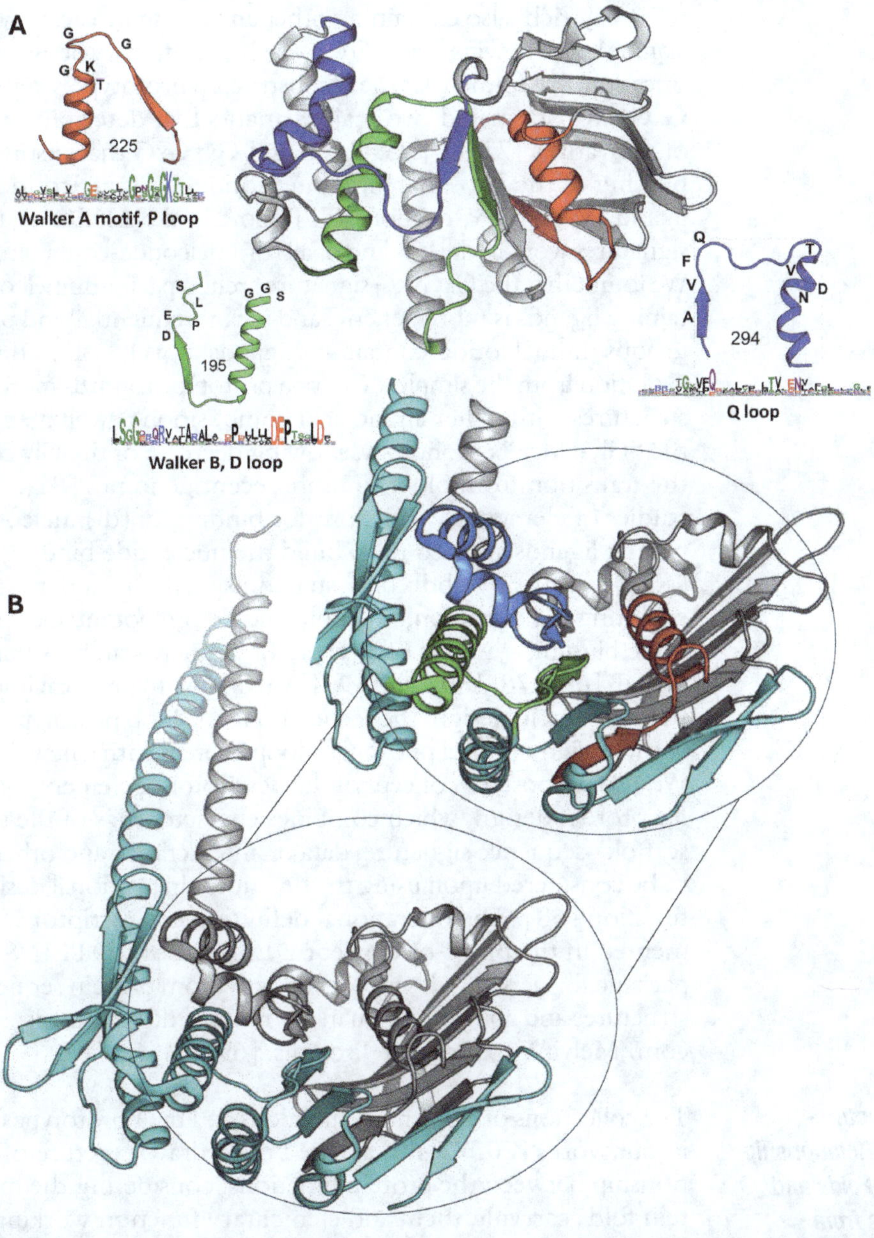

Fig. 4 Continuous (**A**) and discontinuous (**B**) ATP-binding domains and their elementary functional loops with profiles of the prototypes. **A** ATPase domain of the ABC transporter [176]; elementary functional loops and their sequence profiles: P loop (red), D loop (green), Q loop (blue). **B** ATP-binding domain the Mre11-binding Rad50 protein [177]

2.7 Gly-Rich Prototype is an Evolutionary Seed of the Phosphate Binding Function

The elementary function of the phosphate binding is not only present in multiple functions [68, 70, 79, 80, 92, 120, 178–181] and folds [182, 183], but also is one of the most ancient ones [70, 79, 184], giving us a chance to reconstruct the most ancient prototype, from which all constellation of the phosphate-binding functions in modern proteins had started [71]. The reconstructed prototype turned out to be Gly-rich, also containing other ancient amino acids (valine and alanine) and showing versatility towards evolutionary design and diversification. We found that the reconstructed prototype's key signature GxGGxG originated two major variants that determine a diversity of derivatives [71, 80]: GxGxxG and GxxGxG that distinguish the binding of the phosphate in dinucleotide- and nucleotide-binding ligands. With few exceptions (TPP, CoA, and NADP(H)), these two signatures work in the binding of all (di)nucleotide-containing ligands. We found that the GxGxxG signature archetypal for dinucleotide-containing ligands is more generic, and it can frequently bind phosphate groups in nucleotide-containing ligands as well [71]. Of note, the evolution from the simplest Gly-rich prototype towards more complex structures with richer amino acid composition is well manifested in the "Gly-clock" chronology, showing decrease of the Gly content in the transition from older to more recent genomes [82, 185]. The studies of elementary functions for binding of (di)nucleotide-containing ligands allowed us to build the nucleotide binding database (NBDB, https://nbdb.bii.a-star.edu.sg, [80])—a data source for exploring the evolution, function, and design potential of the nucleotide binding—a topic of interest for many researchers that use the NBDB [69, 178, 180, 186–193]. Moving on to practical implications such as protein design, the reconstructed P-loop prototype was used in the design of novel phosphate-loop protein with a new fold [181]. We also proposed a concept of the descriptor of elementary function for protein design, which combines characteristics of the structural scaffold, sequence signature, data on interactions, and other features to be considered upon using the descriptor in a rational design of the function [68]. The operational definition of descriptor was implemented in the proof-of-concept DEFINED-PROTEINS software package for deriving descriptors directly from protein sequences and structures and applying them in the rational design (https://github.com/MelvinYin/Defined_Proteins, [69]).

2.8 Structure-Function Relationship Between Folds and Functions from the Evolutionary Perspective

The collections of elementary functions and their prototypes obtained in our works [70, 77–80] allowed us to draw a picture of the relationship between the protein functions, considering the major protein folds carrying them and elementary function working as their basic units. Figure 5 illustrates a relationship between the functional domains (circles) from conserved domain database (CDD, [101]) via elementary functions (red diamonds), also showing major fold types that provide the structural scaffolds/platforms (marked by corresponding colors designated in the panel of folds on the right). Size of the circle indicates how many families are inside the superfamily. It is easy to see that several common folds (according to SCOP classification [194]), such as hydrolase, α/β-barrel, Rossmann-like, and some others provide scaffolds for many functional families, indicating their high designability [29, 148]. Analyzing relationships between functions and structures via elementary functions, one can obtain previously unseen picture of the presence of descendants of Dayhoff-like peptides of predomain evolution in different folds and functions of modern proteins. It is also possible to follow their evolutionary history and interactions resulting in new structures and functional involvements. Of note, circles with sectors of gray color show an involvement of additional structures into multidomain or oligomeric assemblies required for performing the function of a protein. Figure 5 is used here for illustrating this complex relationship, with many details and relevant discussions presented in our previous works [70, 71, 78, 79]. Accumulating all the data and knowledge, we, finally, proposed in [79] a general view on the emergence and evolution of the protein function. There are still many questions that remain unanswered in the fascinating field of the evolution of protein function, and recent illuminating works only corroborate the complexity of evolutionary relationships and diversity of their determinants [161, 195, 196], calling for further thorough studies. We, however, will move on to considering another aspect of protein function, in which closed loops are also major actor: the dynamics-based regulation of the protein activity— the allosteric regulation.

3 Protein Dynamics and Allostery: Physical Bases and Evolutionary Perspective

What is allostery? Why it is important for protein function and how is it relevant to what we focus on here—basic units of protein structure and function—closed loops? Answering on the first question, we start with a direct quote from the seminal 1963 work "Allosteric Proteins and Cellular Control Systems" [41], postulating that "The biological activity of many proteins is controlled via conformational alteration (allosteric transition) induced in the protein when it binds specific metabolites, which do not interact directly with the substrates or products of reactions. ... Allosteric proteins are assumed to possess at least two stereoscopically different non-overlapping sites: active site binds a substrate and is responsible for biological activity of the protein and allosteric site, upon binding of the effector, initiates reversible alteration of the protein's molecular structure and modifies properties of the active site." Authors also continued later in the work with an assumption that "Allosteric regulation is of such value to living beings that natural selection must have used it to the limit. ... Mutations occurring in an allosteric protein may modify its functional properties in a much larger number of ways." Remarkably, as we already mentioned it in the Introduction, authors of this work Jacques Monod, Jean-Pierre Changeux, and Francois Jacob made above conclusions based on the biochemical analysis of only handful number of proteins. The time, new and powerful methods, high-throughput data including hundreds of thousand structures and even more sequences, as well as opportunity to simulate structural dynamics only corroborated and substantiated their great insights [35, 38, 39, 48, 58, 197].

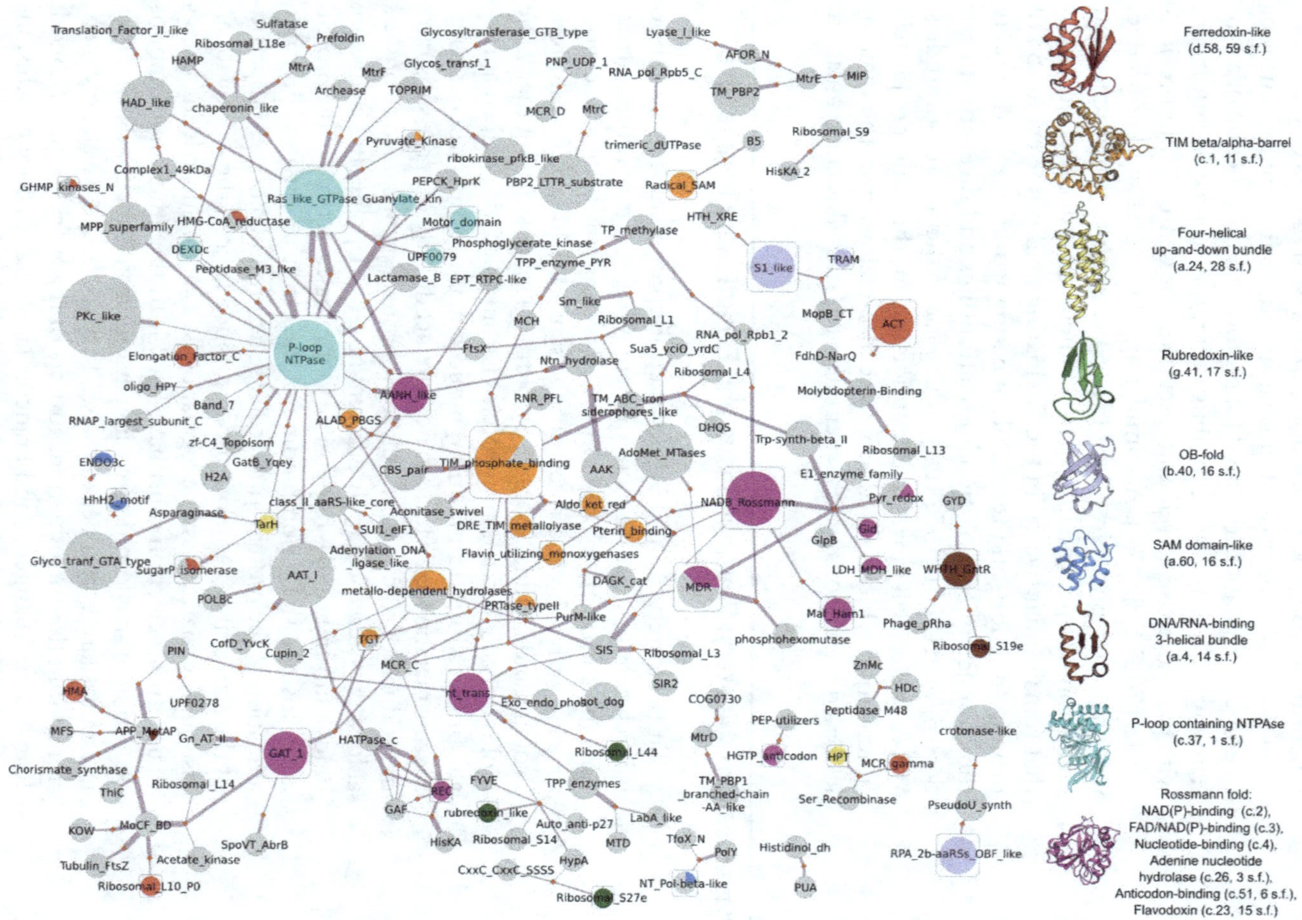

Fig. 5 The map of the structure-function relationships between CDD domains via elementary functions. CDD domains—circles; elementary functions—red rhombuses. Folds (designated according to SCOP, data is from [79]) are shown in squares, gray sectors in colored circles designate multi-fold/domain or complex structure (in the top diagram). Folds dominating in the multi-domain/fold structures are presented by their structures in the bottom diagram.

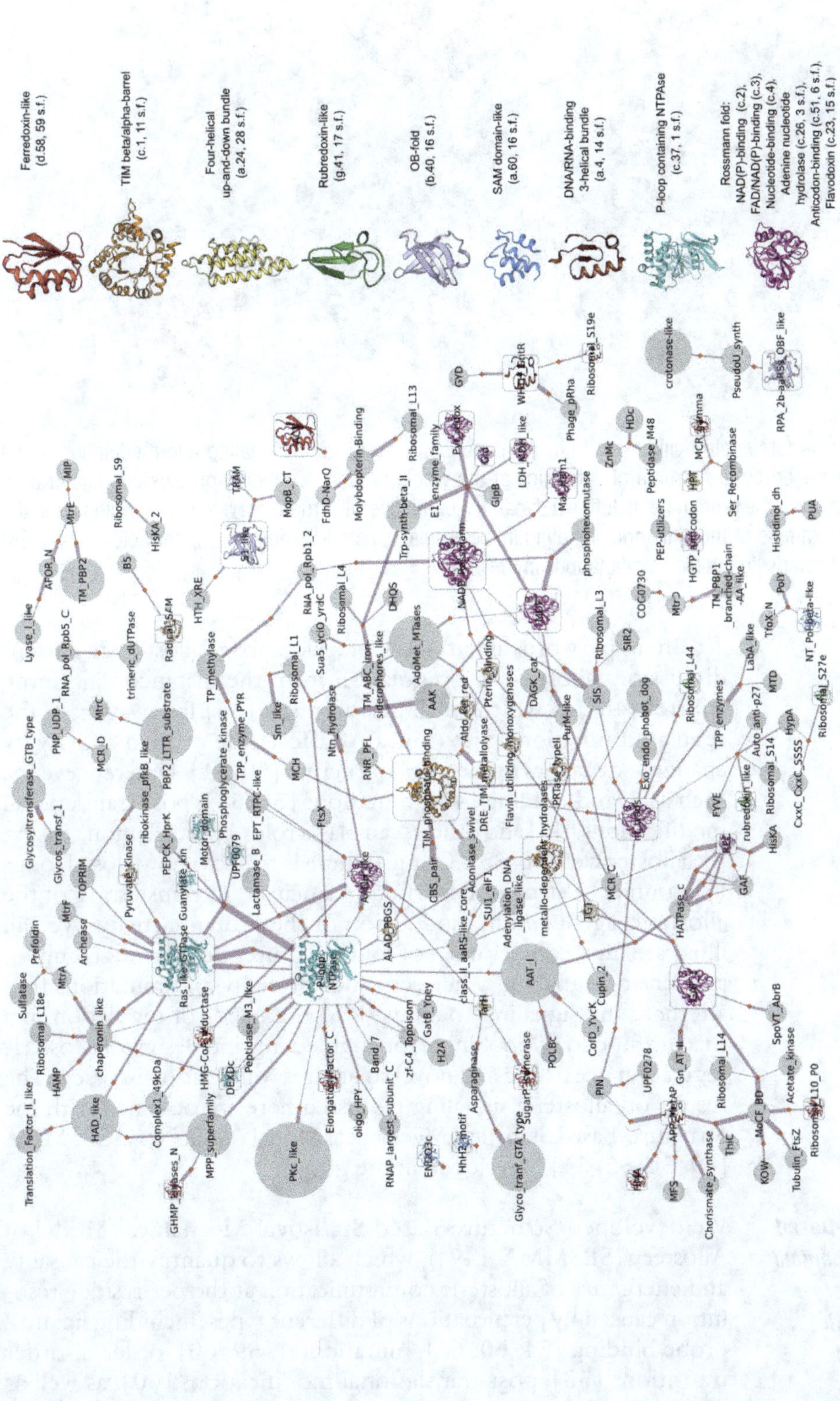

Fig. 5 (continued)

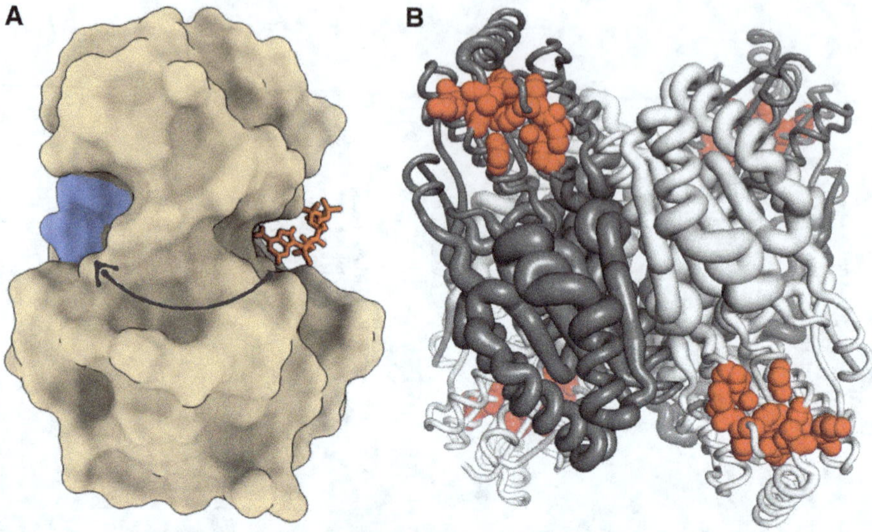

Fig. 6 Allostery—remote control via action from the distance. The allosteric signaling and modulation of the function are determined by the structural dynamics of the whole protein. **A** Binding of allosteric ligand (red, right) affects function of the protein (blue, left) via dynamics of the overall structure, which is changing upon the ligand binding. **B** Example of the allosteric signaling in a classical allosteric protein phosphofructokinase (PFK) upon binding of allosteric ligand (see explanations in the text)

In simple words and to show the relevance of allostery to current discussion: allostery is a regulation from the distance—a remote control [48], in which small perturbation [35] far away from the regulated functional site (Fig. 6A) affects the protein activity by changing dynamics of the whole structure [50, 54]. Different events, such as ligand binding [48], mutations [50, 58], post-translational modifications [47] and others, can play a role of perturbation causing the allosteric effect [35]. Our goal here is to show how closed loops being units of structural scaffolds participate in transferring of the allosteric signal and in modulation of the protein activity. We will illustrate how conservatism of fold structures determines common patterns of signaling, while sequence diversity and mutations may alternate the signaling to adjust it to the needs of regulation in a specific function. A potential for engineering and design of allosteric regulation in existing and novel structures will be also discussed. The results on allosteric signaling discussed here are obtained with the Structure-Based Statistical Mechanical Model of Allostery (SBSMMA [49, 50, 52, 54]) developed in the group.

3.1 Structure-Based Statistical Mechanical Model of Allostery (SBSMMA), a Brief Description and Examples

We developed Structure-Based Statistical Mechanical Model of Allostery (SBSMMA, [49]), which allows to quantify the causality and energetics of allosteric communication at the per-residue resolution caused by perturbations of different types, including ligand/probe binding [54, 60, 66], mutation(s) [59, 60], order-disorder transitions [53], post-translational modifications [60], as well as any combinations thereof. Figure 6B shows an illustration of the communication caused by the binding of allosteric ligand (shown in red) in a classical allosteric protein phosphofructokinase (PFK): the thickness of pipes reflects either negative (thin, negative free energy of allosteric signaling—stabilization of the affected residue/site) or positive (thick, positive free energy of allosteric signaling—potential conformational changes because of the increased dynamics of affected residue/site) allosteric modulation. The most important advantages of the SBSMMA before other models are: (i) it allows one to calculate the free energy of allosteric signal in real energy units—kcal/mol; (ii) it performs calculations in per-residue resolution, allowing to obtain a comprehensive picture of allosteric control over the protein function; (iii) the model can produce high-throughput results or provide detailed consideration, depending on the using the coarse-grained or all-atom input data. Recent sequence-dependent version of the model [51] uses the Myazawa-Jernigan residue-residue potential for the exact consideration of the amino acid types and effects of mutations. The SBSMMA was used for development of the computational framework for comprehensive allosteric control over protein activity [50, 54], including additional original techniques developed in the group, such as reverse perturbation for predicting allosterically linked locations and sites [52, 62], targeted analysis for identifying residues that originate desired allosteric signaling to targeted residues in the functional site [54, 62], and agnostic analysis revealing residues that are extensively allosterically coupled in the protein [54, 62]. Using this framework we explored signaling in many allosteric proteins of interest [35, 49, 50, 56], investigated allosteric effects of mutations, and introduced(based on the analysis of these effects) the notion of allosteric polymorphism [56, 58, 59]. We also showed how order-disorder transitions can trigger and drive the allosteric communication [53], and considered a balance between conserved allosteric signaling determined by the fold structure and its tuning via sequence modifications [57]. On more practical side, we described the allosteric perspective on the mutability and druggability of the SARS-CoV-2 Spike protein [60], proposed a computational approach for gaining comprehensive control of allosteric drugs and actions [62, 63], designed specific allosteric drug candidates for SARS-CoV-2 main protease and K-Ras oncoprotein [63]. We have also implemented the SBSMMA in web-resources (AlloSigMA server [65, 66] and AlloMAPS database [55, 56]), and the latter was recently updated with the high-throughput data on allosteric signaling derived for complete AlphaFold proteomes [55].

3.2 Allosteric Signaling and Role of Closed Loops as its Element

We start here from Allosteric Signaling Fingerprint (ASF, [55, 57]) obtained for the dihydropteroate synthase (DHPS)—shown on the left in the upper panel (Fig. 7). The fingerprint shows an effect of the perturbation of any protein position on all other residues in the

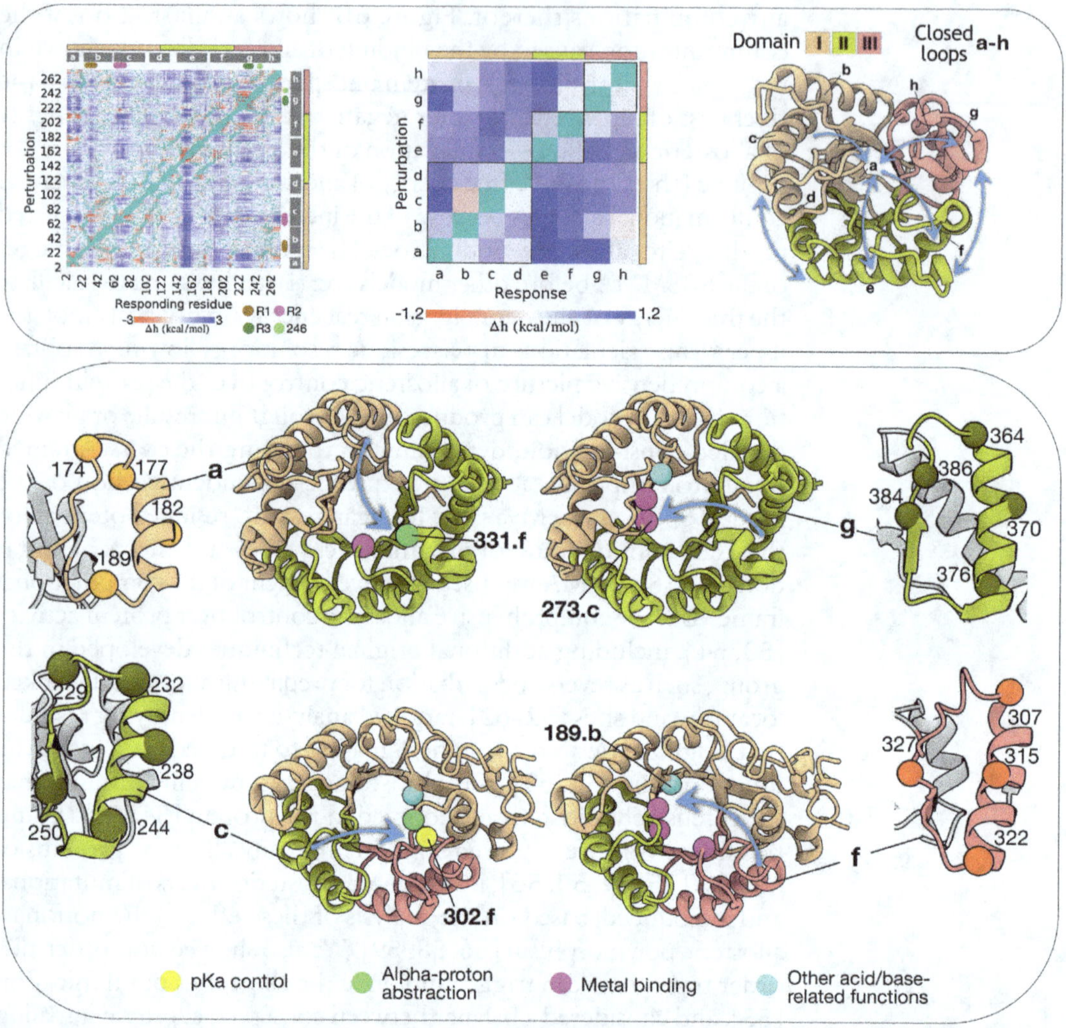

Fig. 7 Allosteric signaling via closed loops and allosteric effects of mutations. **Top panel**. Left: Allosteric Signaling Fingerprint (ASF) of the dihydropteroate synthase (DHPS) in per-residue resolution; Center: ASF of DHSP for closed loops a–h; Right: Example of conserved allosteric signals in αβ-barrel fold. **Bottom panel**. Allosteric effects of mutations on catalytic residues of Methylaspartate ammonia-lyase (MAL, top row) and Rhamonate dehydratase (RD, bottom row) of the Enolase superfamily

structure, providing the exhaustive picture of signaling [50, 55]. The perturbation does not depend on the original residue type on the position, showing the value of allosteric signal on the affected residue upon substitution from the bulky to small amino acid—the allosteric modulation range (for details see [50]). This approach allows us to obtain a compact 2D map of the exhaustive picture of allosteric signaling in the protein [50]. We later complemented it by the same representation for Allosteric Probing Fingerprint (APF, [55]), which represents the allosteric signal on all residues of the protein upon sequential binding of three-residue probe to all residues of the protein. The color gradient from red (negative allosteric free energy) to blue (positive) shows the type and value of signal sent upon perturbation: negative modulation—affected residue is stabilized, positive modulation—the signal produces some work on the residue potentially causing conformational changes, green indicates absence of the allosteric effect [50, 55]. The middle map in the upper panel (Fig. 7) shows the allosteric signaling between closed loops (a–h) that, in turn, form three domains shown on the structure (right, top panel, Fig. 7). We illustrate that α/β-barrel fold reveals conserved preferential signaling between corresponding closed loops and domain determined by the fold architecture (blue arrows), which can be alternated in many different functional (super) families because of their specific functions provided by corresponding changes in sequences. The consistent picture of the structure-based conservatism of allosteric signaling, in which closed loops play a role of elements of allosteric communication, and its alternation based on the sequence divergence was demonstrated for all types of folds, providing a background for the evolutionary-inspired design and engineering of desired allosteric regulation in targets of interest (for details see [57]).

3.3 Allosteric Effects of Mutations

Next, we consider in more details the sequence dependence of allosteric signaling in structurally conserved protein families. The lower panel in Fig. 7 provides an example of the family-dependent signaling in different members of large Enolase superfamily [198]—one of the well-studied mechanistically diverse enzymatic superfamilies, in which enzymes share a common reaction step—is used as the object of this study. According to SFLD [199], Enolase includes more than 10,000 proteins in 20 families, corresponding to different functions. These proteins share a common fold consisting of two domains: the α+β-capping domain in N terminal, which is determinant for the substrate specificity, and the α/β -barrel in C terminal, representing catalytic machinery (Fig. 7, bottom panel)—we focus on the catalytic domain here. The common reaction step is an abstraction of the α-proton of a carboxylate substrate mediated by conserved residues that tend to be in C terminal domain at the ends of the β-strands, constituting the active site. The families are arranged in seven subgroups by positions of residues involved in the conserved reaction step. We illustrate here diversity of signaling determined by specifics of functions in two members of the superfamily—Methylaspartate ammonia-lyase (MAL) and Rhamonate dehydratase (RD). The catalytic residues important for the function (colored according to their roles indicated in the Fig. 7, bottom panel) are delivered to the catalytic site by corresponding closed loops (a–h). Specific locations and interactions of these residues characterize the corresponding family function.

From the perspective of allosteric signaling, the work of catalytic residues can be affected by mutations located in places allosterically connected to catalytic ones, which are archetypal for corresponding functions. For example, the α-proton abstraction carried by the Closed Loop f (CLf) in MAL can be affected by mutations of residues 177 and 189 located in CLa (Fig. 7, top left structure in the bottom panel). At the same time, the α-proton abstraction in RH provided by CLg (Fig. 7, residue 329, green, bottom left structure), and forming the catalytic dyad together with 302 located in CLf, can be tuned by mutations of residues in CLc. Similarly, the Mg^{2+}-binding residue 273 in CLc of MAL is allosterically connected to residues in CLf (Fig. 7, bottom panel, top right), whereas N-terminal electrophilic catalyst residue 189 in CLb in RH receives the allosteric signal from residues in CLf of this protein (Fig. 7, bottom right structure). Referring to other aspects of allosteric effects of mutations, we showed in a series of works that the understanding of them enables the tuning of allosteric biologics [64, 200], prediction of new variants of concern in SARS-CoV-2 infection [60], diagnostic of pathologies cause by allosterically acting nsSNPs [59]. It can also be used in alleviating the drug resistance [58, 62, 63]. All these observations are in a full agreement with the great Monod-Changeux-Jacob's envision on the importance of allosteric mutations [41] dated back to 1963.

3.4 Closed Loop as Basic Elements in the Design of Allosteric Regulation

The understanding of allosteric communication between closed loops and their combinations can be also instrumental in the engineering and design of allosteric signaling in the existing and novel structure [57]. Figure 8 illustrates this potential using example of three proteins: Chemotaxis protein CheY with flavodoxin-like fold, HisF with α/β-barrel, and cytoplasmic leucine-rich repeat of the LRRC8A channel protein. All three structures are built from the same basic element—closed loop of the type β-turn-α-turn. The difference between these folds is in the numbers of loops forming them and in the arrangement in the fold. While eight loops in HisF form the closed relatively rigid barrel, four loops in CheY are arranged in slightly closed and more flexible half of a barrel. At the same time, 15 loops in the leucine-rich repeat are arrange in a pretty rigid and bulky horseshoe-like structure, in which all elements are stacked tightly on top of each other. To see the pattern of signaling and to analyze its persistence and modification, depending on the structure of the overall fold, we start from considering of allosteric signaling in the motif containing three closed loops of the type β-turn-α-turn. Three blue arrows indicate signaling in this motif from extracted different structures (on the left) show that the conserved pattern of signaling is similar for CheY and HisF proteins, but different in the LRR (Fig. 8, left panels with ASFs). However, while this pattern remains mostly unchanged (top row, central panel with ASF) within the CheY fold (Fig. 8, top row, right panel with ASF), it is strongly affected when

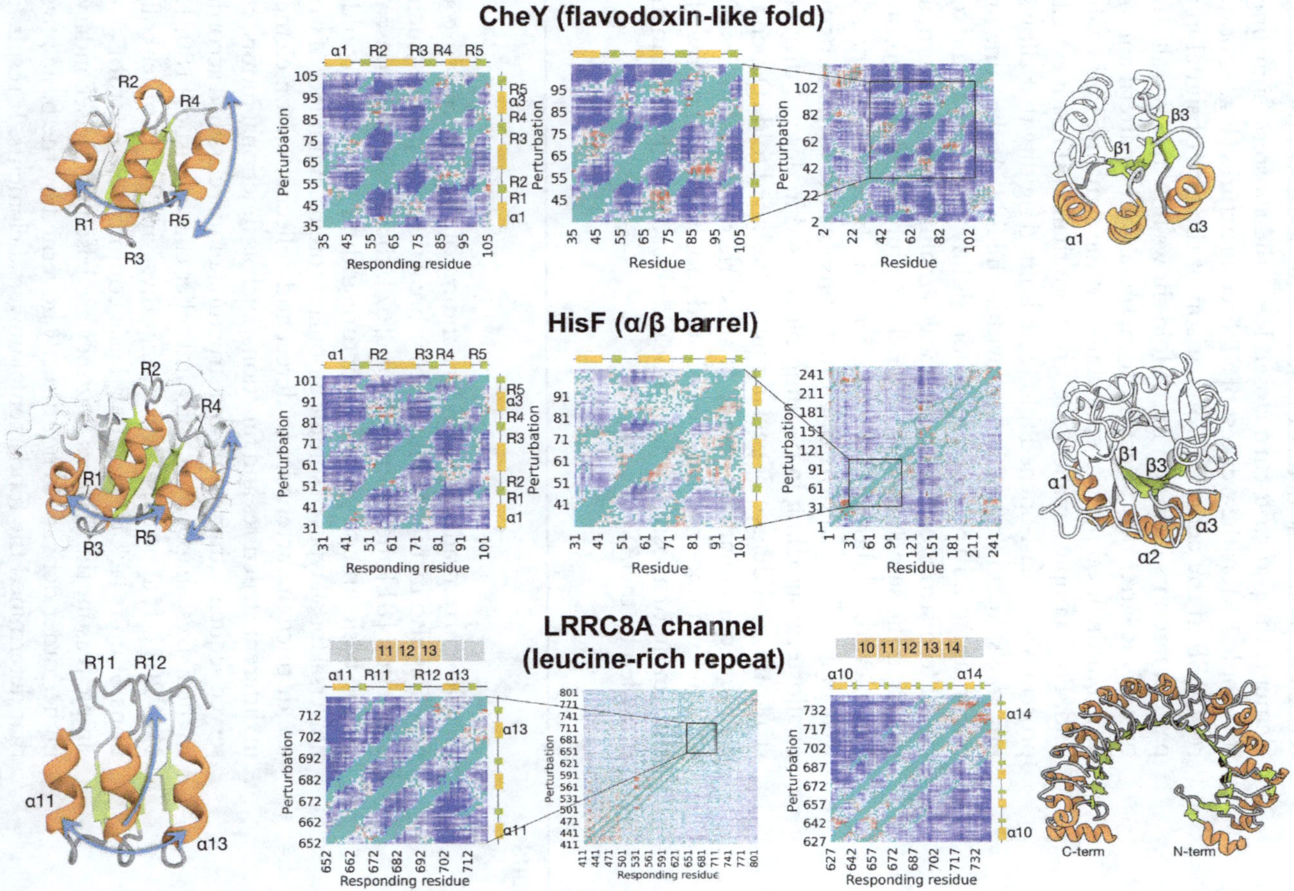

Fig. 8 Towards engineering and design of allosteric regulation. The trade-off between the local and global allosteric signaling in local fragments and in the context of the overall fold. The fragments of three closed loops from flavodoxin-like fold (top row), α/β-barrel (middle row), and leucine-rich repeat (bottom row) are considered

motif considered in the context of the α/β-barrel fold of the HisF protein (Fig. 8, middle row, center). The overall allosteric signaling is also much weaker in the HisF protein (Fig. 8, middle row, right panel), lacking any repetitive or symmetric pattern despite quite regular α/β-barrel structure of the fold. Regardless of the same basic element, the β-turn-α-turn closed loop, the allosteric signaling in the whole LRR is the weakest of all three structures (Fig. 8, bottom row, ASF in the center). Comparison of the fate of signal in the pattern upon its inclusion in the fold shows how the overall structure of the latter may affect the local signal in the pattern and how signal in the local pattern contributes to the signaling in the whole fold. For example, smaller (slightly more than 100 residues) CheY protein has a quasi-opened fold, with limited rigidity that allows to preserve structure of the pattern within the fold. At the same time, larger (about 250 residues) and more tightly packed and rigid α/β-barrel fold stronger affects the local signaling in the pattern, also reducing the signaling in the overall structure of the HisF protein. The picture of allosteric communication obtained here shows us that in any design effort one must consider the original pattern of allosteric signaling in building block used in design, then size, structure, and degree of packing in the designed fold. Depending on the requirement for degree of allosteric regulation in designed structures, an optimal for the task fold should be used followed by the tuning of allosteric communication based on the information provided by the ASFs.

4 Conclusions

Despite current advanced understanding of protein structure and function also manifested in a recent breakthrough in the structure prediction [81] and rational design [67], there is still no consensus on the basic unit of protein structure and function that could be universal for all types of soluble globular proteins. A number of works use distinct characteristics to detect and collect common elements in different structures and functions, such as functional motifs, conserved sequence fragments, and structural motifs classified according to their sizes, shapes, functions and other characteristics [180, 187, 190, 201]. It appears, however, that these and many other collections of units lack to provide a basic common element for a very simple reason: the features that they are based on are secondary or higher order derivatives not directly governed by the basic physics that determined the fundamental universal elements. It results in a challenge in creating "of a cohesive narrative of how proteins emerged and evolved," which was recently recognized for another time elsewhere [202]. I argue, therefore, that consideration of the polymer nature of protein polypeptide backbone allowed us to discover the real basic universal element of globular proteins—closed loops of nearly standard size [87]. Referring to the latter resolves the contradictions between different views and conclusions [21], covering all the aspects of globular proteins from the structural classifications [17, 21, 89, 124], discrete structure of protein domains [125], and co-translational folding [16] to the role of closed loops as structural scaffolds for elementary functions, forming, in turn, basic units of the protein function—elementary functional loops (EFLs, [70, 77]). Combinations of the latter resulted in a diversity of contemporary enzymatic and signaling functions [21, 71, 79], which are also tuned and regulated via allosteric mechanisms using closed loops as elements of protein dynamics and allosteric communication [57]. The other side of this story is protein evolution from the very emergence of first prebiotic peptides with elementary functions to modern proteins as combinations of elementary functional loops—descendants of ancient ring-like "Dayhoff peptides." While physics provided molecular foundation and rules for building biomolecules from simple chemistry, abiogenesis started the biological evolution that picked Dayhoff peptides and used 3.5 billion years to turn them into the protein structure universe with a myriad of individual structures and functions, their complexes, molecular machines, and networks of interactions [203]. Decades of thorough research and dozens of great discoveries have driven us to the state-of-the-art understanding allowing even de novo design [204] of modified and artificial proteins [205–207], receptors [208], sensors [209], and nanomaterials [210, 211]. Nevertheless, Nature can still teach us and help to get better as Emil Fisher proposed it in his Nobel Lecture in 1902: "If we wish to catch up with nature we shall need to use the same methods as she does, and I can foresee a time in which physiological chemistry will not only make greater use of natural enzymes, but will actually resort to creating synthetic ones." And, of course, some of its deep secrets and events like abiogenesis most probably will remain an enigma forever.

Acknowledgments

This work was supported by the Open Frame Individual Research Grant (OF-IRG) MOH-001402-00 to INB from the National Medical Research Council, Ministry of Health, Singapore.

References

1. Kendrew JC, Bodo G, Dintzis HM, Parrish RG, Wyckoff H, Phillips DC (1958) A three dimensional model of the myoglobin molecule obtained by x-ray analysis. Nature 181(4610):662–666. https://doi.org/10.1038/181662a0
2. Perutz MF, Rossmann MG, Cullis AF, Muirhead H, Will G, North AC (1960) Structure of haemoglobin: a three-dimensional Fourier synthesis at 5.5-A. resolution, obtained by X-ray analysis. Nature 185(4711):416–422. https://doi.org/10.1038/185416a0

3. Fisher E (1894) Einfluss der configuration auf die wirkung der enzyme. Ber Dtsch Chem Ges 27:2985–2993
4. Anfinsen CB (1973) Principles that govern the foldingofproteinchains.Science181(4096):223–230. https://doi.org/10.1126/science.181.4096.223
5. Levinthal C (1968) Are there pathways for protein folding? Extrait du J Chim Phys 65:44–45
6. Ivankov DN, Finkelstein AV (2020) Solution of Levinthal's paradox and a physical theory of protein folding times. Biomolecules 10(2). https://doi.org/10.3390/biom10020250
7. Berezovsky IN, Trifonov EN (2002) Loop fold structure of proteins: resolution of Levinthals paradox. J Biomol Struct Dyn 20(1):5–6. https://doi.org/10.1080/07391102.2002.10506817
8. Holtkamp W, Kokic G, Jager M, Mittelstaet J, Komar AA, Rodnina MV (2015) Cotranslational protein folding on the ribosome monitored in real time. Science 350(6264):1104–1107. https://doi.org/10.1126/science.aad0344
9. Thommen M, Holtkamp W, Rodnina MV (2017) Co-translational protein folding: progress and methods. Curr Opin Struct Biol 42:83–89. https://doi.org/10.1016/j.sbi.2016.11.020
10. Bitran A, Jacobs WM, Zhai X, Shakhnovich E (2020) Cotranslational folding allows misfolding-prone proteins to circumvent deep kinetic traps. Proc Natl Acad Sci USA 117(3):1485–1495. https://doi.org/10.1073/pnas.1913207117
11. Bitran A, Jacobs WM, Shakhnovich E (2024) The critical role of co-translational folding: An evolutionary and biophysical perspective. Curr Opin Syst Biol 37:100485
12. Ittah V, Haas E (1995) Nonlocal interactions stabilize long range loops in the initial folding intermediates of reduced bovine pancreatic trypsin inhibitor. Biochemistry 34(13):4493–4506
13. Orevi T, Rahamim G, Hazan G, Amir D, Haas E (2013) The loop hypotesis: contribution of early formed specific non-local intreractions to determination of protein folding pathways. Biophys Rev 5:85–98
14. Orevi T, Rahamim G, Shemesh S, Ben Ishay E, Amir D, Haas E (2014) Fast closure of long loops at the initiation of the folding transition of globular proteins studied by time-resolved FRET-based methods. Bio-Algorithms Med Syst 10(4):169–193
15. Samatova E, Komar AA, Rodnina MV (2024) How the ribosome shapes cotranslational protein folding. Curr Opin Struct Biol 84:102740. https://doi.org/10.1016/j.sbi.2023.102740
16. Berezovsky IN, Kirzhner VM, Kirzhner A, Trifonov EN (2001) Protein folding: looping from hydrophobic nuclei. Proteins 45(4):346–350
17. Berezovsky IN, Trifonov EN (2002) Flowering buds of globular proteins: transpiring simplicity of protein organization. CFG 3(6):525–534. https://doi.org/10.1002/cfg.223
18. Bergasa-Caceres F, Haas E, Rabitz HA (2019) Nature's Shortcut to Protein Folding. J Phys Chem B 123(21):4463–4476. https://doi.org/10.1021/acs.jpcb.8b11634
19. Zhao V, Jacobs WM, Shakhnovich EI (2020) Effect of protein structure on evolution of cotranslational folding. Biophys J 119(6):1123–1134. https://doi.org/10.1016/j.bpj.2020.06.037
20. Berezovsky IN (2011) The diversity of physical forces and mechanisms in intermolecular interactions. Phys Biol 8(3):035002. https://doi.org/10.1088/1478-3975/8/3/035002
21. Berezovsky IN, Guarnera E, Zheng Z (2017) Basic units of protein structure, folding, and function. Prog Biophys Mol Biol 128:85–99. https://doi.org/10.1016/j.pbiomolbio.2016.09.009
22. Shakhnovich E (2006) Protein folding thermodynamics and dynamics: where physics, chemistry, and biology meet. Chem Rev 106(5):1559–1588. https://doi.org/10.1021/cr040425u
23. Berezovskii IN, Esipova NG, Tumanian VG (1997) Isolation of the energy-significant parts of the globule and the hierarchy of the domain structure of the protine macromolecule. Biophysics 42:557–565
24. Berezovskii IN, Esipova NG, Tumanian VG (1998) Spatial Distribution of directed interactions in globular protein structures. Biophysics 43:367–377
25. Berezovskii IN, Esipova NG, Tumanian VG, Namiot VA (1998) A new approach to calculating van der Waals interaction energies in protein mcromolecules. Dielectric constant as a physical parameter. Biophysics 43:909–916
26. Berezovsky IN, Tumanyan VG, Esipova NG (1997) Representation of amino acid sequences in terms of interaction energy in protein globules. FEBS Lett 418(1-2):43–46
27. Berezovsky IN, Zeldovich KB, Shakhnovich EI (2007) Positive and negative design in stability and thermal adaptation of natural proteins. PLoS Comput Biol 3(3):e52. https://doi.org/10.1371/journal.pcbi.0030052
28. Amangeldina A, Tan ZW, Berezovsky IN (2024) Living in trinity of extremes: Genomic

and proteomic signatures of halophilic, thermophilic, and pH adaptation. Curr Res Struct Biol 7:100129. https://doi.org/10.1016/j.crstbi.2024.100129
29. Berezovsky IN, Shakhnovich EI (2005) Physics and evolution of thermophilic adaptation. Proc Natl Acad Sci USA 102(36):12742–12747. https://doi.org/10.1073/pnas.0503890102
30. Goncearenco A, Berezovsky IN (2014) The fundamental tradeoff in genomes and proteomes of prokaryotes established by the genetic code, codon entropy, and physics of nucleic acids and proteins. Biol Direct 9:29. https://doi.org/10.1186/s13062-014-0029-2
31. Goncearenco A, Ma BG, Berezovsky IN (2014) Molecular mechanisms of adaptation emerging from the physics and evolution of nucleic acids and proteins. Nucleic Acids Res 42(5):2879–2892. https://doi.org/10.1093/nar/gkt1336
32. Ma BG, Goncearenco A, Berezovsky IN (2010) Thermophilic adaptation of protein complexes inferred from proteomic homology modeling. Structure 18(7):819–828. https://doi.org/10.1016/j.str.2010.04.004
33. Frauenfelder H, Petsko GA, Tsernoglou D (1979) Temperature-dependent X-ray diffraction as a probe of protein structural dynamics. Nature 280(5723):558–563
34. Frauenfelder H, Sligar SG, Wolynes PG (1991) The energy landscapes and motions of proteins. Science 254(5038):1598–1603
35. Guarnera E, Berezovsky IN (2019) On the perturbation nature of allostery: sites, mutations, and signal modulation. Curr Opin Struct Biol 56:18–27. https://doi.org/10.1016/j.sbi.2018.10.008
36. Jardetzky O (1996) Protein dynamics and conformational transitions in allosteric proteins. Prog Biophys Mol Biol 65(3):171–219
37. Cooper A, Dryden DT (1984) Allostery without conformational change. A plausible model. Eur Biophys J 11(2):103–109
38. Cui Q, Karplus M (2008) Allostery and cooperativity revisited. Protein Sci 17(8):1295–1307. https://doi.org/10.1110/ps.03259908
39. Gunasekaran K, Ma B, Nussinov R (2004) Is allostery an intrinsic property of all dynamic proteins?. Proteins 57(3):433–443. https://doi.org/10.1002/prot.20232
40. Berezovsky IN, Nussinov R (2022) Multiscale allostery: basic mechanisms and versatility in diagnostics and drug design. J Mol Biol: 167751, https://doi.org/10.1016/j.jmb.2022.167751
41. Monod J, Changeux JP, Jacob F (1963) Allosteric proteins and cellular control systems. J Mol Biol 6:306–329
42. Monod J, Wyman J, Changeux JP (1965) On the nature of allosteric transitions: a plausible model. J Mol Biol 12:88–118
43. Koshland Jr. DE, Nemethy G, Filmer D (1966) Comparison of experimental binding data and theoretical models in proteins containing subunits. Biochemistry 5(1):365–385
44. Berezovsky IN (2013) Thermodynamics of allostery paves a way to allosteric drugs. Biochim Biophys Acta 1834(5):830–835. https://doi.org/10.1016/j.bbapap.2013.01.024
45. Goncearenco A, Mitternacht S, Yong T, Eisenhaber B, Eisenhaber F, Berezovsky IN (2013) SPACER: Server for predicting allosteric communication and effects of regulation. Nucleic Acids Res 41:W266–W272. https://doi.org/10.1093/nar/gkt460
46. Mitternacht S, Berezovsky IN (2011) Binding leverage as a molecular basis for allosteric regulation. PLoS Comput Biol 7(9):e1002148. https://doi.org/10.1371/journal.pcbi.1002148
47. Mitternacht S, Berezovsky IN (2011) Coherent conformational degrees of freedom as a structural basis for allosteric communication. PLoS Comput Biol 7(12):e1002301. https://doi.org/10.1371/journal.pcbi.1002301
48. Guarnera E, Berezovsky IN (2016) Allosteric sites: remote control in regulation of protein activity. Curr Opin Struct Biol 37:1–8. https://doi.org/10.1016/j.sbi.2015.10.004
49. Guarnera E, Berezovsky IN (2016) Structure-Based Statistical Mechanical Model Accounts for the Causality and Energetics of Allosteric Communication. PLoS Comput Biol 12(3):e1004678. https://doi.org/10.1371/journal.pcbi.1004678
50. Guarnera E, Berezovsky IN (2019) Toward comprehensive allosteric control over protein activity. Structure 27(5):866–878 e861. https://doi.org/10.1016/j.str.2019.01.014
51. Guarnera E, Berezovsky IN (2023) Sequence-dependent model of allosteric communication. J Chem Phys 158(13):135101. https://doi.org/10.1063/5.013984
52. Tee WV, Guarnera E, Berezovsky IN (2018) Reversing allosteric communication: From detecting allosteric sites to inducing and tuning targeted allosteric response. PLoS Comput Biol 14(6):e1006228. https://doi.org/10.1371/journal.pcbi.1006228
53. Tee WV, Guarnera E, Berezovsky IN (2020) Disorder driven allosteric control of protein activity. Curr Res Struct Biol 2:191–203
54. Tee WV, Tan ZW, Lee K, Guarnera E, Berezovsky IN (2021) Exploring the Allosteric Territory of Protein Function. J Phys Chem B 125(15):3763–3780. https://doi.org/10.1021/acs.jpcb.1c00540

55. Tan ZW, Tee WV, Guarnera E, Berezovsky IN (2023) AlloMAPS 2: allosteric fingerprints of the AlphaFold and Pfam-trRosetta predicted structures for engineering and design. Nucleic Acids Res 51(D1):D345–D351. https://doi.org/10.1093/nar/gkac828
56. Tan ZW, Tee WV, Guarnera E, Booth L, Berezovsky IN (2019) AlloMAPS: allosteric mutation analysis and polymorphism of signaling database. Nucleic Acids Res 47(D1):D265–D270. https://doi.org/10.1093/nar/gky1028
57. Tee WV, Tan ZW, Guarnera E, Berezovsky IN (2022) Conservation and diversity in allosteric fingerprints of proteins for evolutionary-inspired engineering and design. J Mol Biol 167577. https://doi.org/10.1016/j.jmb.2022.167577
58. Guarnera E, Berezovsky IN (2020) Allosteric drugs and mutations: chances, challenges, and necessity. Curr Opin Struct Biol 62:149–157. https://doi.org/10.1016/j.sbi.2020.01.010
59. Tee WV, Guarnera E, Berezovsky IN (2019) On the allosteric effect of nsSNPs and the emerging importance of allosteric polymorphism. J Mol Biol 431(19):3933–3942. https://doi.org/10.1016/j.jmb.2019.07.012
60. Tan ZW, Tee WV, Samsudin F, Guarnera E, Bond PJ, Berezovsky IN (2022) Allosteric perspective on the mutability and druggability of the SARS-CoV-2 Spike protein. Structure 30(4):590–607 e594. https://doi.org/10.1016/j.str.2021.12.011
61. Tan ZW, Tee WV, Berezovsky IN (2022) Learning about allosteric drugs and ways to design them. J Mol Biol 434(17):167692. https://doi.org/10.1016/j.jmb.2022.167692
62. Tee WV, Berezovsky IN (2024) Allosteric drugs: new principles and design approaches. Curr Opin Struct Biol 84:102758. https://doi.org/10.1016/j.sbi.2023.102758
63. Tee WV, Lim SJM, Berezovsky IN (2024) Toward the design of allosteric effectors: gaining comprehensive control of drug properties and actions. J Med Chem 67(19):17191–17206. https://doi.org/10.1021/acs.jmedchem.4c01043
64. Kurochkin IV, Guarnera E, Wong JH, Eisenhaber F, Berezovsky IN (2017) Toward allosterically increased catalytic activity of insulin-degrading enzyme against amyloid peptides. Biochemistry 56(1):228–239. https://doi.org/10.1021/acs.biochem.6b00783
65. Guarnera E, Tan ZW, Zheng Z, Berezovsky IN (2017) AlloSigMA: allosteric signaling and mutation analysis server. Bioinformatics 33(24):3996–3998. https://doi.org/10.1093/bioinformatics/btx430
66. Tan ZW, Guarnera E, Tee WV, Berezovsky IN (2020) AlloSigMA 2: paving the way to designing allosteric effectors and to exploring allosteric effects of mutations. Nucleic Acids Res 48(W1):W116–W124. https://doi.org/10.1093/nar/gkaa338
67. Du Z, Su H, Wang W, Ye L, Wei H, Peng Z, Anishchenko I, Baker D, Yang J (2021) The trRosetta server for fast and accurate protein structure prediction. Nat Protoc 16(12):5634–5651. https://doi.org/10.1038/s41596-021-00628-9
68. Berezovsky IN (2019) Towards descriptor of elementary functions for protein design. Curr Opin Struct Biol 58:159–165. https://doi.org/10.1016/j.sbi.2019.06.010
69. Yin M, Goncearenco A, Berezovsky IN (2021) Deriving and using descriptors of elementary functions in rational protein design. Front Bioinf 1:657529. https://doi.org/10.3389/fbinf.2021.657529
70. Goncearenco A, Berezovsky IN (2010) Prototypesofelementaryfunctionalloopsunravel evolutionary connections between protein functions. Bioinformatics 26(18):i497–i503. https://doi.org/10.1093/bioinformatics/btq374
71. Zheng Z, Goncearenco A, Berezovsky IN (2024) Back in time to the Gly-rich prototype of the phosphate binding elementary function. Curr Res Struct Biol 7:100142. https://doi.org/10.1016/j.crstbi.2024.100142
72. Berezovsky IN, Kirzhner A, Kirzhner VM, Rosenfeld VR, Trifonov EN (2003) Protein sequences yield a proteomic code. J Biomol Struct Dyn 21(3):317–325. https://doi.org/10.1080/07391102.2003.10506928
73. Berezovsky IN, Kirzhner A, Kirzhner VM, Trifonov EN (2003) Spelling protein structure. J Biomol Struct Dyn 21(3):327–339. https://doi.org/10.1080/07391102.2003.10506929
74. Berezovsky IN, Guarnera E, Zheng Z, Eisenhaber B, Eisenhaber F (2017) Protein function machinery: from basic structural units to modulation of activity. Curr Opin Struct Biol 42:67–74. https://doi.org/10.1016/j.sbi.2016.10.021
75. Zeldovich KB, Berezovsky IN, Shakhnovich EI (2007) Protein and DNA sequence determinants of thermophilic adaptation. PLoS Comput Biol 3(1):e5. https://doi.org/10.1371/journal.pcbi.0030005
76. Berezovsky IN, Chen WW, Choi PJ, Shakhnovich EI (2005) Entropic stabilization of proteins and its proteomic consequences. PLoS Comput Biol 1(4):e47
77. Goncearenco A, Berezovsky IN (2011) Computational reconstruction of primordial

prototypes of elementary functional loops in modern proteins. Bioinformatics 27(17): 2368–2375. https://doi.org/10.1093/bioinformatics/btr396
78. Goncearenco A, Berezovsky IN (2012) Exploring the evolution of protein function in Archaea. BMC Evol Biol 12:75. https://doi.org/10.1186/1471-2148-12-75
79. Goncearenco A, Berezovsky IN (2015) Protein function from its emergence to diversity in contemporary proteins. Phys Biol 12(4):045002. https://doi.org/10.1088/1478-3975/12/4/045002
80. Zheng Z, Goncearenco A, Berezovsky IN (2016) Nucleotide binding database NBDB-a collection of sequence motifs with specific protein-ligand interactions. Nucleic Acids Res 44(D1):D301–D307. https://doi.org/10.1093/nar/gkv1124
81. Varadi M, Anyango S, Deshpande M, Nair S, Natassia C, Yordanova G, Yuan D, Stroe O, Wood G, Laydon A, Zidek A, Green T, Tunyasuvunakool K, Petersen S, Jumper J, Clancy E, Green R, Vora A, Lutfi M, Figurnov M, Cowie A, Hobbs N, Kohli P, Kleywegt G, Birney E, Hassabis D, Velankar S (2022) AlphaFold Protein Structure Database: massively expanding the structural coverage of protein-sequence space with high-accuracy models. Nucleic Acids Res 50(D1):D439–D444. https://doi.org/10.1093/nar/gkab1061
82. Trifonov EN (1999) Glycine clock: Eubacteria first, Archaea next, Protoctista, Fungi, Planta and Animalia at last. Gene Ther Mol Biol 4:313–322
83. Trifonov EN (2000) Consensus temporal order of amino acids and evolution of the triplet code. Gene 261(1):139–151. https://doi.org/10.1016/s0378-1119(00)00476-5
84. Trifonov EN, Kirzhner A, Kirzhner VM, Berezovsky IN (2001) Distinct stages of protein evolution as suggested by protein sequence analysis. J Mol Evol 53 (4-5):394–401. https://doi.org/10.1007/s002390010229
85. Trifonov EN, Berezovsky IN (2002) Molecular evolution from abiotic scratch. FEBS Lett 527:1–4
86. Berezovsky IN, Trifonov EN (2001) Protein structure and folding: a new start. J Biomol Struct Dyn 19(3):397–403. https://doi.org/10.1080/07391102.2001.10506749
87. Berezovsky IN, Grosberg AY, Trifonov EN (2000) Closed loops of nearly standard size: common basic element of protein structure. FEBS Lett 466(2-3):283–286
88. Berezovsky IN, Kirzhner VM, Kirzhner A, Rosenfeld VR, Trifonov EN (2002) Closed loops: persistence of the protein chain returns. Protein Eng 15(12):955–957. https://doi.org/10.1093/protein/15.12.955
89. Berezovsky IN, Trifonov EN (2001) Loop fold nature of globular proteins. Protein Eng 14(6):403–407
90. Miller SL (1953) A production of amino acids under possible primitive earth conditions. Science 117(3046):528–529. https://doi.org/10.1126/science.117.3046.528
91. Miller SL, Urey HC (1959) Organic compound synthesis on the primitive earth. Science 130(3370):245–251. https://doi.org/10.1126/science.130.3370.245
92. Romero Romero ML, Rabin A, Tawfik DS (2016) Functional proteins from short peptides: Dayhoff's hypothesis turns 50. Angew Chem Int Ed Engl 55(52):15966–15971. https://doi.org/10.1002/anie.201609977
93. Trifonov EN, Berezovsky IN (2003) Evolutionary aspects of protein structure and folding. Curr Opin Struct Biol 13(1):110–114. https://doi.org/10.1016/s0959-440x(03)00005-8
94. Eigen M, Schuster P (1978) The hypercycle. A principle of natural self-organization. Part C:The realistic hypercycle. Naturwissenschaften 65:341–369
95. Kamtekar S, Schiffer JM, Xiong H, Babik JM, Hecht MH (1993) Protein design by binary patterning of polar and nonpolar amino acids. Science 262(5140):1680–1685. https://doi.org/10.1126/science.8259512
96. Frank-Kamenetskii MD (1990) Energetics of DNA and RNA double helices. In: Landolt-Bornstein series in Biophysics, 1. Springer-Verlag, Heidelberg
97. Ogata Y, Imai E, Honda H, Hatori K, Matsuno K (2000) Hydrothermal circulation of seawater through hot vents and contribution of interface chemistry to prebiotic synthesis. Orig Life Evol Biosph 30(6):527–537. https://doi.org/10.1023/a:1026543825011
98. van der Gulik P, Massar S, Gilis D, Buhrman H, Rooman M (2009) The first peptides: the evolutionary transition between prebiotic amino acids and early proteins. J Theor Biol 261(4):531–539. https://doi.org/10.1016/j.jtbi.2009.09.004
99. Berezovsky IN, Esipova NG, Tumanyan VG, Namiot VA (2000) A new approach for the calculation of the energy of van der Waals interactions in macromolecules of globular proteins. J Biomol Struct Dyn 17(5):799–809. https://doi.org/10.1080/07391102.2000.10506569
100. Berezovsky IN, Namiot VA, Tumanyan VG, Esipova NG (1999) Hierarchy of the interaction energy distribution in the spatial

structure of globular proteins and the problem of domain definition. J Biomol Struct Dyn 17(1):133–155. https://doi.org/10.1080/07391102.1999.10508347
101. Marchler-Bauer A, Anderson JB, Cherukuri PF, DeWeese-Scott C, Geer LY, Gwadz M, He S, Hurwitz DI, Jackson JD, Ke Z, Lanczycki CJ, Liebert CA, Liu C, Lu F, Marchler GH, Mullokandov M, Shoemaker BA, Simonyan V, Song JS, Thiessen PA, Yamashita RA, Yin JJ, Zhang D, Bryant SH (2005) CDD: a Conserved Domain Database for protein classification. Nucleic Acids Res 33:D192–D196. https://doi.org/10.1093/nar/gki069
102. Roy SW, Nosaka M, de Souza SJ, Gilbert W (1999) Centripetal modules and ancient introns. Gene 238(1):85–91
103. Voigt CA, Martinez C, Wang ZG, Mayo SL, Arnold FH (2002) Protein building blocks preserved by recombination. Nat Struct Biol 9(7):553–558. https://doi.org/10.1038/nsb805
104. Yew BK, Chintapalli SV, Upton GG, Reynolds CA (2007) Conservation of closed loops. J Mol Graph 26(3):652–655
105. Qian B, Goldstein RA (2001) Distribution of Indel lengths. Proteins 45(1):102–110
106. Sato Y, Niimura Y, Yura K, Go M (1999) Module-intron correlation and intron sliding in family F/10 xylanase genes. Gene 238(1):93–101
107. Dokholyan NV, Shakhnovich EI (2001) Understanding hierarchical protein evolution from first principles. J Mol Biol 312(1):289–307. https://doi.org/10.1006/jmbi.2001.4949
108. Liu L, Hyeon C (2016) Contact Statistics Highlight Distinct Organizing Principles of Proteins and RNA. Biophys J 110(11):2320–2327. https://doi.org/10.1016/j.bpj.2016.04.020
109. Moret MA, Zebende GF (2007) Amino acid hydrophobicity and accessible surface area. Phys Rev E 75(1 Pt 1):011920. https://doi.org/10.1103/PhysRevE.75.011920
110. Taylor TJ, Vaisman II (2006) Graph theoretic properties of networks formed by the Delaunay tessellation of protein structures. Phys Rev E 73(4 Pt 1):041925. https://doi.org/10.1103/PhysRevE.73.041925
111. Angelov B, Sadoc JF, Jullien R, Soyer A, Mornon JP, Chomilier J (2002) Nonatomic solvent-driven Voronoi tessellation of proteins: an open tool to analyze protein folds. Proteins 49(4):446–456. https://doi.org/10.1002/prot.10220
112. Chintapalli SV, Yew BK, Illingworth CJ, Upton GJ, Reeves PJ, Parkes KE, Snell CR, Reynolds CA (2010) Closed loop folding units from structural alignments: experimental foldons revisited. J Comput Chem 31(15):2689–2701. https://doi.org/10.1002/jcc.21562
113. Fernandez-Fuentes N, Fiser A (2013) A modular perspective of protein structures: application to fragment based loop modeling. Methods Mol Biol 932:141–158. https://doi.org/10.1007/978-1-62703-065-6_9
114. Ben Ishay E, Rahamim G, Orevi T, Hazan G, Amir D, Haas E (2012) Fast subdomain folding prior to the global refolding transition of E. coli adenylate kinase: a double kinetics study. J Mol Biol 423(4):613–623. https://doi.org/10.1016/j.jmb.2012.08.001
115. Caetano-Anolles K, Aziz MF, Mughal F, Caetano-Anolles G (2024) On protein loops, prior molecular states and common ancestors of life. J Mol Evol 92(5):624–646. https://doi.org/10.1007/s00239-024-10167-y
116. Chintapalli SV, Illingworth CJ, Upton GJ, Sacquin-Mora S, Reeves PJ, Mohammedali HS, Reynolds CA (2014) Assessing the effect of dynamics on the closed-loop protein-folding hypothesis. J R Soc Interface R Soc 11(91):20130935. https://doi.org/10.1098/rsif.2013.0935
117. Papandreou N, Berezovsky IN, Lopes A, Eliopoulos E, Chomilier J (2004) Universal positions in globular proteins. Eur J Biochem 271(23-24):4762–4768. https://doi.org/10.1111/j.1432-1033.2004.04440.x
118. Zhang Z, Wayment-Steele HK, Brixi G, Wang H, Kern D, Ovchinnikov S (2024) Protein language models learn evolutionary statistics of interacting sequence motifs. Proc Natl Acad Sci USA 121(45):e2406285121. https://doi.org/10.1073/pnas.2406285121
119. Farias-Rico JA, Schmidt S, Hocker B (2014) Evolutionary relationship of two ancient protein superfolds. Nat Chem Biol 10(9):710–715. https://doi.org/10.1038/nchembio.1579
120. Laurino P, Toth-Petroczy A, Meana-Paneda R, Lin W, Truhlar DG, Tawfik DS (2016) An ancient fingerprint indicates the common ancestry of Rossmann-fold enzymes utilizing different ribose-based cofactors. PLoS Biol 14(3):e1002396. https://doi.org/10.1371/journal.pbio.1002396
121. Newton MS, Guo X, Soderholm A, Nasvall J, Lundstrom P, Andersson DI, Selmer M, Patrick WM (2017) Structural and functional innovations in the real-time evolution of new (betaalpha)8 barrel enzymes. Proc Natl Acad Sci USA 114(18):4727–4732. https://doi.org/10.1073/pnas.1618552114
122. Sterner R, Höcker B (2005) Catalytic versatility, stability, and evolution of the (betaalpha)8-barrel enzyme fold. Chem

Rev 105(11):4038–4055. https://doi.org/10.1021/cr030191z
123. Levitt M, Chothia C (1976) Structural patterns in globular proteins. Nature 261(5561):552–558
124. Berezovsky IN, Trifonov EN (2001) Van der Waals locks: loop-n-lock structure of globular proteins. J Mol Biol 307(5):1419–1426. https://doi.org/10.1006/jmbi.2001.4554
125. Berezovsky IN (2003) Discrete structure of van der Waals domains in globular proteins. Protein Eng 16(3):161–167
126. Koczyk G, Berezovsky IN (2008) Domain Hierarchy and closed Loops (DHcL): a server for exploring hierarchy of protein domain structure. Nucleic Acids Res 36:W239–W245. https://doi.org/10.1093/nar/gkn326
127. Gloge F, Becker AH, Kramer G, Bukau B (2014) Co-translational mechanisms of protein maturation. Curr Opin Struct Biol 24:24–33. https://doi.org/10.1016/j.sbi.2013.11.004
128. Nissen P, Hansen J, Ban N, Moore PB, Steitz TA (2000) The structural basis of ribosome activity in peptide bond synthesis. Science 289(5481):920–930
129. Voss NR, Gerstein M, Steitz TA, Moore PB (2006) The geometry of the ribosomal polypeptide exit tunnel. J Mol Biol 360(4):893–906. https://doi.org/10.1016/j.jmb.2006.05.023
130. Kosolapov A, Deutsch C (2009) Tertiary interactions within the ribosomal exit tunnel. Nat Struct Mol Biol 16(4):405–411. https://doi.org/10.1038/nsmb.1571
131. O'Brien EP, Christodoulou J, Vendruscolo M, Dobson CM (2011) New scenarios of protein folding can occur on the ribosome. J Am Chem Soc 133(3):513–526. https://doi.org/10.1021/ja107863z
132. Berezovsky IN, Zheng Z, Kurotani A, Tokmakov AA, Kurochkin IV (2015) Organization of the multiaminoacyl-tRNA synthetase complex and the cotranslational protein folding. Protein Sci 24(9):1475–1485. https://doi.org/10.1002/pro.2735
133. Choi SI, Kwon S, Son A, Jeong H, Kim KH, Seong BL (2013) Protein folding in vivo revisited. Curr Protein Pept Sci 14(8):721–733
134. Ellis RJ, Hartl FU (1999) Principles of protein folding in the cellular environment. Curr Opin Struct Biol 9(1):102–110
135. Gershenson A, Gierasch LM (2011) Protein folding in the cell: challenges and progress. Curr Opin Struct Biol 21(1):32–41. https://doi.org/10.1016/j.sbi.2010.11.001
136. Kratky O, Prod G (1949) Röntgenuntersuchung gelöster Fadenmoleküle. Recl Trav Chim Pays-Bas 68:1106–1112
137. Kiuhn W (1939) Molekülkonstellation und Kristallitorientierung als Ursachen kautschukähnlicher Elastizität. Kolloid-Z 87:3–12
138. Shimada J, Yamakawa H (1984) Ring-closure probabilities for twisted wormlike chains.Application to DNA. Macromolecules 17:689–698
139. Miller WG, Brant DA, Flory PJ (1967) Random coil configurations of polypeptide copolymers. J Mol Biol 23:67–80
140. Miller WG, Goebel CV (1968) Dimensions of protein random coils. Biochemistry 7(11):3925–3935. https://doi.org/10.1021/bi00851a021
141. Schimmel PR, Flory PJ (1967) Conformational energy and configurational statistics of poly-L-proline. Proc Natl Acad Sci USA 5(1):52–59
142. Brant DA, Flory PJ (1965) The Configuration of Random Polypeptide Chains. I. Experimental Results. J Am Chem Soc 87:2788–2791
143. Nielsen PE, Egholm M, Berg RH, Buchardt O (1991) Sequence-selective recognition of DNA by strand displacement with a thymine-substituted polyamide. Science 254(5037):1497–1500. https://doi.org/10.1126/science.1962210
144. Kuhn H, Demidov VV, Gildea BD, Fiandaca MJ, Coull JC, Frank-Kamenetskii MD (2001) PNA beacons for duplex DNA. Antisense Nucleic Acid Drug Dev 11(4):265–270. https://doi.org/10.1089/108729001317022269
145. Kuhn H, Demidov VV, Coull JM, Fiandaca MJ, Gildea BD, Frank-Kamenetskii MD (2002) Hybridization of DNA and PNA molecular beacons to single-stranded and double-stranded DNA targets. J Am Chem Soc 124(6):1097–1103. https://doi.org/10.1021/ja0041324
146. Eck RV, Dayhoff MO (1966) Evolution of the structure of ferredoxin based on living relics of primitive amino Acid sequences. Science 152(3720):363–366. https://doi.org/10.1126/science.152.3720.363
147. deGennes PG (1990) Minimum number of aminoacids required to build up a specific receptor with a folded polypeptide chain. Introduction to polymer dynamics. Cambridge University Press, Cambridge, pp 17–26
148. Zeldovich KB, Berezovsky IN, Shakhnovich EI (2006) Physical origins of protein superfamilies. J Mol Biol 357(4):1335–1343. https://doi.org/10.1016/j.jmb.2006.01.081
149. Svedberg T (1929) Mass and size of protein molecules. Nature 123(871). https://doi.org/10.1038/123871a0
150. Bresler SE, Talmud DL (1944) On the nature of globular proteins. Compt Rend Acad Sci URSS 43(7):310–314

151. Bresler SE, Talmud DL (1944) On the nature of globular proteins. II A few consequences of the new hypothesis. Compt Rend Acad Sci URSS 43(8):349–350
152. Gerstein M (1998) How representative are the known structures of the proteins in a complete genome? A comprehensive structural census. Fold Des 3(6):497–512
153. Jones S, Stewart M, Michie A, Swindells MB, Orengo C, Thornton JM (1998) Domain assignment for protein structures using a consensus approach: characterization and analysis. Protein Sci 7(2):233–242. https://doi.org/10.1002/pro.5560070202
154. Wheelan SJ, Marchler-Bauer A, Bryant SH (2000) Domain size distributions can predict domain boundaries. Bioinformatics 16(7):613–618
155. Shore D, Baldwin RL (1983) Energetics of DNA twisting. I. Relation between twist and cyclization probability. J Mol Biol 170(4):957–981. https://doi.org/10.1016/s0022-2836(83)80198-3
156. Shore D, Baldwin RL (1983) Energetics of DNA twisting. II. Topoisomer analysis. J Mol Biol 170(4):983–1007. https://doi.org/10.1016/s0022-2836(83)80199-5
157. Shore D, Langowski J, Baldwin RL (1981) DNA flexibility studied by covalent closure of short fragments into circles. Proc Natl Acad Sci USA 78(8):4833–4837. https://doi.org/10.1073/pnas.78.8.4833
158. Aziz MF, Caetano-Anolles K, Caetano-Anolles G (2016) The early history and emergence of molecular functions and modular scale-free network behavior. Sci Rep 6: 25058. https://doi.org/10.1038/srep25058
159. Aziz MF, Mughal F, Caetano-Anolles G (2023) Tracing the birth of structural domains from loops during protein evolution. Sci Rep 13(1):14688. https://doi.org/10.1038/s41598-023-41556-w
160. Chothia C, Gough J, Vogel C, Teichmann SA (2003) Evolution of the protein repertoire. Science 300(5626):1701–1703. https://doi.org/10.1126/science.1085371
161. Nath N, Mitchell JB, Caetano-Anolles G (2014) The natural history of biocatalytic mechanisms. PLoS Comput Biol 10(5):e1003642. https://doi.org/10.1371/journal.pcbi.1003642
162. Noor E, Flamholz AI, Jayaraman V, Ross BL, Cohen Y, Patrick WM, Gruic-Sovulj I, Tawfik DS (2022) Uniform binding and negative catalysis at the origin of enzymes. Protein Sci 31(8):e4381. https://doi.org/10.1002/pro.4381
163. Cai W, Pei J, Grishin NV (2004) Reconstruction of ancestral protein sequences and its applications. BMC Evol Biol 4:33. https://doi.org/10.1186/1471-2148-4-33
164. Harms MJ, Thornton JW (2010) Analyzing protein structure and function using ancestral gene reconstruction. Curr Opin Struct Biol 20(3):360–366. https://doi.org/10.1016/j.sbi.2010.03.005
165. Mirkin BG, Fenner TI, Galperin MY, Koonin EV (2003) Algorithms for computing parsimonious evolutionary scenarios for genome evolution, the last universal common ancestor and dominance of horizontal gene transfer in the evolution of prokaryotes. BMC Evol Biol 3:2. https://doi.org/10.1186/1471-2148-3-2
166. Kullback S, Leibler RA (1951) On information and sufficiency. Ann Math Stat 22(1):142–143
167. Li W, Godzik A (2006) Cd-hit: a fast program for clustering and comparing large sets of protein or nucleotide sequences. Bioinformatics 22(13):1658–1659. https://doi.org/10.1093/bioinformatics/btl158
168. Wootton JC, Federhen S (1996) Analysis of compositionally biased regions in sequence databases. Methods Enzymol 266:554–571
169. Kann MG, Sheetlin SL, Park Y, Bryant SH, Spouge JL (2007) The identification of complete domains within protein sequences using accurate E-values for semi-global alignment. Nucleic Acids Res 35(14):4678–4685. https://doi.org/10.1093/nar/gkm414
170. Altschul SF, Madden TL, Schaffer AA, Zhang J, Zhang Z, Miller W, Lipman DJ (1997) Gapped BLAST and PSI-BLAST: a new generation of protein database search programs. Nucleic Acids Res 25(17):3389–3402. https://doi.org/10.1093/nar/25.17.3389
171. Altschul SF, Gertz EM, Agarwala R, Schaffer AA, Yu YK (2009) PSI-BLAST pseudocounts and the minimum description length principle. Nucleic Acids Res 37(3):815–824. https://doi.org/10.1093/nar/gkn981
172. Atkinson MD (1999) Restricted permutations. Discrete Math 195(1-3):27–38
173. Brenner SE, Koehl P, Levitt M (2000) The ASTRAL compendium for protein structure and sequence analysis. Nucleic Acids Res 28(1):254–256. https://doi.org/10.1093/nar/28.1.254
174. Lo Conte L, Ailey B, Hubbard TJ, Brenner SE, Murzin AG, Chothia C (2000) SCOP: a structural classification of proteins database. Nucleic Acids Res 28(1):257–259. https://doi.org/10.1093/nar/25.1.236
175. Andreeva A, Prlic A, Hubbard TJ, Murzin AG (2007) SISYPHUS-structural alignments for proteins with non-trivial relationships. Nucleic Acids Res 35(Database issue):D253–D259. https://doi.org/10.1093/nar/gkl746

176. Holland IB, Blight MA (1999) ABC-ATPases, adaptable energy generators fuelling transmembrane movement of a variety of molecules in organisms from bacteria to humans. J Mol Biol 293(2):381–399. https://doi.org/10.1006/jmbi.1999.2993
177. Lammens K, Bemeleit DJ, Mockel C, Clausing E, Schele A, Hartung S, Schiller CB, Lucas M, Angermuller C, Soding J, Strasser K, Hopfner KP (2011) The Mre11:Rad50 structure shows an ATP-dependent molecular clamp in DNA double-strand break repair. Cell 145(1):54–66. https://doi.org/10.1016/j.cell.2011.02.038
178. Longo LM, Jablonska J, Vyas P, Kanade M, Kolodny R, Ben-Tal N, Tawfik DS (2020) On the emergence of P-Loop NTPase and Rossmann enzymes from a Beta-Alpha-Beta ancestral fragment. Elife 9. https://doi.org/10.7554/eLife.64415
179. Dym O, Eisenberg D (2001) Sequence-structure analysis of FAD-containing proteins. Protein Sci Publ Protein Soc 10(9):1712–1728. https://doi.org/10.1110/ps.12801
180. Longo LM, Petrovic D, Kamerlin SCL, Tawfik DS (2020) Short and simple sequences favored the emergence of N-helix phospho-ligand binding sites in the first enzymes. Proc Natl Acad Sci USA 117(10):5310–5318. https://doi.org/10.1073/pnas.1911742117
181. Romero Romero ML, Yang F, Lin YR, Toth-Petroczy A, Berezovsky IN, Goncearenco A, Yang W, Wellner A, Kumar-Deshmukh F, Sharon M, Baker D, Varani G, Tawfik DS (2018) Simple yet functional phosphate-loop proteins. Proc Natl Acad Sci USA 115(51):E11943–E11950. https://doi.org/10.1073/pnas.1812400115
182. Denessiouk KA, Rantanen VV, Johnson MS (2001) Adenine recognition: a motif present in ATP-, CoA-, NAD-, NADP-, and FAD-dependent proteins. Proteins 44(3):282–291. https://doi.org/10.1002/prot.1093
183. Brakoulias A, Jackson RM (2004) Towards a structural classification of phosphate binding sites in protein-nucleotide complexes: an automated all-against-all structural comparison using geometric matching. Proteins 56(2):250–260. https://doi.org/10.1002/prot.20123
184. Vyas P, Malitsky S, Itkin M, Tawfik DS (2023) On the origins of enzymes: phosphate-binding polypeptides mediate phosphoryl transfer to synthesize adenosine triphosphate. J Am Chem Soc 145(15):8344–8354. https://doi.org/10.1021/jacs.2c08636
185. Doolittle WF (1997) Fun with genealogy. Proc Natl Acad Sci USA 94(24):12751–12753. https://doi.org/10.1073/pnas.94.24.12751
186. Chu XY, Zhang HY (2020) Cofactors as molecular fossils to trace the origin and evolution of proteins. ChemBioChem 21(22):3161–3168. https://doi.org/10.1002/cbic.202000027
187. Kolodny R (2021) Searching protein space for ancient sub-domain segments. Curr Opin Struct Biol 68:105–112. https://doi.org/10.1016/j.sbi.2020.11.006
188. Romero-Romero S, Kordes S, Michel F, Hocker B (2021) Evolution, folding, and design of TIM barrels and related proteins. Curr Opin Struct Biol 68:94–104. https://doi.org/10.1016/j.sbi.2020.12.007
189. Narunsky A, Kessel A, Solan R, Alva V, Kolodny R, Ben-Tal N (2020) On the evolution of protein-adenine binding. Proc Natl Acad Sci USA 117(9):4701–4709. https://doi.org/10.1073/pnas.1911349117
190. Nepomnyachiy S, Ben-Tal N, Kolodny R (2017) Complex evolutionary footprints revealed in an analysis of reused protein segments of diverse lengths. Proc Natl Acad Sci U S A 114(44):11703–11708. https://doi.org/10.1073/pnas.1707642114
191. Longo LM, Hirai H, McGlynn SE (2022) An evolutionary history of the CoA-binding protein Nat/Ivy. Protein Sci Publ Protein Soc 31(12):e4463. https://doi.org/10.1002/pro.4463
192. Bhagavat R, Srinivasan N, Chandra N (2017) Deciphering common recognition principles of nucleoside mono/di and tri-phosphates binding in diverse proteins via structural matching of their binding sites. Proteins 85(9):1699–1712. https://doi.org/10.1002/prot.25328
193. Heizinger L, Merkl R (2021) Evidence for the preferential reuse of sub-domain motifs in primordial protein folds. Proteins 89(9):1167–1179. https://doi.org/10.1002/prot.26089
194. Murzin AG, Brenner SE, Hubbard T, Chothia C (1995) SCOP: a structural classification of proteins database for the investigation of sequences and structures. J Mol Biol 247(4):536–540. https://doi.org/10.1006/jmbi.1995.0159
195. Koc I, Caetano-Anolles G (2017) The natural history of molecular functions inferred from an extensive phylogenomic analysis of gene ontology data. PLoS One 12(5):e0176129. https://doi.org/10.1371/journal.pone.0176129
196. Schneider G, Neuberger G, Wildpaner M, Tian S, Berezovsky I, Eisenhaber F (2006) Application of a sensitive collection heuristic for very large protein families: evolutionary relationship between adipose triglyceride lipase (ATGL) and classic mammalian lipases. BMC Bioinf 7:164. https://doi.org/10.1186/1471-2105-7-164

197. Wodak SJ, Paci E, Dokholyan NV, Berezovsky IN, Horovitz A, Li J, Hilser VJ, Bahar I, Karanicolas J, Stock G, Hamm P, Stote RH, Eberhardt J, Chebaro Y, Dejaegere A, Cecchini M, Changeux JP, Bolhuis PG, Vreede J, Faccioli P, Orioli S, Ravasio R, Yan L, Brito C, Wyart M, Gkeka P, Rivalta I, Palermo G, McCammon JA, Panecka-Hofman J, Wade RC, Di Pizio A, Niv MY, Nussinov R, Tsai CJ, Jang H, Padhorny D, Kozakov D, McLeish T (2019) Allostery in its many disguises: from theory to applications. Structure 27(4):566–578. https://doi.org/10.1016/j.str.2019.01.003
198. Babbitt PC, Hasson MS, Wedekind JE, Palmer DR, Barrett WC, Reed GH, Rayment I, Ringe D, Kenyon GL, Gerlt JA (1996) The enolase superfamily: a general strategy for enzyme-catalyzed abstraction of the alpha-protons of carboxylic acids. Biochemistry 35(51):16489–16501. https://doi.org/10.1021/bi9616413
199. Pegg SC, Brown SD, Ojha S, Seffernick J, Meng EC, Morris JH, Chang PJ, Huang CC, Ferrin TE, Babbitt PC (2006) Leveraging enzyme structure-function relationships for functional inference and experimental design: the structure-function linkage database. Biochemistry 45(8):2545–2555. https://doi.org/10.1021/bi052101l
200. Kurochkin IV, Guarnera E, Berezovsky IN (2018) Insulin-degrading enzyme in the fight against Alzheimer's disease. Trends Pharmacol Sci 39(1):49–58. https://doi.org/10.1016/j.tips.2017.10.008
201. Alva V, Soding J, Lupas AN (2015) A vocabulary of ancient peptides at the origin of folded proteins. eLife 4:e09410. https://doi.org/10.7554/eLife.09410
202. Ezerzer Y, Frenkel-Pinter M, Kolodny R, Ben-Tal N (2025) A building blocks perspective on protein emergence and evolution. Curr Opin Struct Biol 91:102996. https://doi.org/10.1016/j.sbi.2025.102996
203. Aziz MF, Caetano-Anolles G (2021) Evolution of networks of protein domain organization. Sci Rep 11(1):12075. https://doi.org/10.1038/s41598-021-90498-8
204. Lisanza SL, Gershon JM, Tipps SWK, Sims JN, Arnoldt L, Hendel SJ, Simma MK, Liu G, Yase M, Wu H, Tharp CD, Li X, Kang A, Brackenbrough E, Bera AK, Gerben S, Wittmann BJ, McShan AC, Baker D (2024) Multistate and functional protein design using RoseTTAFold sequence space diffusion. Nat Biotechnol. https://doi.org/10.1038/s41587-024-02395-w
205. Lauko A, Pellock SJ, Sumida KH, Anishchenko I, Juergens D, Ahern W, Jeung J, Shida A, Hunt A, Kalvet I, Norn C, Humphreys IR, Jamieson C, Krishna R, Kipnis Y, Kang A, Brackenbrough E, Bera AK, Sankaran B, Houk KN, Baker D (2025) Computational design of serine hydrolases. Science eadu2454. https://doi.org/10.1126/science.adu2454
206. Vazquez Torres S, Benard Valle M, Mackessy SP, Menzies SK, Casewell NR, Ahmadi S, Burlet NJ, Muratspahic E, Sappington I, Overath MD, Rivera-de-Torre E, Ledergerber J, Laustsen AH, Boddum K, Bera AK, Kang A, Brackenbrough E, Cardoso IA, Crittenden EP, Edge RJ, Decarreau J, Ragotte RJ, Pillai AS, Abedi M, Han HL, Gerben SR, Murray A, Skotheim R, Stuart L, Stewart L, Fryer TJA, Jenkins TP, Baker D (2025) De novo designed proteins neutralize lethal snake venom toxins. Nature. https://doi.org/10.1038/s41586-024-08393-x
207. Ward T, Kalvet I, Lozhkin B, Morris E, Zhang KC, D., Ernst MZ, X., Baker D (2025) De Novo Design and Evolution of an Artificial Metathase for Cytoplasmic Olefin Metathesis. https://doi.org/10.21203/rs.3.rs-5849532/v1
208. Piraner DI, Abedi MH, Duran Gonzalez MJ, Chazin-Gray A, Lin A, Zhu I, Ravindran PT, Schlichthaerle T, Huang B, Bearchild TH, Lee D, Wyman S, Jun YW, Baker D, Roybal KT (2025) Engineered receptors for soluble cellular communication and disease sensing. Nature 638(8051):805–813. https://doi.org/10.1038/s41586-024-08366-0
209. Zhang JZ, Nguyen WH, Greenwood N, Rose JC, Ong SE, Maly DJ, Baker D (2024) Computationally designed sensors detect endogenous Ras activity and signaling effectors at subcellular resolution. Nat Biotechnol 42(12):1888–1898. https://doi.org/10.1038/s41587-023-02107-w
210. Dowling QM, Park YJ, Fries CN, Gerstenmaier NC, Ols S, Yang EC, Wargacki AJ, Dosey A, Hsia Y, Ravichandran R, Walkey CD, Burrell AL, Veesler D, Baker D, King NP (2025) Hierarchical design of pseudosymmetric protein nanocages. Nature 638(8050):553–561. https://doi.org/10.1038/s41586-024-08360-6
211. Lee S, Kibler RD, Ahn G, Hsia Y, Borst AJ, Philomin A, Kennedy MA, Huang B, Stoddard B, Baker D (2025) Four-component protein nanocages designed by programmed symmetry breaking. Nature 638(8050):546–552. https://doi.org/10.1038/s41586-024-07814-1

Chapter 8

Methods to Characterize Mutant Spectra of RNA Viruses: The Case of SARS-CoV-2

Celia Perales, María Eugenia Soria, Ana Isabel de Ávila, Brenda Martínez-González, Isabel Gallego, Carlos García-Crespo, Pilar Somovilla, Pablo Mínguez, Ignacio Gadea and Esteban Domingo

Abstract

Viral isolates consist of complex mixtures of many variants that are termed mutant spectra, distributions, clouds or swarms. They include information on virus behavior that is not captured by consensus sequences. Here, we describe experimental procedures, a bioinformatics pipeline, and calculations to characterize the mutant spectrum of RNA viruses. The data are obtained through the high-resolution MiSeq Illumina ultra-deep sequencing platform. We describe protocols for SARS-CoV-2 patients' isolates and laboratory populations, which can be adapted to other RNA viral pathogens. Precautions for sample handling to avoid cross-contaminations and controls for mutation and deletion detection reliability are also outlined.

Keywords Ultra-deep sequencing, Viral quasispecies, Mutation, Deletion, RNA quantification, cDNA amplification, MiSeq Illumina, COVID-19

1 Introduction

The recent recognition of an extreme diversity of the cellular and sub-cellular world poses new challenges regarding its molecular characterization and biological significance. In the ample and complex virosphere and virus discovery projections [1, 2], there is a largely unexplored, additional level of complexity. It consists in multitude of variations found among the individual genomic molecules whose ensemble we refer to as a viral isolate. The set of genomes within an isolate is termed mutant spectrum, distribution, cloud, or swarm. Such intra-host genetic and functional heterogeneity goes beyond the classic concept of genetic polymorphism since it does not describe permanent variant forms, but, rather, complex swarms with a fleeting existence [3–5].

Viral quasispecies is used to mean the ensemble of genomes present in a population of viruses at a given time. It does not refer to

Gustavo Caetano-Anollés (ed.), *Evolutionary Genomics*, Methods in Molecular Biology, vol. 2981, https://doi.org/10.1007/978-1-0716-4836-0_8,

differences among consensus sequences of different isolates (those that are often deposited in sequence data banks). "Population of viruses" may mean either the viral genomes present in a sample from an individual infected cell, from a set of cells, from a tissue, organ or an entire organism. It may also mean the genomes in a laboratory preparation of virus of any origin (a field isolate grown in cell culture, samples from experimental evolution designs, etc.). In contrast to the mutant spectrum, the consensus sequence of the same population is a weighted average of the different sequences included in the sample. In a consensus sequence each genomic position is occupied by the nucleotide which is most represented in the corresponding position when the sequences of the individual genomes are aligned. The consensus sequence does not inform of the different minority genomes that compose the population under study.

Important biomedical information embodied in the mutant spectrum of a viral isolate is the possible presence at low frequency of antiviral-, antibody-, or cytotoxic T cell-escape mutants, host cell tropism variants, interferon (IFN)-inducing or IFN-responsive genomes, or variants prone to establish persistent infections, among other phenotypes (reviewed in [3]). Traits expressed by individual genomes (when they increase in frequency) may differ from the traits expressed by the ensemble where they were located. Such minority, individual genomes can shift towards dominance either by positive selection (i.e., selection of antiviral-resistant mutants as a cause of treatment failure) or by bottleneck events.

Mutant spectra can also provide an alternative and more realistic assessment of residue conservation than consensus sequences. A recent trend in antiviral research is to achieve universal vaccines or pan-genomic antiviral agents. Such designs rely largely on the targeting of conserved genomic residues or domains, often in their encoded proteins. Residue conservation based on alignment of consensus sequences is less stringent than residue conservation based on mutant spectra [6]. Thus, mutant spectra provide a far more solid base than consensus sequences to design universal vaccines or antiviral compounds.

More difficult to detect (but increasingly reported in the literature) are interactions of cooperation, interference, or complementation among viruses from the same swarm ([7–9], among other examples). As the identification of sequences capable of such intra-population interactions increases, mutant spectrum composition offers a valuable information source for potential cross-influences. Another asset that has been unveiled with SARS-CoV-2 research is the possibility that mutant spectra recapitulate the range of sequences that may become epidemiologically prominent at later times ([10, 11], and references therein). That is, there may be limits to the incessant tendency of the virus to explore new sequences, and mutant spectra may inform of such limits.

Despite its large (29.9 Kb) genome, and its encoding a 3′–5′ exonuclease activity (ExoN) which is active in vitro, SARS-CoV-2 isolates contain complex mutant spectra, and the virus participates of the biomedical consequences of quasispecies dynamics ([12–14] and other studies). Experimental evolution designs with SARS-CoV-2 (and other viral pathogens that can infect cells in culture) intended to investigate the response to selective agents, replicative fitness variations, etc. may benefit of initiating the experiments with a single infectious genome. Therefore, we describe a protocol for SARS-CoV-2 isolation of a single infectious genome by plaque isolation in cell culture. We use Vero E6 but the procedures can be extended to other cells that are productively infected by SARS-CoV-2. From the available evidence, it is likely that future emergent RNA viral pathogens will also share quasispecies implications. Thus, a presentation of ultra-deep sequencing data processing protocols, and ensuing calculations aimed at dissecting SARS-CoV-2 mutant spectra with high resolution seems justified.

Some obvious limitations should be stated. Unavoidably, as technology stands at the moment, the ultra-deep sequencing results can only provide an approximate picture of the genome composition of a viral population. This is due to sampling effects, and to technical limitations in copying accuracy during nucleic acid amplification reactions, and nucleotide discrimination in the ultra-deep sequencing instruments. The sampling effect means that the isolate that reaches the laboratory is only a part of the total original biological sample, and, in addition, in the processes that follow (RNA extraction, copying into cDNA, amplification, purification, and sequencing) often only a part of the total sample enters the next step. The overall technical accuracy can be evaluated by quality scores calculated from internal controls included in the sequencing platforms (for example, the Q score in the Illumina MiSeq platform). In the Methods section we describe this and other controls performed to evaluate the reliability of ultra-deep sequencing of SARS-CoV-2 genomic regions.

2 Materials

Use disposable gloves throughout.

2.1 SARS-CoV-2 Cell Culture System

2.1.1 Cell Culture and Maintenance

1. Frozen vials of Vero E6 cells [Ref: CRL-1586, RRID: CVCL_0574; ATCC] (*see* **Note 1**).
2. Dulbecco's modified Eagle's medium (DMEM) supplemented with 1 mM sodium pyruvate, 1% nonessential amino acids, 4 mM L-glutamine, 50 mg/ml gentamicin, 0.2 mg/ml antimycotic, and 10% fetal bovine serum (FBS).

3. EDTA (0.02%).
4. Trypsin-EDTA (0.25%; 0.02%).
5. Glass bottles.
6. 100 mm tissue culture dishes.
7. Serological pipettes.
8. Aspirating pipettes.
9. Surface disinfectant Suredis VT-1, 70% ethanol.
10. 37°C water bath.
11. Cell culture incubator at 37°C, 5% CO_2.
12. Class IIA biosafety cabin.

2.1.2 Preparation of SARS-CoV-2 Stock Virus

1. Vero E6 cells (cultured to confluence) in a 25 cm^2 tissue culture flask (25 cm^2 canted neck cell culture flask with non-vented screw cap) (*see* **Note 2**).
2. SARS-CoV-2 reference virus (*see* **Notes 3 and 4**).
3. PBS 1×.
4. DMEM supplemented with 2% fetal bovine serum (FBS) and 25 mM [4-(2-Hydroxyethyl)piperazine-1-ethanesulfonic acid] (HEPES) (*see* **Note 5**).
5. Conical tubes.
6. Serological pipettes.
7. Aspirating pipettes.
8. Micropipettes (20 μl, 200 μl and 1000 μl).
9. Micropipette sterile filter tips.
10. 2 ml externally threaded sterile cryogenic storage tubes with rubber seals.
11. Cell culture incubator at 37 °C, 0% CO_2.
12. Class IIA biosafety cabin.
13. Freezer at −80 °C.
14. Virucidal decontaminant Virkon.
15. Perasafe.
16. 70% ethanol.

2.1.3 SARS-CoV-2 Infection

1. Vero E6 cells (cultured to confluence) in a 12.5 cm^2 tissue culture flasks or multiwell plates (it depends on the experiment) (*see* **Note 2**).
2. SARS-CoV-2 reference virus (*see* **Notes 3 and 4**).
3. PBS 1×.
4. DMEM supplemented with 2% fetal bovine serum (FBS) and 25 mM [4-(2-Hydroxyethyl)piperazine-1-ethanesulfonic acid] (HEPES) (*see* **Note 5**).

5. Conical tubes.
6. Serological pipettes.
7. Aspirating pipettes.
8. Micropipettes (20 μl, 200 μl, and 1000 μl).
9. Micropipette sterile filter tips.
10. 2 ml externally threaded sterile cryogenic storage tubes with rubber seals.
11. Cell culture incubator at 37 °C, 0% CO_2.
12. Class IIA biosafety cabin.
13. Freezer at −80 °C.
14. Virucidal decontaminant Virkon.
15. Perasafe.
16. 70% ethanol.

2.1.4 Titration

1. Vero E6 cells (cultured to confluence) in 6-well plates (*see* **Note 2**).
2. SARS-CoV-2 reference virus (*see* **Notes 3 and 4**).
3. PBS 1×.
4. DMEM supplemented with 2% fetal bovine serum (FBS) and 25 mM [4-(2-Hydroxyethyl)piperazine-1-ethanesulfonic acid] (HEPES) (see Note 5), for dilutions.
5. DMEM 2×.
6. Agar 1%.
7. Fetal bovine serum (FBS).
8. DEAE-Dextran.
9. Conical tubes.
10. Serological pipettes.
11. Aspirating pipettes.
12. Micropipettes (20 μl, 200 μl, and 1000 μl).
13. Micropipette sterile filter tips.
14. 2 ml externally threaded sterile cryogenic storage tubes with rubber seals.
15. Cell culture incubator at 37°C, 0% CO_2.
16. Class IIA biosafety cabin.
17. Freezer at −80°C.
18. Virucidal decontaminant Virkon.
19. Perasafe.
20. 70% ethanol.

2.1.5 Biological Cloning

1. Vero E6 cells (cultured to confluence) in 6-well plates (*see* **Note 2**).
2. SARS-CoV-2 reference virus (*see* **Notes 3 and 4**).
3. PBS 1×.
4. DMEM supplemented with 2% fetal bovine serum (FBS) and 25 mM [4-(2-Hydroxyethyl)piperazine-1-ethanesulfonic acid] (HEPES) (*see* **Note 5**), for dilutions.
5. Sodium deoxycholate.
6. DMEM 2×.
7. Agar 1%.
8. Fetal bovine serum (FBS).
9. DEAE-Dextran.
10. Conical tubes.
11. Serological pipettes.
12. Aspirating pipettes.
13. Plastic Pasteur pipettes.
14. Automatic pipettor.
15. Micropipettes (20 μl, 200 μl and 1000 μl).
16. Micropipette sterile filter tips.
17. 2 ml externally threaded sterile cryogenic storage tubes with rubber seals.
18. Cell culture incubator at 37 °C, 0% CO_2.
19. Class IIA biosafety cabin.
20. Freezer at −80 °C.
21. Virucidal decontaminant Virkon.
22. Perasafe.
23. 70% ethanol.

2.1.6 Virus and Drugs

1. Vero E6 cells (cultured to confluence) in 24-well plates (*see* **Note 2**).
2. SARS-CoV-2 reference virus (*see* **Notes 3 and 4**), or biological samples from COVID-19 patients (*see* **Note 6**).
3. Drugs at required concentrations according to CC_{50} and IC_{50} values; for example, ribavirin stocks 0.001 M and 0.01 M.
4. PBS 1×.
5. DMEM supplemented with 2% fetal bovine serum (FBS) and 25 mM [4-(2-Hydroxyethyl)piperazine-1-ethanesulfonic acid] (HEPES) (*see* **Note 5**).
6. Conical tubes.
7. Serological pipettes.

8. Aspirating pipettes.
9. Micropipettes (20 μl, 200 μl, and 1000 μl).
10. Micropipette sterile filter tips.
11. 2 ml externally threaded sterile cryogenic storage tubes with rubber seals.
12. Cell culture incubator at 37 °C, 0% CO_2.
13. Class IIA biosafety cabin.
14. Freezer at −80 °C.
15. Virucidal decontaminant Virkon.
16. Perasafe.
17. 70% ethanol.

2.1.7 RNA Extraction

1. Viral RNA with QIAamp Viral RNA Mini Kit (Qiagen).
2. 96–100% ethanol.
3. Purified (RNAse-free) water.
4. 15 ml conical tubes.
5. Sterile, RNase-free 1.5 ml low-binding tubes.
6. 2 ml externally threaded sterile cryogenic storage tubes with rubber seals.
7. Micropipettes (200 μl and 1000 μl).
8. Vortex mixer.
9. Centrifuge.
10. Low-binding filter pipette tips.
11. Class IIA biosafety cabin.
12. Freezer at −80 °C.
13. Virucidal decontaminant Virkon.
14. Perasafe.
15. 70% ethanol.

2.2 SARS-CoV-2 RNA Quantification and Amplification

2.2.1 SARS-CoV-2 Standard Curve for RNA Quantification by Real-Time RT-PCR

1. USA-WA1/2020 viral RNA previously extracted (see Sect. 2.1.7).
2. Transcriptor One Step RT-PCR kit (Roche).
3. QIAquick gel extraction kit (Qiagen).
4. Agarose MP (Roche).
5. TAE 1× (dilution from TAE 10 X; Roche).
6. DNA dye Midori Green Advance (Thermo Scientific).

7. GeneRuler 1 kb Plus DNA Ladder (Thermo Scientific).
8. TriTrack DNA Loading Dye (6X) (Thermo Scientific).
9. Qubit® dsDNA HS Assay Kit (ThermoFisher).
10. Qubit® RNA BR Assay Kit (ThermoFisher).
11. Qubit® Assay tubes (ThermoFisher).
12. GoTaq Flexi DNA Polymerase (Promega).
13. pGEM-T vector (Promega).
14. Illustra TempliPhi DNA Sequencing Template Amplification Kit (GE Healthcare).
15. Wizard® Plus SV Minipreps DNA Purification System (Promega).
16. Wizard® PCR Preps DNA Purification Resin (5.4 M Guanidine Thiocyanate) (Promega).
17. Wizard® Minicolumns (Promega).
18. T7 RiboMAX Express Large Scale RNA Production system (Promega).
19. Mini Quick Spin RNA Columns (Roche).
20. Pair of oligonucleotides previously designed for amplification of nsp12-coding region at 50 ng/μl (nsp12-CoV2-u14511 5'-GGATGTAAACTTACATAGCTCTA-3' and nsp12-CoV2-d15717 5'-AGAGAGTATCATCATTGAGAAATG-3').
21. Isopropanol (2-propanol).
22. Plasmid pGEM3zf+.
23. DH5α *Escherichia coli.*
24. LB + Ampicillin (100 μg/ml) plates.
25. LB medium without antibiotic.
26. X-gal (20 mg/ml) and IPTG (20%).
27. Sterile glycerol.
28. PCR cabin free of genetic material.
29. Balance.
30. Microwave.
31. Test tube and bottles without autoclaving.
32. Mold and comb for gels. Electrophoresis cuvette and power supply.
33. LED light transilluminator, plastic knives and protective glasses.
34. Qubit® fluorometer.
35. Bath at 37 °C.
36. Heater at 37 °C.

37. Glass beads for seeding plates.
38. Timer.
39. Vortex.
40. Microcentrifuge.
41. Thermocycler.
42. Thermoblock.
43. Automatic pipettes.
44. Sterile RNase-free low-binding tips with filter.
45. Sterile RNase-free 1.5 ml and 0.2 ml low-binding tubes.
46. 15/50 ml conical tubes.
47. 10 ml polystyrene tube.

2.2.2 Detection and Quantification of SARS-CoV-2 RNA by Real-Time RT-PCR

1. SARS-CoV-2 viral RNA previously extracted (see Sect. 2.1.7).
2. LightCycler RNA Master SYBR Green I (Roche): LightCycler RNA Master SYBR Green l [contains Tth DNA Polymerase, reaction buffer, dNTP mix (with dUTP instead of dTTP), and SYBR Green I], $Mn(OAc)_2$ 50 mM stock solution, PCR-grade water.
3. PCR cabin free of genetic material.
4. Pair of oligonucleotides at 3 μM designed for amplification of an nsp12-coding region (nsp12-CoV2-u14890: 5'-GATGGTGGCTGTATTAATGCT-3' and nsp12-CoV2-d15320: 5'-GGGATGGCTCTATCACATTTA-3').
5. LightCycler® System (Roche).
6. LightCycler® Capillaries (20 μl) (Roche).
7. LightCycler® Centrifuge Adapters (Roche).
8. Automatic pipettes.
9. Sterile RNase-free low-binding tips with filter.
10. Sterile RNase-free 1.5 ml low-binding tubes.

2.2.3 Amplification by RT-PCR

1. SARS-CoV-2 viral RNA previously extracted (see Sect. 2.1.7).
2. Transcriptor One Step RT-PCR kit (Roche): transcriptor enzyme mix (Transcriptor Reverse Transcriptase, Protector RNase Inhibitor, and a DNA polymerase enzyme blend, consisting of Taq DNA polymerase and a proofreading polymerase), RT-PCR reaction buffer 5× (Tris, $MgCl_2$, sodium salts of dNTPs 1.5 mM each), PCR-grade water.
3. PCR cabin genetic material free.
4. Pair of oligonucleotides previously designed for the amplifications of nsp12- and S-coding regions at 50 ng/μl.
5. Vortex.

6. Thermocycler.
7. Automatic pipettes.
8. Sterile RNase-free tips with filter.
9. Sterile RNase-free 1.5 ml and 0.2 ml low-binding tubes.

2.2.4 Gel Electrophoresis

1. Agarose MP (Roche).
2. TAE 1× (dilution form TAE 10×; Roche).
3. DNA dye Midori Green Advance (Thermo Scientific).
4. GeneRuler 1 kb Plus DNA Ladder (Thermo Scientific).
5. TriTrack DNA Loading Dye (6×) (Thermo Scientific).
6. Balance.
7. Microwave.
8. Vortex.
9. Test tube and bottles without autoclaving.
10. Mold and comb for gels.
11. Electrophoresis cuvette and power supply.
12. LED light transilluminator and protective glasses.
13. Automatic pipettes.
14. Sterile RNase-free low-binding tips with filter.
15. Sterile RNase-free 1.5 ml low-binding tubes.

2.2.5 DNA Purification and Quantification

1. QIAquick gel extraction kit (Qiagen).
2. Isopropanol (2-propanol).
3. Balance.
4. Vortex.
5. LED light transilluminator, plastic knifes and protective glasses.
6. Automatic pipettes.
7. Sterile RNase-free low-binding tips with filter.
8. Sterile RNase-free 1.5 ml low-binding tubes.
9. Sterile RNase-free 2 ml tubes.
10. 15/50 ml conical tubes.
11. Qubit® fluorometer.
12. Qubit® dsDNA HS Assay kit (ThermoFisher): Qubit® dsDNA HS Reagent and Qubit® dsDNA HS Buffer.
13. Qubit® Assay tubes (Thermofisher).

2.3 SARS-CoV-2 Amplicon Ultra-Deep Sequencing

2.3.1 Amplicon Normalization and Purification

1. KAPA Pure Beads (30 ml) (Kapa).
2. Thermoblock.
3. Centrifuge.
4. Vortex.
5. Automatic pipettes.
6. Sterile RNase-free low-binding tips with filter.
7. Sterile RNase-free 1.5 ml low-binding tubes.
8. MiliQ water.
9. Absolute ethanol.
10. Elution Buffer (EB): 10 mM Tris-HCl, pH 8.5.
11. Magnetic rack for 1.5 ml tubes.

2.3.2 Library Preparation

1. KAPA Hyper Prep (Kapa).
2. SeqCap Adapter Kit A/B (Roche).
3. KAPA Pure Beads (30 ml) (Kapa).
4. Thermoblock.
5. Thermal cycler.
6. Centrifuge.
7. The materials that follow are identical to those listed in Sect. 2.3.1.

2.3.3 Sequencing with MiSeq Platform

1. MiSeq® platform Illumina.
2. MiSeq® Reagent Kit v3 (600 cycle) (Illumina).
3. PhiX Control v3 kit (Illumina).
4. Thermoblock.
5. Vortex.
6. Automatic pipettes.
7. Ice.
8. Tween 20 (Sigma Aldrich).
9. 10 mM Tris-HCl, pH 8.0–8.5.
10. 200 nM Tris-HCl, pH 7.0–7.8.
11. NaOH 2N.
12. MiliQ water.
13. Distilled water.
14. Fiber-free tissues.

15. Sterile RNase-free 1.5 ml low-binding tubes.
16. Sterile RNase-free low-binding tips with filter.

2.3.4 Bioinformatics Analysis with SeekDeep

SeekDeep is freely available under the GNU Lesser General Public License v3.0 and is actively maintained on GitHub (https://github.com/bailey-lab/SeekDeep). Information regarding its usage and program details can be found on the SeekDeep website (http://baileylab.umassmed.edu/SeekDeep/). It is recommended to run SeekDeep analysis in a computer with linux-based OS. Our setup includes a HPC cluster with 560GB RAM and 280 cores available. Follow SeekDeep specifications for setup limitations.

2.3.5 Coverage and Frequency Cut-off

Mutations and deletions can be detected with a 0.1% frequency cut-off, contingent upon clean read coverage (see Sect. 3.3.5).

3 Methods

3.1 SARS-CoV-2 Cell Culture System

3.1.1 Cell Culture and Maintenance

1. Vero E6 cells are periodically thawed from a large frozen stock and passaged a maximum of 30 times at a split ratio of 1:6 (3 days) to 1:8 (4 days).
2. When cells are nearly confluent, aspirate growth media and wash the cells once with 2 ml of EDTA (0.02%).
3. Add 1 ml of 0.25% trypsin-EDTA to a p100 dish and incubate at 37°C, 5% CO_2 for 3–4 min, until cells begin to detach.
4. Add 1 ml of DMEM 10% FBS to the plate and gently pipet up and down several times to disaggregate clumps.
5. Add cells back to a tissue culture plate to the desired sub-cultivation ratio with a final volume of 10 ml for a new plate.
6. Move the plate with cross movements to avoid accumulation of cells towards the plate center.
7. Place the cells in the incubator at 37°C, 5% CO_2, and culture until near confluence (3–4 days for Vero E6 cells).

3.1.2 Preparation of SARS-CoV-2 Stock Virus

1. Seed 3×10^6 Vero E6 cells in a 25 cm^2 tissue culture flask (*see* **Note 7**).
2. Introduce the flask in a biosafety level 3 laboratory (BSL-3) to continue the experiment.
3. Introduce in the cabin the SARS-CoV-2 reference virus (*see* **Note 4**), and let it thaw.

4. Prepare the inoculum with 20 µl of SARS-CoV-2 reference virus in DMEM supplemented with 2% FBS and 25 mM HEPES.
5. Aspirate growth media from Vero E6 cells with an aspirating pipette.
6. Add the inoculum onto the cells very carefully and tightly close the flask cap. Swirl gently the flask to cover all the monolayer. Decontaminate the flask external surface with Virkon.
7. Place the flask with the infected cells inside an airtight container. Decontaminate the flask exterior with Virkon.
8. Place the container in a cell culture incubator at 37 °C.
9. Swirl gently the container every 15 min for 1 h for virus adsorption.
10. After 1 h, place the flask again into the biosafety cabin, and remove the initial inoculum with an aspirating pipette connected to a vacuum bomb that transfers the liquid to a Virkon solution.
11. Wash the monolayer once with PBS 1×.
12. Add 5 ml of DMEM supplemented with 2% FBS and 25 mM HEPES. Repeat Steps 7 and 8.
13. Incubate the flask for 48 h until cytopathic effect appears.
14. Transfer the supernatant to a 15 ml conical tube.
15. Aliquot the supernatant into 2 ml externally threaded sterile cryogenic storage tubes.
16. Store virus aliquots at −80 °C.
17. To control for the absence of contamination, a mock-infected flask is maintained in parallel.

3.1.3 SARS-CoV-2 Infection

1. Seed Vero E6 cells in a multiwell tissue culture plate or flask (*see* **Note** 7).
2. Introduce the plate/flask in a BSL-3 laboratory to continue the experiment.
3. Introduce in the cabin the SARS-CoV-2, and let it thaw.
4. Prepare the inoculum to reach a multiplicity of infection (MOI) of 0.001 plaque-forming units (PFU)/cell in DMEM supplemented with 2% FBS and 25 mM HEPES.
5. Aspirate growth media from Vero E6 cells with an aspirating pipette.
6. Add the inoculum onto the cells very carefully and tightly close the flask cap or plate. Gently swirl the plate/flask to cover all the monolayer. Decontaminate the plate/flask external surface with Virkon.

7. Place the plate/flask with the infected cells inside an airtight container. Decontaminate the flask external surface with Virkon.
8. Place the container in a cell culture incubator at 37 °C.
9. Swirl gently the container every 15 min for 1 h for virus adsorption.
10. After 1 h, place the plate/flask again into the biosafety cabin and remove the initial inoculum with an aspirating pipette connected to a vacuum bomb that transfers the liquid to a Virkon solution.
11. Wash the monolayer once with PBS 1×.
12. Add the amount of DMEM supplemented with 2% FBS and 25 mM HEPES according to the plate/flask used. Repeat Steps 7 and 8.
13. Incubate the plate/flask for 48–72 h until cytopathic effect appears (*see* **Note 8**).
14. Transfer the supernatant to 2 ml externally threaded sterile cryogenic storage tubes.
15. Store virus aliquots at −80 °C.
16. To control for the absence of contamination, a mock-infected plate/flask is maintained in parallel.

3.1.4 Titration

1. Seed 5×10^5 cells per well of Vero E6 cells in 6-well plates with DMEM 1× supplemented with 10% FBS (2 ml/well) (*see* **Note 7**).
2. Introduce the plate/flask in a BSL-3 laboratory to continue the experiment.
3. Introduce in the cabin the SARS-CoV-2, and let it thaw.
4. Prepare ten-fold serial dilutions of the virus samples in DMEM supplemented with 2% FBS and 25 mM HEPES.
5. Aspirate growth media from Vero E6 cells with an aspirating pipette.
6. Add 200 μl inoculum of each dilution onto the cells very carefully, and gently swirl the plates to cover the monolayers. Decontaminate the plates external surface with Virkon.
7. Place the plates with the infected cells inside an airtight container. Decontaminate the container external surface with Virkon.
8. Place the container in a cell culture incubator at 37°C.
9. Swirl gently the container every 15 min for 1 h for virus adsorption.
10. Prepare plaque medium just before use: half volume of DMEM 2× supplemented with 25 mM HEPES, and the

other half volume of 1% agar. Add 1% FBS and 1% DEAE-Dextran to the final mixture.

11. After 1 h, place the plates again into the biosafety cabin and remove the initial inoculum with an aspirating pipette connected to a vacuum bomb that transfers the liquid to a Virkon solution.
12. Wash the monolayer once with PBS 1×.
13. Add 2 ml of plaque medium to each well (*see* **Note 9**). Repeat Steps 7 and 8.
14. Incubate the plates for 72 h.
15. After that time, fix the cells with 10% formaldehyde for 1 h.
16. Remove the formaldehyde and stain the cells with 2% crystal violet staining solution for 10 min. Remove the crystal violet, wash with water, and let it dry.
17. Quantify the plaques as PFU/ml.

3.1.5 Biological Cloning

1. Prepare a stock of 1% sodium deoxycholate in culture medium (50 mg sodium deoxycholate in 5 ml of DMEM supplemented with 10% FBS at 37°C, and mix with vortex). Once dissolved, pass through a 0.45 μM filter, prepare aliquots, and store at −80 °C.
2. Seed 5×10^5 cells per well of Vero E6 cells in 6-well plates with DMEM 1× supplemented with 10% FBS (2 ml/well) (*see* **Note 2**).
3. Introduce the plates in a BSL-3 laboratory to continue the experiment.
4. Introduce into the cabin the SARS-CoV-2 virus preparation and let it thaw.
5. Incubate 1000 PFUs of virus with 0.01% sodium deoxycholate during 10 min at room temperature in the cabin (*see* **Note 10**).
6. Dilute 1/10 in DMEM supplemented with 2% FBS and 25 mM HEPES.
7. Aspirate growth media from Vero E6 cells with an aspirating pipette.
8. Add 200 μl of inoculum and of a 1/10 dilution onto the cells very carefully, and gently swirl the plates to cover all the monolayer. Decontaminate the plates external surface with Virkon.
9. Place the plates with the infected cells inside an airtight container. Decontaminate the outside surface with Virkon.
10. Place the container in a cell culture incubator at 37°C.

11. Swirl gently the container every 15 min for 1 h for virus adsorption.
12. After 1 h, place the plates again into the biosafety cabin and remove the initial inoculum with an aspirating pipette connected to a vacuum bomb that transfers the liquid to a Virkon solution.
13. Wash the monolayer once with PBS 1×.
14. Add 2 ml of plaque medium to each well (*see* **Note 9**). Repeat Steps 9 and 10.
15. Incubate the plates for 72 h.
16. After that time, monitor the formation of lysis plaques under an optical microscope and mark the plaques on the plate.
17. Inside the cabin, puncture labeled, well-isolated lysis plaque with a plastic Pasteur pipette and carefully bring to 500 μl DMEM supplemented with 2% FBS and 25 mM HEPES into 2 ml externally threaded sterile cryogenic storage tubes.

3.1.6 Virus and Drugs

1. Seed 1×10^5 Vero E6 cells per well in 24-well plates with DMEM 1× supplemented with 10% FBS (1 ml/well) (*see* **Note 2**).
2. Pretreat cells with the corresponding concentrations of the drugs for the desired length of time (i.e., 16 h for nucleoside analogs). Remove the medium, and add 500 μl of DMEM supplemented with 10% FBS, and the corresponding drug solution, to reach the required concentration.
3. Introduce the plate and the drug stocks in the BSL-3 laboratory to continue the experiment.
4. Introduce into the cabin the SARS-CoV-2 virus preparation and let it thaw.
5. Prepare the inoculum to reach a MOI of 0.001 PFU/cell in DMEM supplemented with 2% FBS and 25 mM HEPES.
6. Aspirate growth media from Vero E6 cells with an aspirating pipette.
7. Add the inoculum onto the cells very carefully and tightly close the plate. Gently swirl the plate to cover all the monolayer. Decontaminate the plate external surface with Virkon.
8. Place the plate with the infected cells inside an airtight container. Decontaminate the external surface with Virkon.
9. Place the container in a cell culture incubator at 37°C.
10. Swirl gently the container every 15 min for 1 h for virus adsorption.

11. After 1 h, place the plate again into the biosafety cabin and remove the initial inoculum with an aspirating pipette connected to a vacuum bomb that transfers the liquid to a Virkon solution.
12. Wash the monolayer once with PBS 1×.
13. Add the 500 μl of DMEM supplemented with 2% FBS, 25 mM HEPES and the corresponding drug concentration. Repeat Steps 8 and 9.
14. Incubate the plate for 48 h until cytopathic effect appears.
15. Transfer supernatant into 2 ml externally threaded sterile cryogenic storage tubes.
16. Store tubes with virus at −80 °C.

3.1.7 RNA Extraction

1. Prepare aliquots of the following reagents:

   AVL buffer: Aliquots in 15 ml conical tubes. If precipitates are present, incubate at 80°C without exceeding 5 min.
   Carrier RNA-AVE: Add the AVE buffer to the carrier RNA (*see* **Note 11**). Add 310 μl of AVE buffer to the tube containing 310 μg of lyophilized carrier RNA to obtain a solution of 1 μg RNA/μl. Dissolve completely and divide in aliquots. Store at −20°C.
   AW1 buffer: Add 25 ml of 96–100% ethanol to the buffer. Aliquot in 15 ml conical tubes (*see* **Note 12**).
   AW2 buffer: Add 30 ml of 96–100% ethanol to the buffer. Aliquot in 15 ml conical tubes (*see* **Note 12**).
2. Prepare a mixture of 0.56 ml of AVL buffer + 5.6 μl of carrier RNA-AVE and 560 μl of 100% ethanol per sample (see Table 1). Introduce both preparations into the BSL-3 laboratory to continue the experiment.
3. Transfer the virus stock tubes from the freezer inside an airtight container, and place it inside the safety cabin (*see* **Note 13**).
4. Pipette 560 μl of the mix (AVL buffer + carrier RNA-AVE) into externally threaded tubes with a rubber seal.
5. Add 140 μl of virus supernatant to each tube. For the negative control, add 140 μl of sterile water.
6. Vortex vigorously for 15 s.
7. Incubate at room temperature inside the cabin for 10 min.
8. Tap the tubes gently to make sure any liquid in the cap drops down, always inside the cabin.
9. Add 560 μl of 100% ethanol and mix vigorously with a vortex for 15 s.
10. Repeat the gentle tapping to ensure any liquid in the cap drops to the bottom of the tube, inside the cabin.

Table 1
Mix required to use QIAamp Viral RNA Mini protocol

Number of samples	Buffer AVL (ml)	Carrier RNA-AVE (μl)	Number of samples	Buffer AVL (ml)	Carrier RNA-AVE (μl)
1	0.56	5.6	13	7.28	72.8
2	1.12	11.2	14	7.84	78.4
3	1.68	16.8	15	8.40	84.0
4	2.24	22.4	16	8.96	89.6
5	2.80	28.0	17	9.52	95.2
6	3.36	33.6	18	10.08	100.8
7	3.92	39.2	19	10.64	106.4
8	4.48	44.8	20	11.20	112.0
9	5.04	50.4	21	11.76	117.6
10	5.60	56.0	22	12.32	123.2
11	6.16	61.6	23	12.88	128.8
12	6.72	67.2	24	13.44	134.4

11. Decontaminate the tubes one by one with a paper towel impregnated with Virkon. Then, wrap the tubes with paper soaked in PeraSafe. Place the tubes in a transport box that has been previously disinfected with Virkon. Impregnate gloves and sleeves with Virkon.
12. Wait for 10 min, change gloves, remove the box with the tubes, and place it inside the SAS. Once the SAS can be opened, place the tubes to a rack and transfer them to the standard laboratory.
13. Return the stock virus tubes to the freezer, clean everything as per the manual, and exit the BSL-3 laboratory following the manual's instructions (*see* **Note 14**).
14. Spin the tubes briefly to ensure any droplets on the walls or cap are collected.
15. Transfer 630 μl of the solution to a column.
16. Centrifuge for 1 min at 8000 r.p.m. After centrifugation, empty the collection tube.
17. Repeat Steps 15 and 16 until all the solution has passed through the column.
18. Add 500 μl of AW1 buffer.
19. Centrifuge for 1 min at 8000 r.p.m. and empty the collection tube.

20. Add 500 μl of AW2 buffer.
21. Centrifuge for 3 min at 14,000 r.p.m. Change the collection tube.
22. Centrifuge for 1 min at 14,000 r.p.m. to dry the membrane as much as possible.
23. Place the column in a clean 1.5 ml tube (Eppendorf).
24. Add 60 μl of AVE buffer (if the samples are from cell culture) or 40 μl (if the samples are from patients, such as nasopharyngeal swab or serum) (*see* **Note 15**). Incubate at room temperature for 1 min.
25. Centrifuge for 1 min at 8000 r.p.m. Store the samples at −80°C (*see* **Note 16**).

3.2 SARS-CoV-2 RNA Quantification and Amplification

3.2.1 SARS-CoV-2 Standard Curve for RNA Quantification by Real-Time RT-PCR

1. RT-PCR amplification reaction to obtain the insert with Transcriptor One Step RT-PCR kit. Keep reagents on ice.
   (a) Calculate RT-PCR mix necessary for the reaction (see Table 2).
   (b) Irradiate with ultra-violet for at least 15 min, prior to its use, and take out the reagents from the freezer and put them on ice: PCR-grade water, oligonucleotides, and buffer 5×.
   (c) Label a 0.2 ml tube.
   (d) Using the PCR cabin, prepare RT-PCR mix in a 1.5 ml low-binding tube and mix with vortex. The Transcriptor enzyme mix will be added at the end.
   (e) Distribute the mix in previously labeled 0.2 ml tubes.

Table 2
RT-PCR reaction mix

Reagent	Volume (1 reaction)
PCR-grade water (vial 3)	30 μl
RT-PCR reaction buffer 5× (vial 2)	10 μl
Nsp12-CoV2-u14511 (50 ng/μl)	2 μl
Nsp12-CoV2-d15717 (50 ng/μl)	2 μl
Transcriptor enzyme mix (vial 1)	1 μl
Final volume	45 μl

(f) Outside the PCR cabin, add 5 μl of RNA to each 0.2 ml tube and mix with the pipette.

(g) Put the tubes on the thermocycler and configure the program (see Table 3).

(h) Once the amplification is finished, store samples at 4°C for the agarose gel electrophoresis analysis if they are going to be used in 1–2 h time, or at −20°C for a longer of time.

2. Agarose gel electrophoresis.

(a) Weigh the amount of Agarose MP necessary for the preparation of gel at a concentration of 1% (see Table 4 for a 60 ml agarose gel).

(b) Put in a 250 ml bottle the agarose and the required volume of TAE.

(c) Heat the mix in the microwave until the agarose is completely dissolved.

(d) Cool down the bottle in a container with water.

(e) Add Midori Green volume (see Table 4) and put the contents of the bottle in a mold for gels with the comb. Wait until the gel has hardened.

(f) Prepare the RT-PCR product. Add TriTrack DNA Loading Dye (6×) to the reaction mixture to obtain a 1× concentration (10 μl of TriTrack DNA Loading Dye for 50 μl of reaction).

Table 3
RT-PCR program to amplify SARS-CoV-2 RNA

Step	Temperature (°C)	Time	Cycles
Retrotranscription	50°C	30 min	1
First denaturation	94°C	7 min	1
Denaturation	94°C	10 s	
Annealing	48°C	30 s	35
Elongation	68°C	40 s	
Final elongation	68°C	7 min	1

Table 4
Agarose gel

Gel volume (ml)	Percentage of agarose (%)	Agarose MP (gr)	TAE 1× volume (ml)	Midori Green volume (μl)
60	1	0.6	60	4–5

(g) Prepare in a 1.5 ml low-binding tube the molecular weight marker: 5 μl of 1 Kb Plus DNA ladder + 5 μl of miliQ water + 2 μl of TriTrack DNA Loading Dye. Mix with the pipette.

(h) Put the gel without the comb in an electrophoresis cuvette and cover the gel with TAE 1×.

(i) Load the molecular weight marker and the RT-PCR product in a different well. Switch on the power supply to start the electrophoresis (80–90 V for 30–40 min).

(j) Once the electrophoresis is finished, remove the gel from the cuvette, clean with ethanol the LED transilluminator, and examine the gel.

3. DNA purification with QIAquick gel extraction kit.

(a) Cut the band with a plastic knife and place the piece of agarose in a labeled 2 ml tube (check that the negative control has not produced an amplification band).

(b) Weigh the band on a balance and add 3 volumes of QG buffer.

(c) Incubate the mix at 50°C for 10 min in the thermoblock (or until the agarose has melted).

(d) Vortex 15 s. Add one volume of isopropanol. Vortex 15 s.

(e) Label a column and add 600 μl of the sample, centrifuge 1 min at 13,000 r.p.m. and discard the supernatant. Repeat this step with the rest of the sample.

(f) Add 500 μl of QG buffer and centrifuge 1 min 13,000 r.p.m. and discard the supernatant.

(g) Add 690 μl of PE buffer, centrifuge 1 min at 13,000 r.p.m and discard the supernatant.

(h) Place the column in a new collecting tube and centrifuge 1 min at 14,000 r.p.m. to dry the membrane.

(i) Put the column in a new low-binding labeled tube of 1.5 ml.

(j) Add 30 μl of EB buffer and incubate 1 min at room temperature.

(k) Centrifuge 1 min at 13,000 r.p.m. discard the column and store the purified DNA solution at 4°C for a short period of time (1–5 h), or at −20°C for a long period of time.

4. Quantification of DNA by fluorimetry (Qubit®) with Qubit® dsDNA HS Assay kit.

(a) Label a 0.5 ml tube for the quantified DNA, and two additional 0.5 ml tubes (Qubit® Assay tubes) that will be used for the preparation of the standards (*see* **Note 17**).

(b) Prepare the “working solution” in a Falcon tube, as required for the total number of tests (prepare a small excess) (*see* Table 5 and **Note 18**). Vortex.

(c) Add 190 µl of the working solution to each of the standard tubes.

(d) Add 198 µl of the working solution to the tube of the test sample.

(e) Add 10 µl of each one of the standards to its corresponding tube.

(f) Add 2 µl of the test sample to its corresponding tube.

(g) Vortex the mixture (important) and incubate for at least 5 min at room temperature (*see* **Note 19**).

(h) Start the Qubit® fluorometer by touching the screen, and in the main screen of the equipment select the “dsDNA” option.

(i) Select the kit used for quantification (Qubit® dsDNA HS Assay Kit) and select “Read standards”.

(j) Lift the lid of the Qubit® and place the tube with the standard number 1. Lower the cover and select “Read standard”. Once the reading is done, the equipment will ask you to insert standard 2. Lift the lid and replace the tube with standard 1 by the tube with standard 2. Lower the cover and select “Read standard”.

(k) Select “Read samples” and select the sample volume used (2 µl) for the dilution with the working solution and the units in which we want the measurements (ng/µl). Remove the tube with the standard number 2 and place the test sample to be quantified. Select “Read tube”.

(l) Once all the samples have been analyzed, select “Done”. Connect a pen-drive to the equipment and select the quantifications to be saved with the box on the left. Select “Export” and remove the pen-drive from the instrument and the last tube that has been quantified.

Table 5
Working solution preparation

Number of samples	Number of standards	Total number of reactions	Reagent volume (1 µl/reaction)	Buffer volume (199 µl/reaction)
1	2	3	4 µl	796 µl

5. A-Tailing to the insert for subsequent ligation to pGEM-T vector with GoTaq Flexi DNA Polymerase. Keep reagents on ice.
   (a) Calculate A-Tailing mix necessary for the reaction (see Table 6).
   (b) Incubate in thermalcycler marking the final 10 µl volume (see Table 7).
6. Prepare the adapter ligation reaction. Keep reagents on ice.

Box 1. DNA insert calculation. We use the 1:3 ratio of our insert. The pGEM-T vector has a size of 3000 bp and a concentration of 50 ng/µl.

$$\frac{\text{Vector (ng)} \times \text{Insert length (bp)}}{\text{Vector length (bp)}} \times 3 = \text{Insert quantity (ng)}$$

   (a) Prepare the mix (in a small excess) (see Table 8). We use a positive ligation control with a DNA control [c (+) ligation], and a negative ligation control that will indicate if the vector has been ligated without insert [c (−) ligation].
   (b) Incubate at 4°C overnight.

Table 6
A-Tailing reaction mix

Reagent	Volume (1 reaction)
DNA insert	5 µl
Taq buffer (5× GoTaq)	1 µl
25 mM $MgCl_2$	0.6 µl
1 mM dATP	2 µl
PCR-grade water	0.4 µl
Taq	1 µl
Final volume	10 µl

Table 7
A-tailing reaction program

Step	Temperature	Duration
A-tailing	70°C	40 min
Hold	10°C	∞

Table 8
Ligation reaction mix

Reagent	Volume (1 reaction)	C(+) ligation	C(−) ligation
2× rapid ligation buffer	5 μl	5 μl	5 μl
pGEM-T vector	1 μl	1 μl	1 μl
DNA insert	**See Box 1**	2 μl	–
T4 DNA ligase	1 μl	1 μl	1 μl
PCR-grade water	Up to 10 μl	1 μl	3 μl
Final volume	**10 μl**	**10 μl**	**10 μl**

7. Transformation into DH5α *Escherichia coli.*
   (a) Thaw DH5α *Escherichia coli* on ice quickly. Stir gently.
   (b) Distribute 100 μl into previously chilled 1.5 ml tubes on ice.
   (c) Add the plasmid solution [add 5 μl of the ligation, 0.5 μl of the pGEM3zf+ plasmid that we use as a control (+), and a negative transformation control without plasmid].
   (d) Incubate on ice for 30 min.
   (e) Incubate 30 s at 42°C in thermoblock.
   (f) Incubate on ice 5 min.
   (g) Add bacteria to 0.4 ml of LB medium (without antibiotic) in 10 ml polystyrene tube.
   (h) Incubate at 37°C with shaking 45 min in bath. During this incubation we prepare the LB + Ampicillin (100 mg/ml) plates to seed the bacteria. We add to each plate 50 μl of X-gal (20 mg/ml) and 5 μl of IPTG (20%). Prepare a mix according to the number of plates, and distribute 55 μl of the mix per plate.
   (i) Seed bacteria on LB plates with ampicillin. In the case of controls, seed 50 μl of bacterial suspension. In the case of samples, seed 4 plates with 20 μl, 50 μl, 100 μl, and the rest, respectively.
   (j) Incubate at 37°C overnight in stove.
   (k) Next day observe results:

   Negative ligation control: there should be few blue colonies.

   Positive ligation control: many white colonies; there might be a few blue colonies.

   Negative transformation control: no colonies.

   Positive transformation control: a lawn of colonies.

   Observe the plates. Positive colonies (with the vector plasmid and insert) appear white.

8. Spike colonies and streak.
   (a) Choose an adequate number of colonics and make a streak on a new plate of LB + Ampicillin (100 μg/ml) + X-gal + IPTG.
   (b) Incubate at 37°C overnight in stove.
   (c) Confirm the next day that the colonies are white.
9. Use Illustra TempliPhi DNA Sequencing Template Amplification Kit.
   (a) Prepare as many 0.2 ml tubes as colonies to be spiked.
   (b) Add 5 μl of TDB to each tube.
   (c) Pipette the colony with pipette tip (*see* **Note 20**).
   (d) Denature the sample at 95°C for 3 min in thermocycler.
   (e) Add 5 μl of Premix T to each tube (*see* **Note 21**).
   (f) Incubate at 30°C 16–18 h in thermocycler, and inactivate the enzyme at 65°C for 10 min, and cool to 4°C.
   (g) The amplified DNA does not require purification.
10. Sanger Sequencing.
    (a) Send the complete Templiphi product for sequencing (10 μl).
    (b) Sequencing with primers T7 and SP6 to check the insert and its orientation.
11. Grow the chosen colony.
    (a) Grow the chosen colony (colony containing our insert in the correct orientation) in 10 ml of LB + Ampicillin (100 μg/ml). Grow in a 50 ml flask.
    (b) In parallel put a negative control under the same conditions.
    (c) Grow at 37°C overnight in a shaking bath.
12. Wizard® Plus SV Minipreps DNA Purification System.
    (a) From 10 ml of culture, prepare an aliquot for storage in glycerol: with 750 μl de culture + 250 μl of sterile glycerol (store at −80°C) and divide the rest of the culture into 2 ml tubes with ~2 ml each.
    (b) Harvest 1–5 ml (for high-copy-number plasmid) or 10 ml (for low-copy-number plasmid) of bacterial culture by centrifugation for 5 min at 10,000 × *g*. Remove the supernatant and the excess medium by inverting the tube on a paper towel.
    (c) Add 250 μl of Cell Resuspension Solution and resuspend the cell pellet by vortexing or pipetting. Transfer the resuspended cells to a sterile 1.5 ml microcentrifuge tube(s) (*see* **Note 22**).

(d) Add 250 µl of Cell Lysis Solution and mix by inverting the tube 4× (do not vortex). Incubate ~1–5 min or until the cell suspension clears (*see* **Note 23**).

(e) Add 10 µl of Alkaline Protease Solution and mix again by inverting the tube 4 times. Incubate for 5 min at room temperature. Do not exceed 5 min of incubation with Alkaline Protease Solution, as nicking of the plasmid DNA may occur.

(f) Add 350 µl of Neutralization Solution. Immediately, mix by inverting the tube 4× (do not vortex).

(g) Centrifuge the bacterial lysate at maximum speed (around 14,000 × *g*) in a microcentrifuge for 10 min at room temperature.

(h) Prepare plasmid DNA purification units. Insert one Spin Column into one 2 ml Collection Tube for each sample.

(i) Transfer the cleared lysate (~850 µl, Step 12.7) to the prepared Spin Column by decanting. Avoid disturbing or transferring any of the white precipitate with the supernatant (*see* **Note 24**).

(j) Centrifuge the supernatant at maximum speed in a microcentrifuge for 1 min at room temperature. Remove the Spin Column from the tube. Discard the flow through from the Collection Tube and reinsert the Spin Column into the Collection Tube.

(k) Add 750 µl of Column Wash Solution to the Spin Column (previously diluted with 95% ethanol).

(l) Centrifuge at maximum speed in a microcentrifuge for 1 min at room temperature. Remove the Spin Column from the tube. Discard the flow through and reinsert the Spin Column into the Collection Tube.

(m) Repeat the washing procedure but, in this step, use 250 µl of Column Wash Solution.

(n) Centrifuge at maximum speed in a microcentrifuge (around 14,000 × *g*) for 2 min at room temperature.

(o) Transfer the Spin Column to a new, sterile 1.5 ml microcentrifuge tube. If the Spin Column has Column Wash Solution associated with it, centrifuge again for 1 min at maximum speed.

(p) Transfer the Spin Column to a new, sterile 1.5 ml microcentrifuge tube.

(q) Elute the plasmid DNA by adding 50 µl of Nuclease-Free Water to the Spin Column. Centrifuge at maximum speed for 1 min at room temperature in a microcentrifuge. Discard the Spin Column.

13. Quantification of DNA by fluorimetry (Qubit®) with Qubit® dsDNA HS Assay kit (see point 4 of Sect. 3.2.1.).
14. Linearization of template DNA.
    (a) Calculate Linearization mix necessary for the reaction (see Table 9).
    (b) Incubate in thermoblock at 37°C overnight.
    (c) Spin the tubes and incubate in thermoblock at 65°C for 10 min to inactivate the enzyme.
    (d) Reserve 2 µl to analyze, together with the sample after purification, by agarose gel electrophoresis.
    (e) Freeze at −20°C if not proceeding to the next step immediately.
15. DNA purification with Wizard® PCR Preps DNA Purification Resin and Wizard® Minicolumns.
    (a) Incubate in the thermoblock at 55°C an aliquot of H_2O for the final elution.
    (b) Add 1 ml of Wizard DNA Clean-up Resin (shake well) to each linearization mixture. Mix well for 1 min by inversion without vortex.
    (c) Insert a 5 ml syringe (with previously removed plunger) into the minicolumn, and the latter into a 5 ml polystyrene tube. Add the DNA-resin mixture onto the syringe, and press the plunger slowly.
    (d) Remove the syringe from the minicolumn.
    (e) Remove the plunger from the syringe, reinsert the syringe into the minicolumn, and add 2 ml of 80% isopropanol (freshly prepared). Slowly squeeze the plunger on the syringe.
    (f) Transfer the column to a new 1.5 ml low-binding tube.
    (g) Centrifuge at 12,000 r.p.m. for 2 min to dry thoroughly.

Table 9
Linearization reaction mix

Reagent	Volume (1 reaction)
Miniprep product	X µl (2 µg)
Buffer 2 (10×)	10 µl
BSA (100×)	1 µl
Nde I	2 µl
PCR-grade water	Up to 100 µl
Final volume	**100 µl**

(h) Transfer to a new 1.5 ml low-binding tube and add 30 μl H_2O preheated to 55°C over the column. Incubate for 1 min, and elute the DNA by centrifuging at 12,000 r.p.m. for 1 min.

16. Quantification of DNA by fluorimetry (Qubit®) with Qubit® dsDNA HS Assay kit (see point 4 of Sect. 3.2.1).

17. Agarose gel electrophoresis. Check plasmid linearization by 1% agarose gel electrophoresis (50 ml) (pre-purification and the post-purification sample, 2 μl of each) (see point 2 of Sect. 3.2.1).

18. *In vitro* transcription with T7 RiboMAX Express Large Scale RNA Production system.

    (a) Calculate transcription mix necessary for the reaction (see Table 10).

    (b) Incubate in thermoblock at 37°C 30 min.

    (c) Add 1 μl of DNAse RQ1.

    (d) Incubate in thermoblock at 37°C 30 min.

    (e) Reserve 5 μl to analyze, together with the sample after purification, by agarose gel electrophoresis.

19. Purification of transcribed RNAs with Mini Quick Spin RNA Columns:

    (a) Remove the cap from the top of the column and then break the end from under the column and insert into a 1.5 ml tube.

    (b) Centrifuge the column at 1000 × *g* for 1 min to remove the residual buffer.

    (c) Insert the column into a new clean 1.5 ml tube.

    (d) Add the sample from the *in vitro* transcription onto the center of the column.

    (e) Centrifuge at 1000 × *g* for 4 min.

    (f) Store the eluate with the purified RNA at −80°C.

Table 10
Transcription reaction mix

Reagent	Volume (1 reaction)
RiboMAX Express T7 Buffer (2×)	10 μl
T7 Mix	2 μl
Linearized DNA	X μl (1 μg)
PCR-grade water	Up to 20 μl
Final volume	**20 μl**

(g) Reserve 3 μl to analyze, together with the sample before purification, by agarose gel electrophoresis.

20. Quantify RNA in triplicate by fluorimetry (Qubit®) with the Qubit RNA BR Assay Kit (follow point 4 of Sect. 3.2.1).
21. Agarose gel electrophoresis (see point 2 of Sect. 3.2.1).
    (a) Clean the cuvette, mold and comb with soap and water before use. Leave in distilled H_2O with a H_2O_2 (1:30 in volume) for 10–20 min and blot dry with paper.
    (b) Check the integrity of the RNA in the pre-purification sample (5 μl) and the post-purification sample (3 μl) by agarose gel electrophoresis (50 ml 1% agarose; 80–90 V for 30–40 min).
22. Once the transcribed RNA has been checked by agarose gel electrophoresis and quantified (Qubit®), calculate the number of RNA molecules/μl.
23. From the calculations of the previous step, make serial dilutions with at least 5 points to be used to produce the standard curve for Real-Time RT-PCR quantifications.

3.2.2 Detection and Quantification of SARS-CoV-2 RNA by Real-Time RT-PCR

1. Keep RT-PCR reagents on ice.
2. Prepare the mix corresponding to the number of samples to be quantified, including 5 for the standard curve and 1 as negative control (prepare a small excess) in a PCR cabin (where no biological samples are handled) (see Table 11).
3. Prepare as many capillaries in the LightCycler Centrifuge Adapters as samples to be quantified.
4. Distribute 19 μl of the mix to each capillary.
5. Add 1 μl H_2O to the negative control in the first capillary and close it.
6. Add 1 μl of the RNA from the standard curve to capillaries 2–6 (10^4, 10^5, 10^6, 10^7, 10^8 molecules, respectively) and close it.

Table 11
RT-PCR reaction mix

Reagent	Volume (1 reaction)
PCR-grade water (vial 3)	6.2 μl
$Mn(OAc)_2$ stock solution, 50 mM (vial 2)	1.3 μl
Oligonucleotide nsp12-CoV2-u14890 (3 μM)	2 μl
Oligonucleotide nsp12-CoV2-d15320 (3 μM)	2 μl
LC RNA Master SYBR Green I (vial 1)	7.5 μl
Final volume	**19 μl**

Table 12
Light Cycler® RT-PCR program

Step	Temperature	Hold	Acquisition	Cycles
RT	61°C	20 min	None	1
Denaturalization	95°C	30 s	None	1
Quantification	95°C	5 s	None	45
	48°C	10 s	None	
	72°C	17 s	Single	
Melting	95°C	0 s	None	1
	65°C	15 s	None	
	95°C	0 s	Continuous (slope 0.1°C/s)	
Cooling	40°C	30 s	None	1

7. Add 5 μl of the RNA to be quantified to each capillary and close it.
8. Centrifuge at 900 × g to capillaries for 5 s.
9. Load the capillaries into the carrousel of the Light Cycler® in order and insert them into the unit.
10. Turn on the computer and the Light Cycler® equipment.
11. Open the Light Cycler application and click "Run".
12. Click on "Open Exp File".
13. Select the program and check each step of the program (see Table 12).
14. Press "Run" and save the experiment.
15. Mark the number of samples and press "Enter samples later".
16. When the program starts, press "Edit samples" and label the samples.
17. Click "Done" and let it run. It takes about 1 h.

3.2.3 Amplification by RT-PCR

1. Calculate RT-PCR mix necessary for each reaction. For the amplification of one sample: 30 μl of PCR-grade water, 10 μl of reaction buffer 5×, 2 μl of oligonucleotide forward (50 ng/μl) and oligonucleotide reverse (50 ng/μl), 1 μl Transcriptor enzyme mix (*see* **Note 25**).
2. Irradiate with ultra-violet for at least 15 min, prior to its use.
3. Take the reagents from the freezer and place them on ice: PCR-grade water, oligonucleotides, and buffer 5×.

4. Label as many 0.2 ml tubes as samples to amplify.
5. Using the PCR cabin (where no biological samples are handled), prepare RT-PCR mix for each amplicon in a 1.5 ml tube and shake with vortex. The Transcriptor enzyme mix will be added at the end. Finally, distribute the mix in 0.2 ml previously labeled tubes.
6. Outside of the PCR cabin, add 5 μl of RNA to each 0.2 ml tube, and mix with the pipette.
7. Place on the tubes in the thermocycler and configure the program (see Table 13).
8. Once the amplification has been completed, we proceed with the gel electrophoresis analysis. Samples can be stored at 4°C until the gel electrophoresis if they are going to be used in a short period of time (1–2 h), or at −20°C for longer times.

3.2.4 Gel Electrophoresis

1. Weigh the amount of Agarose MP necessary for the preparation of gel at a concentration of 2% (see Table 14 for an example of a 60 ml agarose gel).
2. Put in a 250 ml bottle the agarose and the required volume of TAE.
3. Heat the mix in the microwave oven until the agarose is completely melted. Cool down the bottle in a recipient with water.

Table 13
RT-PCR program to amplify SARS-CoV-2

Step	Temperature (°C)[a]	Time	Cycles
Retrotranscription	50°C	30 min	1
First denaturation	94°C	7 min	1
Denaturation	94°C	10 seg	
Annealing	46°C/48°C	30 seg	35/45
Elongation	68°C	40 seg	
Final elongation	68°C	7 min	1

[a]Annealing temperature varies depending on the amplicon.

Table 14
Amount of each component for the preparation of 2% agarose gel of 60 ml

Gel volume (ml)	Percentage of agarose (%)	Agarose MP (gr)	TAE 1x volume (ml)	Midori Green volume (μl)
60	2	1.2	60	4–5

4. Add Midori Green volume (see Table 14) and put the contents of the bottle in a mold for gels and place the comb until the agarose hardens.
5. Prepare the samples from amplification in Step 3.2.2 adding loading buffer for 50 μl of reaction to obtain a 1× concentration (10 μl of loading buffer for 50 μl of reaction mixture).
6. Prepare in a 1.5 ml tube the molecular weight marker: 5 μl of 1 Kb Plus DNA ladder + 5 μl of miliQ water + 2 μl of loading buffer. Mix with the pipette.
7. Place the gel without the comb in an electrophoresis cuvette, covering the gel with TAE 1×.
8. Load each amplification mixture prepared in Step 5 and the marker prepared in Step 6 in a separate well. Switch on the power supply to start the electrophoresis (80–90 V for 30–40 min).
9. Once the electrophoresis has finished, remove the gel form the cuvette, clean with ethanol the transilluminator, and examine the gel.

3.2.5 DNA Purification and Quantification

1. If we observe the corresponding DNA band for the amplicon, and if the negative control does not show any amplification band, cut each band with a plastic knife and place each of them in labeled 2 ml tubes.
2. The next step is the purification with QIAquick Gel Extration Kit. First, weigh the material with each band on the balance, and calculate the corresponding amount of buffer QC multiplying the weight for three. Incubate the mixture during 10 min at 50°C in the thermoblock (or until the agarose is melted). Vortex 15 s. Add a volume of isopropanol. Vortex 15 s.
3. Label as many columns as samples to be processed. Add 600 μl of the sample, centrifuge 1 min at 13,000 r.p.m. and discard the supernatant. Repeat this step with the rest of the samples.
4. Add 500 μl of QG buffer and centrifuge 1 min 13,000 r.p.m. Discard the supernatant.
5. Add 690 μl of PE buffer, centrifuge 1 min at 13,000 r.p.m. Discard the supernatant and centrifuge again at 14,000 r.p.m. to dry the membrane.
6. Label as many low-binding tubes of 1.5 ml as samples we have.
7. Put each column in its corresponding low-binding tube. Add 30 μl of EB buffer. Incubate 1 min. Centrifuge 1 min at 13,000 r.p.m. Discard the column and store the purified DNA at 4°C for a short period of time (hours) or at −20°C.

8. Quantification of DNA by fluorimetry (Qubit®) with Qubit® dsDNA HS Assay kit (see point 4 of Sect. 3.2.1).

3.3 SARS-CoV-2 Amplicon Ultra-Deep Sequencing

3.3.1 Amplicon Normalization and Purification

1. Adjust the concentration of each amplicon separately to 4 $\times 10^9$ molecules/μl with EB in a new sterile 1.5 ml low-binding tube (*see* **Note 26**).
2. Mix 20 μl of each amplicon solution that belongs to the same sample in a labeled low-binding tube. (We call this mixture "pool").
3. Remove the Kapa Pure Beads from the refrigerator, and allow them to reach room temperature before use.
4. Determine the total volume of sample to be purified (volume of each Step 2 pool).
5. Prepare and label as many 1.5 ml low-binding tubes as samples to be purified.
6. Use a DNA:bead ratio of 1:1.
7. Vortex the bottle of Kapa Beads vigorously to homogenize.
8. Add the volume of Kapa Beads calculated in Step 4 into each tube.
9. Turn on and set the thermoblock at 1400 r.p.m., 10 min, 25°C.
10. Add the corresponding DNA sample to each tube and mix gently by pipetting (*see* **Note 27**).
11. Incubate the tubes in the thermoblock under the conditions detailed in Step 9.
12. Prepare 50 ml of 80% ethanol by diluting with MiliQ water (*see* **Note 28**).
13. Place the tubes in the magnetic rack and allow the bead pellet to form.
14. Raise the temperature of the thermoblock to 37°C.
15. Remove the supernatant with p1000 tips and discard it without removing the tube from the rack.
16. Add 1 ml of 80% ethanol to each tube (not directly on the pellet). Leave for 1 min.
17. Remove the supernatant with p1000 tips without touching the beads (*see* **Note 29**).
18. Repeat Steps 16 and 17.
19. Spin all tubes and place them back in the magnetic rack. Allow the bead pellet to form.

20. Using the p200 tips, remove any remaining ethanol without touching the beads.
21. Allow the tubes to dry (open) in the magnetic rack, making sure that the pellet does not dry out completely (~2 min).
22. Remove the tubes from the magnetic rack.
23. Add 50 μl EB to each tube to re-suspend the samples. Mix gently by pipetting until the pellet is well mixed.
24. Place the tubes in the thermoblock and set at 1400 r.p.m., 5 min and 37°C.
25. Place the tubes in the magnetic rack. Allow pellet to form.
26. Prepare and label as many 1.5 ml low-binding tubes as the number of samples.
27. Using the p200 tips, remove the supernatant without touching the beads, and store it in its corresponding tube.
28. Quantify each of the pools by fluorimetry (Qubit®) with Qubit® dsDNA HS Assay kit (see point 4 of Sect. 3.2.1).

3.3.2 DNA Library Preparation

1. Standardize the concentration of each pool separately to 1.5 ng DNA/μl with EB in a new sterile 0.2 ml low-binding tube (*see* **Note 30**).
2. Prepare the End repair and A-tailing reaction. Keep reagents on ice.
   (a) Prepare the mix corresponding to the number of pools (prepare a small excess) (see Table 15).
   (b) Distribute 10 μl of the mix to each pool. Mix by pipetting and spin.

Table 15
End Repair and A-tailing reaction mix

Reagent	Volume (1 reaction)
End Repair and A-tailing Buffer	7 μl
End Repair and A-tailing Enzyme Mix	3 μl

Table 16
End Repair and A-tailing reaction program

Step	Temperature	Duration
End repair	20°C	30 min
A-tailing	65°C	30 min
Hold	4°C	∞

   (c) Incubate in thermal cycler marking the final 60 μl volume (*see Table* 16 and **Note 31**).
   (d) Proceed directly to the next step; storage at −20 °C may result in loss of performance.
3. Prepare the adapter ligation reaction. Keep reagents on ice.
   (a) Prepare the mix corresponding to the number of pools (prepare a small excess) (see Table 17).
   (b) Add 2.5 μl of each adapter (stock 15 μM) to each pool (*see* **Note 32**). The adapter-to-insert ratio should be 50:1.
   (c) Distribute 47.5 μl of the mix to each tube. Mix by pipetting and spin.
   (d) Incubate in the thermal cycler at 20°C for 15 min (*see* **Note 33**).
4. Purify each of the pools with the KAPA Pure Beads kit.
   (a) Prepare and label as many 1.5 ml low binding tubes as we have pools, and mix the pool with KAPA Pure Beads (see Table 18).
   (b) Mix by pipetting.

Tablc 17
Ligation reaction mix

Reagent	Volume (1 reaction)
Ligation buffer	30 μl
DNA ligase	10 μl
PCR-grade water	7.5 μl

Table 18
Ratio DNA:bead 1:0.8

Reagent	Volume (1 reaction)
Adapter ligation reaction product	110 μl
KAPA pure beads	88 μl

Table 19
PCR reaction mix

Reagent	Volume (1 reaction)
KAPA HiFi HotStart Ready Mix (2×)	25 μl
KAPA Library Amplification Primer Mix (10×)	5 μl

Table 20
PCR program

Step	Temperature	Duration	Cycles
Initial denaturation	98°C	45 s	1
Denaturation	98°C	15 sec	
Annealing	60°C	30 s	5
Elongation	72°C	30 s	
Final elongation	72°C	1 min	1
Hold	4°C	∞	

(c) Incubate in a thermoblock at 25 °C for 10 min with agitation.

(d) Incubate in magnetic rack until pellet is formed (until supernatant is clear).

(e) Remove and discard the supernatant. Keep the tube in the magnetic rack.

(f) Add 1 ml of freshly prepared 80% ethanol (*see* **Note 28**).

(g) Incubate in magnetic rack at room temperature for 30 s.

(h) Remove and discard supernatant. Keep the tube in the magnetic rack.

(i) Repeat Steps 4.6, 4.7, and 4.8.

(j) Spin tubes and incubate on magnetic rack until pellet is formed. Remove as much of the supernatant as possible using a fine-tipped pipette.

(k) Open the tubes and allow the beads to dry at room temperature for 5 min, until remaining ethanol has evaporated (*see* **Note 34**).

(l) Remove the tubes from the magnetic rack and re-suspend the beads in 25 μl EB.

(m) Incubate in thermoblock at 37 °C for 5 min with shaking to elute DNA from beads.

(n) Incubate on magnetic rack until pellet is formed (until the supernatant is clear).

(o) Transfer 20 μl of the supernatant to a new sterile 0.2 ml low-binding tube.

5. Prepare the library amplification reaction. Keep reagents on ice.

   (a) Prepare the mix corresponding to the number of pools (prepare a small excess) (see Table 19).

Table 21
Ratio DNA:bead 1:0.8

Reagent	Volume (1 reaction)
Adapter ligation reaction product	50 µl
KAPA pure beads	40 µl

(b) Distribute 30 µl of the mix to each tube. Mix by pipetting and spin.

(c) Incubate in thermal cycler marking the final 50 µl volume (*see* Table 20 and **Note 31**).

(d) Proceed directly to the next step; storage at −20°C may result in loss of performance.

6. Purify each of the pools with the KAPA Pure Beads kit.

(a) Prepare and label as many 1.5 ml Low-Binding tubes as the number of pools, and mix the pool with KAPA Pure Beads (see Table 21):

(b) Mix by pipetting.

(c) Incubate in a thermoblock at 25°C for 10 min with agitation.

(d) Incubate in magnetic rack until pellet is formed (until supernatant is clear).

(e) Remove and discard the supernatant. Keep the tube in the magnetic rack.

(f) Add 1 ml of freshly prepared 80% ethanol (*see* **Note 28**).

(g) Incubate in magnetic rack at room temperature for 30 s.

(h) Remove and discard supernatant. Keep the tube in the magnetic rack.

(i) Repeat Steps 6.6, 6.7, and 6.8.

(j) Spin tubes and incubate on magnetic rack until pellet is formed. Remove as much of the supernatant as possible using a fine-tipped pipette.

(k) Open the tubes and allow the beads to dry at room temperature for 5 min, until remaining ethanol has evaporated (*see* **Note 34**).

(l) Remove the tubes from the magnetic rack and re-suspend the beads in 25 µl EB.

(m) Incubate in thermoblock at 37°C for 5 min with shaking to elute DNA from beads.

(n) Incubate on magnetic rack until pellet is formed (incubate until the supernatant is clear).

(o) Transfer the supernatant (as much as possible) to a new, labeled 1.5 ml low-binding tube.

7. Quantify each of the pools by fluorimetry (Qubit®) with Qubit® dsDNA HS Assay kit (see Step 4 of Sect. 3.2.1).
8. Standardize the concentration of each pool separately to 4 nM with EB in a new sterile 1.5 ml low-binding tube.
9. Mix in a new sterile 1.5 ml low-binding tube the corresponding volume of each of the pools (in order to include 2 μl for each amplicon).
10. Determine the amount (by quantitative techniques such as quantitative PCR or PicoGreen) and quality (by bioanalyzer) of the final DNA pool generated in the previous step.

3.3.3 Sequencing with MiSeq Platform

1. Remove the MiSeq® Reagent Kit v3 cartridge from the freezer.
2. Take the tube containing the HT1 Buffer and allow to thaw in a container with ice.
3. Slowly thaw the cartridge in a pan of water at room temperature (*see* **Note 35**).
4. Thaw 2 μl of PhiX 10 nM in a container with ice.
5. Place on ice the library solution at 4 nM of DNA that has been previously quantified.
6. If the library quantification is close to the expected theoretical value (4 nM), mix in a 1.5 ml low-binding tube: 5 μl of the library solution + 5 μl of NaOH 0.2 N. In another 1.5 ml low-binding tube, mix: 2 μl of PhiX 10 nM + 3 μl of Tris-HCl 10 mM pH 8.0–8.5 + 5 μl of NaOH 0.2 N (*see* **Note 36**).
7. Vortex the two tubes and incubate 5 min at room temperature.
8. To obtain a final concentration of 20 pM, add 990 μl of HT1 Buffer to each of the tubes and place tubes on ice (*see* **Note 37**).
9. Determine the dilution of the library and PhiX to be loaded into the MiSeq (it depends on the density of clusters we want to obtain, and on previous experience in sequencing libraries).
10. Calculate the percentage of PhiX to be mixed with the library to obtain the final pool. In our runs we usually add 20% PhiX.
11. Mix the two volumes in a new low-binding tube.
12. Mix by vortex, spin the tube, and keep on ice.
13. Incubate the new pool solution (library +PhiX) for 2 min in thermoblock at 95°C.
14. Invert the reagent cartridge 10× (dry the cartridge beforehand with fiber-free tissues).
15. Pierce with a pipette tip the position of the cartridge where the sample is loaded.

16. Load the 600 µl of the pool solution with a new pipette tip.
17. Load the cartridge into the MiSeq along with the remaining reagents and flowcell following the manufacturer's instructions.

3.3.4 Bioinformatics Analysis with SeekDeep

1. For SeekDeep and dependencies installation follow the recipe at https://seekdeep.brown.edu/installing/installingSeekDeep_Ubuntu.html.
2. Get path with fastqs available (FASTQPATH), R1 and R2 sequences should be available. In the fastq file, the first string before "_" should be the name of the sample.
3. Locate virus target sequences' fasta files (TARGETFASTAPATH).
4. Make a config file with oligonucleotide and sample matches (sampleNames.tab) with a 3 columns format "#target\tsample\trep1", where rep1 can be omitted.
5. Locate oligonucleotides' sequences file (idFile.tab) with a 3 columns format "target\tforward\treverse".
6. Locate oligonucleotide status configuration file (overlapStatusFnp.txt) with format "target\tstatus".
7. Run SeekDeep setupTarAmpAnalysis command to prepare analysis. See https://seekdeep.brown.edu/usages/setupTarAmpAnalysis_usage.html.

Command line:

```
SeekDeep setupTarAmpAnalysis --numThreads 80 --refSe-
qsDir $TARGETFASTAPATH --conservativePopClus --sam-
ples sampleNames.tab --outDir $outDir --inputDir
$FASTQPATH --idFile idFile.tab --overlapStatusFnp ../
overlapStatusFnp.txt "--extraExtractorCmds=--check-
RevComplementForPrimers --primerNumOfMismatches
3" "--extraProcessClusterCmds=--fracCutOff 0 -res
cueExcludedOneOffLowFreqHaplotypes".
```

8. SeekDeep generates $outDir and creates all needed files to obtain final clusters, including ./runAnalysis.sh script, a pipeline with commands for:
   (a) SeekDeep extractorPairedEnd. Demultiplexing and filtering.
   (b) SeekDeep qluster. Clustering of reads.
   (c) SeekDeep processClusters. Curate and refine haplotypes.
   (d) SeekDeep genProjectConfig tasks in that order. Results generator.
9. Run runAnalysis.sh script, adapting the number of threads to your systems configuration.
10. To extract sequences per amplicon and samples we apply this command:

```
for file in popClustering/*/analysis/final/*gz; do
amplicon=$(echo "$file" | tr '\/' '\n'| head -n2 |
tail -n1); name="$(basename -- $file)"; echo $ampli-
con; gunzip -c $file | s
ed -n '1~4s/^@/>/p;2~4p' > fastas_final/${ampli-
con}_${name}.fasta; done
```

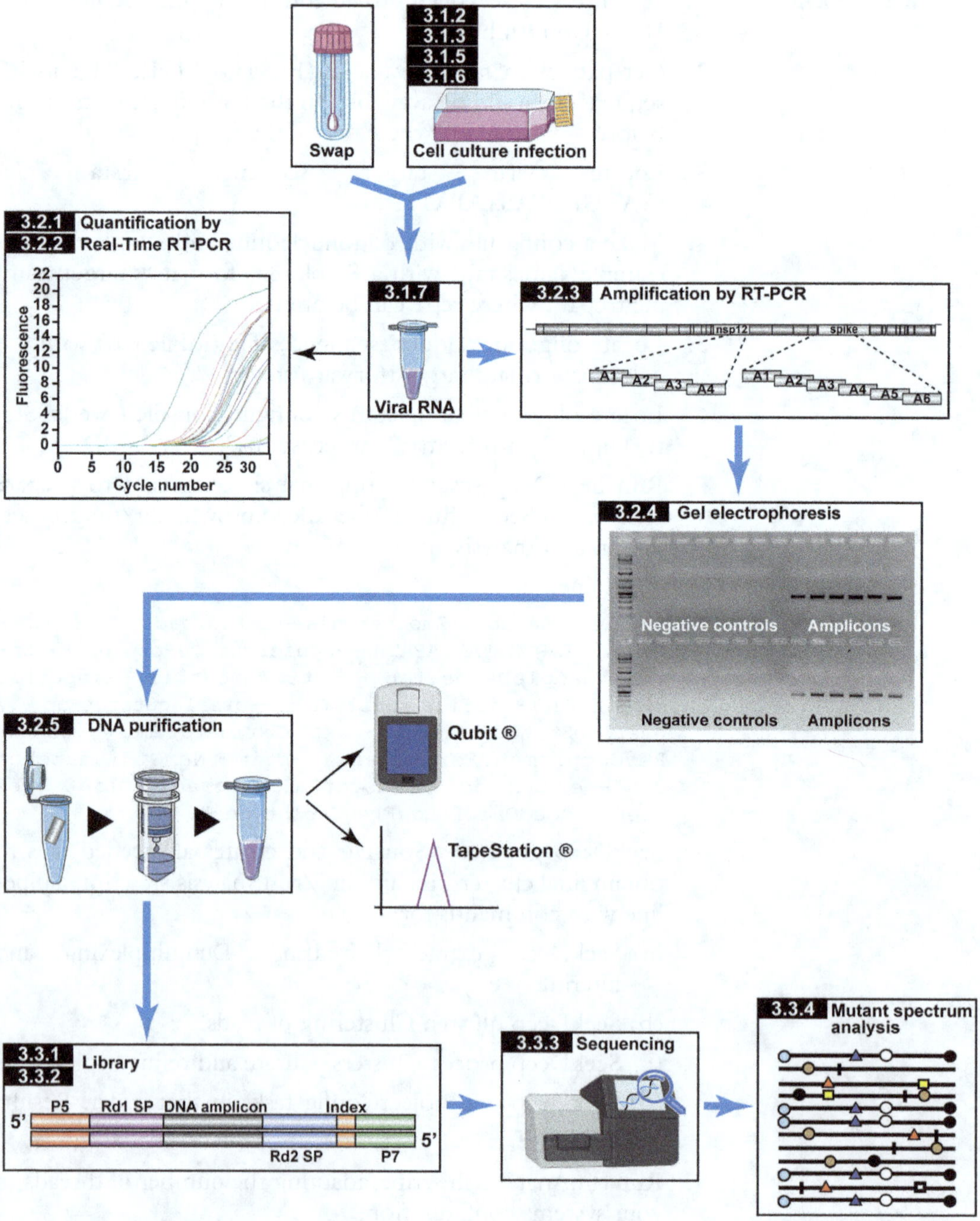

Fig. 1 Workflow of the major steps from virus preparation to the interpretation of the mutant spectrum. Only the order of the main procedures described in Sect. 3 are indicated. Section numbers in filled rectangles correspond to those detailed in the text

3.3.5 Coverage and Frequency Cut-off

The average clean read coverage we reach with the present protocol is 110,074 (range 89,201 to 129,807). This sets the basis for reliability of the 0.1% mutation and deletion frequency cut-off attained. Reliability of this resolution is further supported by: (i) high (>80%) residues with a quality score Q > 30; (ii) the great majority of mutations and their relative abundance in the three codon positions, as well as the representation of amino acids in the GISAID data bank, are all very similar with 0.1% and 0.5% frequency cut-off; (iii) the most abundant low frequency mutations are not those expected from technical mistakes in the Illumina system used; (iv) duplicate analyses of the same sample give similar mutant spectrum composition (for details see [10, 13, 15]) (*see also* **Note 38**).

The major steps in Sect. 3 are depicted as a schematic workflow in Fig. 1.

4 Notes

1. All work with uninfected cells should be carried out in a Biosafety level 2 laboratory in Class II biosafety cabins, using good practice protocols/techniques and sterile materials.
2. For isolation, propagation, and titration of USA-WA1/2020 SARS-CoV-2, we used Vero E6 cells because the initial working stock of this virus was obtained in these cells [16].
3. All manipulations with SARS-CoV-2 should be carried out in a Biosafety level 3 laboratory in Class IIA biosafety cabins, using good practice protocols/techniques and sterile materials.
4. SARS-CoV-2 reference virus [strain USA-WA1/2020, NR-52281 (deposited by the Centers for Disease Control and Prevention and obtained through BEI Resources, NIAID, NIH) (SRA accession no. NR-52281_70036318)].
5. We work with totally sealed containers in which CO_2-mediated pH control is not possible. Therefore, HEPES is included to maintain the pH of the culture medium.
6. Care should be taken to avoid cross-contamination among samples at the time of their collection (usually nasopharyngeal swabs for COVID-19 diagnosis) at the Hospital or Health Center, their transport to the BSL-3 facility, and their treatment in the laboratory. We handle each sample separately in high containment Class II, type A biological safety cabins that comply with European regulation EN-12469. Cabins are maintained under UV irradiation for 30 min after use. As a control, one open tube containing water was maintained during several successive extractions of nasopharyngeal swabs

(except during UV irradiation periods). Then, the contents of the tube were subjected to the sequential procedure for RT-PCR amplification (Sects. 3.1.7, 3.2.3, and 3.2.4). No DNA band was obtained. This and other controls are recommended to ensure absence of cross-contamination at any step from sample collection to preparation of amplicons for ultra-deep sequencing. This is critical for high-resolution mutant spectrum analyses [13].

7. At the day of infection, Vero E6 cells are at 90–100% confluence.
8. Time of infection until reaching cytopathic effects depends on the SARS-CoV-2 variant.
9. Take care not to move the plates until the medium is solidified.
10. A control without sodium deoxycholate should be performed in parallel.
11. Keep the AVE buffer at room temperature.
12. Indicate the preparation date, as the expiration is one year.
13. In case of consecutive extractions, perform each one separately in different cabins, including a negative control in each process.
14. For consecutive extractions: Remove all materials from the previous extraction before starting the next one. Clean between extractions with bleach and ethanol. Change gloves. The final tubes from each extraction should be removed at the end of each process. The negative controls for each extraction should be handled last at each step.
15. To obtain a more concentrated final eluate.
16. Once the final elution is performed, keep the tubes cold until they are stored at −80°C.
17. Do not write on the side of the tubes, only write on the cap.
18. Never prepare the working solution in a glass container.
19. The mixture is stable up to 3 h.
20. Do not transfer too much material or agar.
21. Premix always on ice, and tubes with premix should always be kept on ice.
22. To prevent shearing of chromosomal DNA, do not vortex after Step 12.3. Mix only by inverting the tubes.
23. It is important to observe partial clearing of the lysate before proceeding to addition of the Alkaline Protease Solution (Step 12.5); however, do not incubate for longer than 5 min.
24. If the white precipitate is accidentally transferred to the Spin Column, pour the Spin Column contents back into a sterile 1.5 ml microcentrifuge tube and centrifuge for another 5–10 min at maximum speed. Transfer the resulting supernatant

into the same Spin Column that was used initially for this sample. The Spin Column can be reused only for this sample.

25. Include a negative control for each amplicon mix. To ensure that the amount of template RNA did not represent a limitation, amplification reactions were carried out with dilutions 1:10, 1:100 and 1:1000 of each RNA preparation; we continued with the sequencing reaction using the undiluted RNA sample only when dilution 1:1000 yielded a visible DNA band in the electrophoresis analysis.
26. The final volume should be >20 μl.
27. It is important not to vortex in this case.
28. It is important that the 80% ethanol used in this protocol is freshly made.
29. It is not necessary to remove the total supernatant volume in this case.
30. The final dilution volume is 50 μl.
31. Thermal cycler lid at 85 °C (if possible).
32. If using the adapters for the first time, resuspend the freeze-dried sample in 33.3 μl of PCR-grade water.
33. Leave the lid of the thermal cycler open or turn off the temperature of the lid. Do not heat the lid of the thermal cycler under any circumstances.
34. Do not allow the beads to dry excessively, as this will reduce the yield.
35. Do not exceed the water level specified on the cartridge.
36. If the library has a value <3.3 nM, dilute the library to a concentration of 2 nM, and adjust the volumes described in point 6 of Sect. 3.3.3.
37. In the case of the 2 nM option, add 10 μl of 200 nM Tris-HCl pH 7–8 to neutralize the excess NaOH, and add 970 μl of HT1 Buffer to obtain the 20 pM concentration.
38. A variety of experimental and bioinformatics procedures have been described in the literature to ensure deep sequencing reliability. Their inclusion is out of the scope of the present article. Here we have described protocols that, in our hands, have provided reproducible results.

Acknowledgments

This work was supported by the Ministerio de Ciencia e Innovación, grants PID2020-113888RB-I00/AEI/10.13039/501100011033 and 202220I116, by the Ministerio de Ciencia, Innovación y

Universidades (MICIU), grant PID2023-146622OB-I00 financed by MICIU/AEI/10.13039/501100011033 and by FEDER, UE, through the "Severo Ochoa" Programmes for Centres of Excellence in R&D [CEX2021-001154-S and CEX2023-001386-S]. The work was also supported by the European Commission-Next Generation EU (regulation EU 2020/2024) through the CSIC's Global Health Platform (PTI Salud Global). The work was also funded by grants PI21/00139 funding from Instituto de Salud Carlos III (ISCIII), co-funded by the European Union, CSIC-COV19-014 from Agencia Estatal Consejo Superior de Investigaciones Científicas (CSIC) and project 525/C/2021 from Fundació La Marató de TV3, grants 202136-30 and 202136-31. We also acknowledge the projects TEC-2024/BIO-66 (SALAINDEC-CM from Comunidad de Madrid/FEDER) and S2018/BAA-4370 (PLATESA2 from Comunidad de Madrid/FEDER). Institutional grants from the Fundación Ramón Areces and Banco Santander to the CBMSO are also acknowledged. The team at CBMSO belongs to the Global Virus Network (GVN), and to the Biomedicine Unit of Universidad de Castilla La Mancha (UCLM), associated with CSIC. BM-G is supported by predoctoral contract PFIS FI19/00119 from Instituto de Salud Carlos III cofinanced by Fondo Social Europeo (FSE), "El FSE invierte en tu futuro". PS is supported by postdoctoral contract Margarita Salas, CA1/RSUE/2021 from MCIU. PM was supported by the Miguel Servet program of the Instituto de Salud Carlos III (CPII21/00015), cofinanced by the European Regional Development Fund (ERDF).

References

1. Anthony SJ, Epstein JH, Murray KA, Navarrete-Macias I, Zambrana-Torrelio CM, Solovyov A, Ojeda-Flores R, Arrigo NC, Islam A, Ali Khan S, Hosseini P, Bogich TL, Olival KJ, Sanchez-Leon MD, Karesh WB, Goldstein T, Luby SP, Morse SS, Mazet JA, Daszak P, Lipkin WI (2013) A strategy to estimate unknown viral diversity in mammals. mBio 4:e00598–13
2. Hou X, He Y, Fang P, Mei SQ, Xu Z, Wu WC, Tian JH, Zhang S, Zeng ZY, Gou QY, Xin GY, Le SJ, Xia YY, Zhou YL, Hui FM, Pan YF, Eden JS, Yang ZH, Han C, Shu YL, Guo D, Li J, Holmes EC, Li ZR, Shi M (2024) Using artificial intelligence to document the hidden RNA virosphere. Cell 187:6929–6942
3. Domingo E (2020) Virus as populations, 2nd ed. Academic Press, Elsevier, Amsterdam
4. Domingo E, Martínez-González B, Somovilla P, García-Crespo C, Soria ME, de Ávila AI, Gadea I, Perales C (2025) A general and biomedical perspective of viral quasispecies. RNA 31:1–16
5. Sardanyés J, Perales C, Domingo E, Elena SF (2024) Quasispecies theory and emerging viruses: challenges and applications. NPJ Viruses 2:54
6. García-Crespo C, Soria ME, Gallego I, Avila AI, Martinez-Gonzalez B, Vazquez-Sirvent L, Gomez J, Briones C, Gregori J, Quer J, Perales C, Domingo E (2020) Dissimilar conservation pattern in Hepatitis C virus mutant spectra, consensus sequences, and data banks. J Clin Med 9:3450
7. Shirogane Y, Watanabe S, Yanagi Y (2016) Cooperative interaction within RNA virus mutant spectra. Curr Top Microbiol Immunol 392:219–229
8. Shirogane Y, Harada H, Hirai Y, Takemoto R, Suzuki T, Hashiguchi T, Yanagi Y (2023) Collective fusion activity determines neurotropism of an en bloc transmitted enveloped virus. Sci Adv 9:eadf3731
9. Yousaf I, Hannon WW, Donohue RC, Pfaller CK, Yadav K, Dikdan RJ, Tyagi S, Schroeder DC, Shieh WJ, Rota PA, Feder AF, Cattaneo

R (2023) Brain tropism acquisition: the spatial dynamics and evolution of a measles virus collective infectious unit that drove lethal subacute sclerosing panencephalitis. PLoS Pathog 19:e1011817

10. Martínez-González B, Soria ME, Mínguez P, Lorenzo-Redondo R, Salar-Vidal L, López-García A, Esteban-Muñoz M, Durán-Pastor A, Somovilla P, García-Crespo C, de Ávila AI, Gómez J, Esteban J, Fernández-Roblas R, Gadea I, Domingo E, Perales C (2024) SARS-CoV-2 mutant spectra as variant of concern nurseries: endless variation? Front Microbiol 15:1358258
11. De Ávila AI, Soria ME, Martinez-Gonzalez B, Somovilla P, Minguez P, Salar-Vidal L, Esteban-Muñoz M, Martin-Garcia M, Zuñiga S, Sola I, Enjuanes L, Gadea I, Perales C, Domingo E (2025) SARS-CoV-2 biological clones are genetically heterogeneous, and include clade-discordant residues. J Virol 99(5):e0225024
12. Martínez-González B, Soria ME, Vázquez-Sirvent L, Ferrer-Orta C, Lobo-Vega R, Mínguez P, de la Fuente L, Llorens C, Soriano B, Ramos-Ruíz R, Corton M, López-Rodríguez R, García-Crespo C, Somovilla P, Durán-Pastor A, Gallego I, de Avila AI, Delgado S, Morán F, López-Galíndez C, Gómez J, Enjuanes L, Salar-Vidal L, Esteban-Muñoz M, Esteban J, Fernández-Roblas R, Gadea I, Ayuso C, Ruíz-Hornillos J, Verdaguer N, Domingo E, Perales C (2022) SARS-CoV-2 mutant spectra at different depth levels reveal an overwhelming abundance of low frequency mutations. Pathogens 11:662
13. Delgado S, Somovilla P, Ferrer-Orta C, Martinez-Gonzalez B, Vazquez-Monteagudo S, Munoz-Flores J, Soria ME, Garcia-Crespo C, de Avila AI, Duran-Pastor A, Gadea I, Lopez-Galindez C, Moran F, Lorenzo-Redondo R, Verdaguer N, Perales C, Domingo E (2024) Incipient functional SARS-CoV-2 diversification identified through neural network haplotype maps. Proc Natl Acad Sci USA 121:e2317851121
14. Domingo E, Martinez-Gonzalez B, Garcia-Crespo C, Somovilla P, de Avila AI, Soria ME, Duran-Pastor A, Perales C (2023) Puzzles, challenges, and information reservoir of SARS-CoV-2 quasispecies. J Virol 97:e0151123
15. Schirmer M, D'Amore R, Ijaz UZ, Hall N, Quince C (2016) Illumina error profiles: resolving fine-scale variation in metagenomic sequencing data. BMC Bioinf 17:125
16. Harcourt J, Tamin A, Lu X, Kamili S, Sakthivel SK, Murray J, Queen K, Tao Y, Paden CR, Zhang J, Li Y, Uehara A, Wang H, Goldsmith C, Bullock HA, Wang L, Whitaker B, Lynch B, Gautam R, Schindewolf C, Lokugamage KG, Scharton D, Plante JA, Mirchandani D, Widen SG, Narayanan K, Makino S, Ksiazek TG, Plante KS, Weaver SC, Lindstrom S, Tong S, Menachery VD, Thornburg NJ (2020) Severe acute respiratory syndrome coronavirus 2 from patient with coronavirus disease, United States. Emerg Infect Dis 26:1266–1273

Chapter 9

Geographic and Temporal Dynamics of Viral Proteome Sequences

Tre Tomaszewski

Abstract

Long-term tracking and widespread sequencing provide ample amounts of data to study the mutations of viral proteomes leading to the emergence of variants along a temporal dimension. Described here is a process for identifying the seasonal trajectories of mutational patterns via both temporal and geographic dynamics.

Keywords Proteome, Seasonality, Mutation, Epidemiology, Bioinformatics

1 Introduction

During the SARS-CoV-2 pandemic, large-scale collection and sequencing provided ample data to study the progression of the virus through surveillance of proteomic mutations as they accumulate and coalesce into distinct haplotypes over time. Haplotypes are sets of mutations that tend to be inherited together in successful viral variants. Understanding these dynamics is essential for unveiling the mechanisms behind variant emergence and for anticipating evolutionary trajectories by identifying how mutations emerge, spread, and coalesce into stable haplotypes.

Recent studies have highlighted how climatic factors and latitude-specific behaviors contribute to the diversification of SARS-CoV-2 over time [1–3]. While some variants achieve global prominence and persist over seasons, others remain geographically constrained. Contextualized in an evolutionary framework, such fluctuations suggest fitness depends not only on typical mechanisms such as infectivity or immune escape, but also on greater environmental features such as climate.

Figure 1 illustrates some different dynamics of accumulation of SARS-CoV-2 variants and haplotypes in different latitudes of the world during the pandemic. Accumulation plots of the first two

Gustavo Caetano-Anollés (ed.), *Evolutionary Genomics*, Methods in Molecular Biology, vol. 2981,
https://doi.org/10.1007/978-1-0716-4836-0_9,

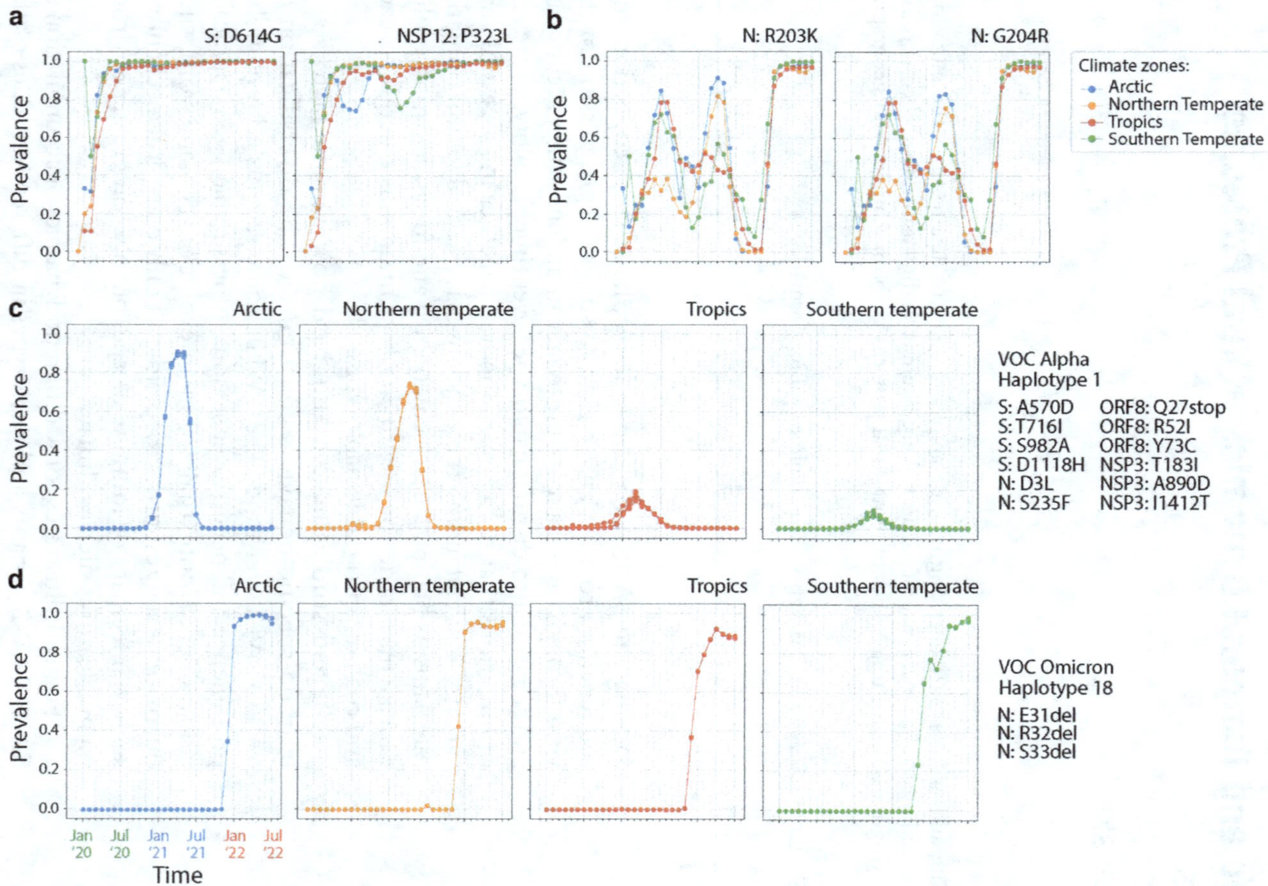

Fig. 1 Accumulation plots illustrating the emergence of haplotypes during the first two and a half years of the COVID-19 pandemic. Accumulation plots of the two mutants that make up the first and second haplotypes to appear during the pandemic, Haplotypes 5 and 2, are provided in panels (**a**) and (**b**), respectively. Accumulation plot overlaps of the 12 mutant markers that define Haplotype 1 of VOC Alpha and Haplotype 18 of VOC Omicron in different climatic zones are provided in panels (**c**) and (**d**), respectively

haplotypes that appeared worldwide showed distinct behaviors. Haplotype 5 involved four mutations, two of which were nonsynonymous, the well-known D614G amino acid substitution of the spike (S) protein and the P323L substitution of the NSP12 polymerase protein (Fig. 1a). The haplotype accumulated early achieving high prevalence levels that maintained throughout the pandemic in all climatic regions, with haplotype decoupling (different accumulation patterns in mutants) manifesting especially in Southern temperate regions. Haplotype 2 involved two mutations in the nucleocapsid (N) protein, R203K and G204R. Both mutations did not achieve the high levels of prevalence of Haplotype 5 until they became part of the late appearing variant of concern (VOC) Omicron. The mutations exhibited significant variabilities in different climatic regions with prevalence levels producing two peaks of accumulation prior to their recruitment into VOC Omicron (Fig. 1b). Accumulation plots of Haplotype 1 of VOC Alpha (encompassing 12 mutations) and Haplotype 18 of VOC Omicron (encompassing 3 mutations) show their monolithic behavior typical of coalescence processes (Fig. 1c, d). However, Haplotype 1 achieved significant prevalence in Arctic and Northern temperate regions but not in Southern temperate and tropical regions, while Haplotype 18 did not show that behavior but revealed mutant decoupling in individual regions.

The computational process described here provides a detailed methodological approach to track and explore mutation dynamics in the viral proteome across different climatic zones and temporal periods. This approach can help distinguish the interplay between viral genetics and external pressures, offering predictive insight when identifying mutations that may enhance transmissibility or evade immune responses. While the focus of the analysis in the original study centered on SARS-CoV-2, the approach detailed here may be applicable to analyses on the evolutionary dynamics of other widespread and/or persistent RNA viruses.

2 Materials

As examples, all scripts and data used for replicating the original analysis can be found on the project GitHub (https://github.com/gcalab/files/tree/master/F1000Research) and archived on Zenodo (https://zenodo.org/records/7636393) (*see* **Note 1**). The project includes sequence metadata and files featuring a list of countries along with coordinates and climate zones.

2.1 Datasets and Metadata

SARS-CoV-2 Genomic Metadata obtained from the GISAID EpiCoV™ repository (link). Accessing the data requires registration. Under the EpiCoV™ tab, click the "Downloads" link at the top. In the window that appears, find "metadata" under this tab.

Alternatively, conducting a search under the "Search" tab will allow prefiltering of datasets, although there are caveats to this method (*see* **Note 2**).

The retrieved dataset must contain:

- Accession IDs for identification
- A list of amino acid substitutions for each sequence ("AA substitutions") (*see* **Note 3**)
- The geographic location of sequence collection, and
- The sequence *collection* date (not to be confused with the *submission* date)
- The host type for filtering (optional)

The geographic dataset must include

- Country names
- Latitude

The data should be organized in a tabular format. Files with data organized as tab-separated (*.txt or *.tsv) or comma-separated values (*.csv) are recommended. GISAID metadata is downloaded in this format, though it may be compressed (in a format such as ZIP, TAR, or XZ) and require decompression prior to use.

2.2 Software and Tools

The analysis used the following software:

1. Python (v3.8.*) (https://www.python.org/)
2. Matplotlib (v3.5.1) (https://matplotlib.org/) for visualization [4]

Most programming or scripting languages (such as R or Python) can accomplish the data processing process. Depending on the size of the dataset, a system with 16 GB RAM, *minimum*, is recommended. The processing steps, particularly one-hot encoding of substitutions, may require significant memory.

Visualization can be accomplished via any plotting library or package.

3 Methods

3.1 Data Preprocessing

1. Load the SARS-CoV-2 metadata from the field-delimited GISAID file into the programming environment as a tabular data format (*see* **Note 4**).
2. Filter the sequences to retain only those relevant for the analysis to minimize the amount of data processed (*see* **Note 5**).

3. Discard sequences that do not meet these criteria. Sequences with missing, corrupted, or uninformative (e.g., featuring only a year) collection dates should be excluded or assigned an approximate date derived from other metadata to ensure temporal accuracy.
4. Remove features not applicable to the analysis should also be discarded.
5. Remove fields used to filter the sequences, if they are no longer necessary to limit the amount of data loaded during the following steps.
6. Save results to CSV file.

3.2 Geographic Climate Zones

1. Assign a climate zone to each sequence by mapping the normalized country names to their corresponding latitude using the country coordinate dataset.
2. Based on the latitude, classify each sequence into one of the following climate zones (*see* **Note 6**):
   (a) Tropics (30°S to 30°N),
   (b) Northern Temperate (30°N to 60°N),
   (c) Southern Temperate (30°S to 60°S), or
   (d) Arctic (60°N to 90°N),
   (e) Antarctic (60°S to 90°S).
3. In the case that the genomic metadata does not feature consistent country names, normalize country names in the sequence metadata using the geographic dataset as lookup table.

3.3 Parsing Amino Acid Substitutions

1. Extract the amino acid substitution data from the "AA substitutions" field in the metadata. As per the nomenclature [5] from the Human Genome Variation Society, GISAID substitution metadata are represented as a list. Each substitution label is a single string constructed from:
   (a) the protein name,
   (b) reference amino acid,
   (c) position, and
   (d) substituted amino acid
2. Use regular expressions to parse each substitution into these components (*see* **Note 7**).
3. Aggregate all unique amino acid substitutions across the entire dataset.
4. Store each unique substitution along with a count of the sequences that contain it in preparation for further analysis.

3.4 Data Transformation and Filtering

1. For each sequence, convert amino acid substitutions into a one-hot encoded format indicating the absence or presence as "0" or "1", respectively. Only existing substitutions need be represented as features.
2. Optionally, reduce the dataset to the most significant substitutions by applying a "relevancy threshold" function. This function may be a formally predefined value, or a heuristic based on preliminary data exploration and should be adjusted based on the dataset's distribution characteristics (*see* **Note 8**). This adjustment will affect the number of retained substitutions and reduce disk space and memory requirements.
3. Store the one-hot encoded format as a CSV file.

3.5 Temporal Grouping and Aggregation

1. Convert the collection date of each sequence into a temporal index (e.g., by month), with the earliest date as index #0, to facilitating temporal analysis of substitution trends.
2. Group the sequences by their assigned climate zone *and* monthly index to produce a compound "climate-time" value.
3. For each group, conduct a simple count of the total number of sequences and the number of occurrences for each amino acid substitution.
4. Normalize the counts into the incidence rate (between 0 and 1) of each amino acid substitution for each climate zone and chosen period.
5. Store this as a CSV file.

3.6 Visualization

1. Generate temporal line plots for high-variance substitutions in each climate zone to visualize the trends of each substitution.
2. (Optional) Save as vector files (*.svg).
3. (Optional) Edit as needed in vector editing program as desired.

3.7 Downstream Analysis

1. Explore patterns of accumulation of different variants and identify through overlap those that are consistent enough to warrant inclusion into an haplotype.
2. Map mutants and haplotypes along the amino acid sequence of individual proteins.
3. Visualize their possible impact in protein atomic structure.
4. Based on results, consider dividing mutants into finer geographical locales.

4 Notes

1. The repository scripts also are responsible for file size and memory management due to the size of the data used to obtain results. Depending on the size of the dataset being analyzed, these additional steps may or may not be necessary.
2. Alternatively, use the "Search" tab to access metadata for specific, prefiltered sequences. For search, the "Host" field dropdown should be set to "Human". Tick the "Complete" and "High Coverage," and "Collection Date Complete" checkboxes on the right-hand side. Adjust the "Collection" dates as needed. A drawback to this method is that only 10,000 sequences to be downloaded at a time, which can be subject to user error.
3. The metadata contains a field listing all amino acid substitutions, referenced against the "hCoV-19/Wuhan/WIV04/2019" sequence (GISAID accession ID: EPI_ISL_402124, GenBank accession version: MN908947.3) [6].
4. Encoding many unique substitutions can be memory-intensive, to reduce memory usage implement batch processing and/or use sparse matrices for numerical data.
5. If the entire metadata dataset was downloaded (not prefiltered in the download stage), the genomic metadata may contain incomplete or low-quality sequences. Filter these during preprocessing to maintain data quality. In the example analysis, only sequences obtained from human hosts and labeled as "complete" or "high-coverage" were used.
6. Large countries like China, Russia, or the USA span multiple climate zones, which may reduce the accuracy of classification. If more granular mapping is required, these countries should be divided into regions as applicable.
7. The regular expression to decompose the substitutions is

```
(?P<gene>[A-Z][A-Za-z]*[0-9]{,2}[ab]?)_(?P<ref>.+?)
(?P<loc>[0-9]+?)(?P<sub>[A-Za-z].*)
```

   This is the Python style of regular expressions, though any comparable expression in another language will work.

8. The heuristic used in the original analysis calculated the mean and standard deviation for the frequency of each substitution across all sequences. From these only sequences with a frequency greater than or equal to one standard deviation above the mean were retained; substitutions with low variance (indicating few sequences featured the substitution) were removed.

5 Acknowledgments

I gratefully acknowledge all data contributors, including the authors and their originating laboratories responsible for obtaining the specimens, and their submitting laboratories for generating the genetic sequence and metadata and sharing via the GISAID Initiative. I also acknowledge support by the Office of Research and Office of International Programs, College of Agricultural, Consumer and Environmental Sciences of the University of Illinois at Urbana-Champaign.

References

1. Caetano-Anollés K, Hernandez N, Mughal F et al (2021) The seasonal behaviour of COVID-19 and its galectin-like culprit of the viral spike. Methods Microbiol 57:27–81. https://doi.org/10.1016/bs.mim.2021.10.002s
2. Tomaszewski T, Gurtler V, Caetano-Anollés K, Caetano-Anollés G (2022) The emergence of SARS-CoV-2 variants of concern in Australia by haplotype coalescence reveals a continental link to COVID-19 seasonality. Methods Microbiol 50:233–268. https://doi.org/10.1016/bs.mim.2022.03.003
3. Tomaszewski T, Ali MA, Caetano-Anollés K, Caetano-Anollés G (2023) Seasonal effects decouple SARS-CoV-2 haplotypes worldwide. F1000Res 12:267. https://doi.org/10.12688/f1000research.131522.1
4. Hunter JD (2007) Matplotlib: a 2D graphics environment. Comput Sci Eng 9:90–95. https://doi.org/10.1109/MCSE.2007.55
5. Dunnen JT, den Antonarakis SE (2000) Mutation nomenclature extensions and suggestions to describe complex mutations: a discussion. Hum Mutat 15:7–12. https://doi.org/10.1002/(SICI)1098-1004(200001)15:1<7::AID-HUMU4>3.0.CO;2-N
6. Zhou P, Yang X-L, Wang X-G et al (2020) A pneumonia outbreak associated with a new coronavirus of probable bat origin. Nature 579:270–273. https://doi.org/10.1038/s41586-020-2012-7

Chapter 10

Proteome-wide Ab Initio Structural Analysis of Viral Evolution

Muhammad Asif Ali and Gustavo Caetano-Anollés

Abstract

The rapid advancement of sequencing technologies has outpaced experimental protein structure determination. AI-driven tools like AlphaFold now enable the prediction of near-experimental-quality atomic structures. However, certain protein regions—such as intrinsically disordered segments or those with transient or context-dependent conformations—may not be fully captured by experimental methods like NMR or cryo-EM. Here, we use AlphaFold to model a reference structure and bridge gaps in experimental models within a comparative framework for assessing structural deviations in sequences yet to be experimentally resolved. To achieve an accurate depiction of changes in protein structure, these regions are analyzed for their flexibility and biochemical properties. This approach provides a systematic strategy for refining structural interpretations of proteins in evolving viral populations, aiding in functional annotation, and improving our understanding of interactions and evolution of viral proteins.

Keywords Proteome, Seasonality, Mutation, Epidemiology, Bioinformatics

1 Introduction

The COVID-19 pandemic revealed a seasonal pattern of haplotype diversification of the SARS-CoV-2 virus at both the protein sequence and structure levels [1–3]. The accumulation of specific mutations indicated that as the virus navigated the mutational landscape, different regions and positions along its sequence and structure exhibited varying degrees of mutational entropy [4]. Favorable mutations were more frequently retained (entropy fixation at a position), while less beneficial mutations were selectively eliminated, allowing alternative mutations to be tested at the same position (entropy reversal at a position). Just as successful mutations were inherited together from haplotypes, forming larger mutant constellations characteristic of variants of concern (VOCs), similar patterns were observed in regions of structural deviation [2, 3].

Independent regions, despite their distinct mutational profiles, often adopted the same or similar substructures over the course of the pandemic, carrying these features from earlier haplotypes

Gustavo Caetano-Anollés (ed.), *Evolutionary Genomics*, Methods in Molecular Biology, vol. 2981,
https://doi.org/10.1007/978-1-0716-4836-0_10,

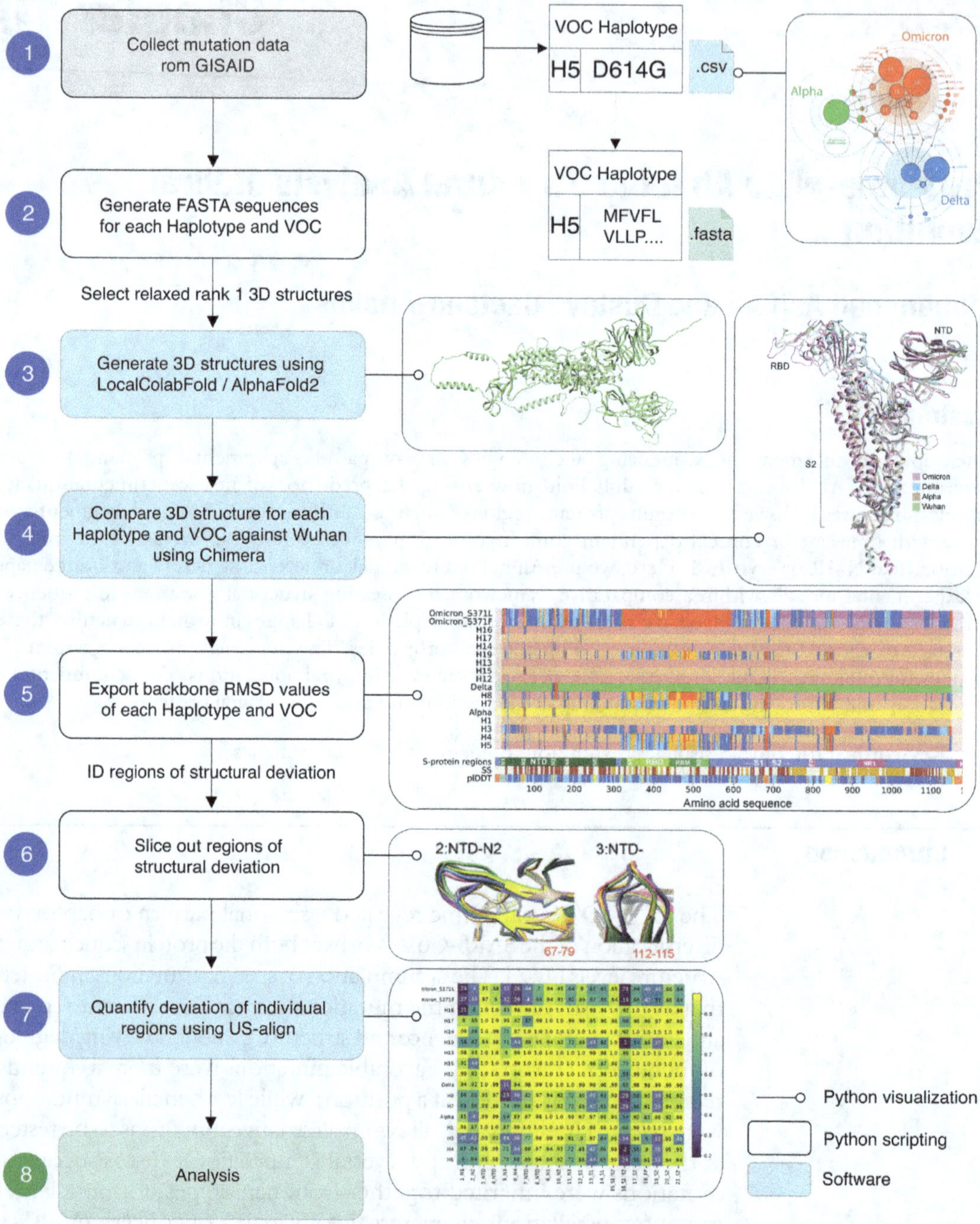

Fig. 1 Structural analysis pipeline. A flowchart that provides a step-by-step breakdown of the steps and software used to analyze protein structures

into later haplotypes and VOCs. Different levels of cooperativity were observed. For example, the S-protein VOCs Alpha and Delta exhibited antagonistic behavior, counteracting the effects of their respective haplotypes, whereas VOC Omicron acted synergistically, amplifying the impact of individual haplotypes. The N-protein exhibited similar patterns of structural recruitments across haplotypes and VOCs. Additionally, protein disorder analysis revealed a tendency to decrease disorder over time, accompanied by changes in binding capability at specific regions. While normal mode analysis (NMA) remained largely consistent, except for a few specific areas, visualizing the normal modes provided deeper insight by illustrating overall shifts in protein movement behavior. These properties demonstrate varying degrees of recruitment, which is particularly interesting from an evolutionary perspective, as it allows the breakdown of evolutionary patterns into specific regions, structures, and their associated biochemical properties and behavior [2, 3].

Here, we use the evolving populations of the SARS-CoV-2 virus to demonstrate how proteome-level ab initio structure prediction can help dissect the structural effects of viral evolution. Figure 1 outlines a workflow for analyzing protein sequences in comparison to a reference sequence and structure. By generating structural models at atomic resolution using AlphaFold [5] for newly sequenced mutants or older sequences lacking experimentally determined structures, we can perform structural comparisons using tools such as Chimera [6] and US-align [7] to analyze whole proteins and specific regions [8].

Structural changes should be documented over time, by location, or both (such as in the case of haplotypes). Further biochemical and protein disorder analyses can then complement these structural comparisons. This approach can be applied to any virus or protein sequence, offering an effective strategy to explore the evolutionary landscape using recent mutant sequences grouped into haplotypes defined by time and location, with the potential to reveal seasonal patterns [1]. However, further study is needed to link these structural changes to functional properties, such as immune evasion or protein stability. In the case of SARS-CoV-2, many regions of structural deviation coincided with known epitopes, which align with the expectation that the virus evolves to evade immune responses, particularly in regions that interact with the immune system. Therefore, regions of interest should be further explored for their specific properties through a relevant literature review.

2 Materials

All modeled three-dimensional (3D) structures can be accessed on the model archive repository for the M-protein, S-protein, and N-protein [1–3, 9–11].

2.1 Datasets

The Wuhan reference FASTA sequence (EPI_ISL_402124) was obtained from the online web-service GISAID [12]. VOC and haplotype definitions were retrieved from the data curated by Tomaszewski et al [1]. Table 1 describes the order of progression of VOCs and haplotypes and their associated mutations.

Table 1

Emergence of VOCs and haplotypes and their mutant constellations. Time of origin is given in a scale of relative appearance. Data from ref. [1]

Haplotype/VOC	Mutations	Time of origin
Omicron	T19I, L24-, P25-, P26-, A27S, A67V, H69-, V70-, T95I, G142D, V143-, Y144-, Y145-, N211-, L212I, V213G, + 214EPE, G339D, R346K, S371F, S373P, S375F, T376A, D405N, R408S, K417N, N440K, G446S, L452R, S477N, T478K, E484A, Q493R, Q493R, G496S, N501Y, Y505H, T547K, D614G, H655Y, N679K, P681H, N764K, D796Y, N856K, Q954H, N969K, L981F	16
H16	L24-, P25-, P26-, A27S, V213G, S371F, T376A, D405N, R408S	15
H17	N211-, L212I, S371L	14
H14	A67V, V143-, Y145-, G496S, T547K, N856K, L981F	13
H19	S373P, S375F, K417N, S477N	12
H13	E484A, Q493R, Y505H	11
H15	N764K	10
H12	G339D, H655Y, N679K, P681H, D796Y, Q954H, N969K	9
Delta	T19R, T95I, G142D, E156G, F157-, R158-, A222V, L452R, T478K, D614G, P681R, D950N	8
H8	E156G, R158-, F157-	7
H7	T19R, P681R	6
Alpha	H69-, V70-, Y144-, N501Y, A570D, D614G, P681H, T716I, S982A, D1118H	5
H1	A570D, T716I, S982A, D1118H	4
H3	H69-, V70-, Y144-	3
H4	N501Y	2
H5	D614G	1
Wuhan	–	0

2.2 Software and Tools

The analysis uses the following software:

1. Local Colab Fold (version 1.4.0) [13] (Can be replaced with the latest version of AlphaFold)
2. UCSF Chimera (version 1.16 linux_x86_64) [6]
3. AS2TS server used in CASP [14]
4. USAlign [7]
5. IUPred2A [15]
6. Bio3D package in R [16]

3 Methods

3.1 AlphaFold Modeling

1. Obtain the sequence of the protein structure you will use as reference for validation either from the NCBI data repository, another relevant database, or directly from PDB. Ensure that the full-length sequence, including all amino acids (not just the resolved ones), is used (*see* **Note 1**).
2. Prepare mutation data as CSV files, with each file corresponding to a specific mutant. Each entry should list amino acid changes in the following format: "D614G" (Original Amino Acid—Position in Protein Sequence—New Amino Acid).
3. Using Python, generate new FASTA sequences by introducing the specified mutations into the reference sequence. Replace the original amino acids with the corresponding mutated residues at the specified positions, then save the modified sequences as new FASTA files (*see* **Note 2**).
4. Input each FASTA file into the offline version of AlphaFold2 [5], called LocalColabFold [13], to generate 3D structural models. Select the highest-ranking structure (relaxed_rank_1) for downstream analysis. Key parameters include: --num-recycle 3 and --use-gpu-relax. Proteins with highly disordered regions may benefit from additional recycles (e.g., 6 or 12), though diminishing returns occur beyond this point. Example command:

```
colabfold_batch --amber --random-seed 42 --templates --num-recycle
3 --use-gpu-relax $fasta_file output/
```

3.2 Structure Validation

1. To validate the reference AlphaFold structure against an experimentally determined structure (cryo-EM or NMR), use the AS2TS server [17, 18], which was employed in CASP14 to assess structural similarity (*see* **Note 3**).
2. Set the superposition parameters to: -4 -o2 -gdc -lga_m -stral -d:4.0
3. AS2TS provides three input options for structural comparisons; use the second option to upload PDB files. For "Molecule 1 to rotate," click the browse option to select the reference PDB file from AlphaFold. For "Molecule 2 no change," select the experimentally determined structure PDB file. Click START to begin processing. This step may take some time, and an intermediate webpage will display your results.
4. Save the intermediate result as a text file or directly copy it for the last step.
5. Open a new tab for the AS2TS webpage and set the GDT_TS parameters to: -3 -o2 -gdc -lga_m -stral -d:4.0 -al

6. In the fourth input field on the AS2TS webpage, paste the intermediate result from the previous step, ensuring no molecules are selected, and click START.
7. The output will include the GDT_TS score. For more accurate assessment, multiply the final GDT_TS score by the fraction of superimposed alpha carbons (CA) over the total number of alpha carbons: GDT_TS × (superimposed CA/Total CA).
8. If the final score exceeds 70, the structure is considered sufficiently accurate for downstream analysis.

3.3 Mutant Structure Comparison

1. Each mutant 3D structure is compared to the reference 3D structure (all reference structures hereafter refer to the AlphaFold-generated structure) using Chimera's [6] "Structure Comparison" features found in the "Tools" menu. Open all 3D models in a single Chimera session. Use the "Match Maker" option, setting the Wuhan structure (or your reference PDB file) as the "Reference Structure" and all other structures as "Structures to Match." Enable "show pairwise alignments," keeping all other options as default, and click Apply.
2. To create a multiple structural alignment, use the "Match Align" feature with default settings. This allows for visual inspection of regions of interest.
3. Using the pairwise alignments, extract RMSD (Root Mean Square Deviation) differences between each residue of the reference and mutant structures. In the "Multi Align Viewer" window (which appears after running "Match Maker"), select "RMSD-backbone" from the "Headers" drop-down menu to obtain RMSD differences between backbone atoms. Save the results as a text file for each structure.
4. Use Python to parse the saved header files and extract all amino acid positions/regions where structural deviation exceeds 3Å. Store the start and end positions of these regions in a CSV file, recording deviations relative to the Wuhan structure.
5. Visualize these high-deviation regions as a heatmap using Matplotlib in Python. The original study used a stacked bar graph, with each bar at height one uniquely colored per position, allowing a lollipop plot of mutations to be overlayed. Alternatively, a simple heatmap can be generated using Seaborn in Python.
6. Extract regions exceeding the 3Å threshold for independent structural analysis. Use the ResSelect function in Python's Bio.PDB package [19] to isolate (slice out) and save these regions

from the 3D structures as separate PDB files. Example Python code to save individual regions:

```
s = bpdb.PDBParser().get_structure('temp', og_pdb)
    class ResSelect(bpdb.Select):
            def accept_residue(self, res):
            if res.id[1] >= start_res and res.id[1]
<= end_res and res.parent.id == chain_id:
                    return True
                  else:
                    return False
            io = bpdb.PDBIO()
            io.set_structure(s)
            new_name = name + ".pdb"
              io.save(out + region + '/' + new_name,
ResSelect())
```

7. US-align [7] to compare the isolated regions of mutant structures against the Wuhan reference. Analyze the structures with and without further superposition (-se option). Further superposition helps determine whether structural differences arise from morphological changes rather than simple shifts in 3D space. Example US-align commands:

```
USalign -se -dir1 path/to/pdbs/mutant.pdb path/to/
pdbs/file/list.txt path/to/pdbs/reference.pdb >
output_folder/results_se.txt
USalign -dir1 path/to/pdbs/mutant.pdb path/to/
pdbs/file/list.txt path/to/pdbs/reference.pdb >
output_folder/results_no_se.txt
```

8. Visualize TM scores for each region from US-align as a heatmap using Matplotlib in Python.

3.4 Mapping Intrinsic Disorder and Binding Capability

1. For proteins with high intrinsic disorder alongside structured domains, further analysis of the disordered regions is necessary, as standard structural analysis may not yield meaningful results. Use the IUPred2A online server [15] to assess intrinsic disorder (IUPred scores) and binding capability (Anchor2 scores). Combine all mutant sequences into one multi-FASTA file and submit it to the server with default settings.
2. Parse the TXT result file using Python. Visualize these scores as line graphs in Python, plotting them against the reference sequence. Ensure the Wuhan sequence is represented with a distinct line color to highlight differences.
3. Another approach to evaluating protein flexibility at the residue level involves using the Bio3D package in R [16, 20]. Use NMA to calculate residue fluctuations and save them to a CSV file for visualization in Python using *modes <- nma(pdb)* and

then *fluctuations <- modes$fluct*. Example R script to extract fluctuations from bio3d.nma:

```
library('bio3d')
pdbfiles <- list.files(pattern="*.pdb", full.names=
TRUE)
for (x in pdbfiles){
     pdb <- read.pdb(x)
     modes <- nma(pdb)
     fluctuations <- modes$fluct
     name <- gsub("[./pdb]", "", x)
     n <- paste("out/",name,".csv", sep="")
      write.csv(fluctuations, file = n, row.names =
FALSE)}
```

4. Comparing normal modes of different mutants against the reference structure can reveal differences in protein flexibility and movement.

3.5 Downstream Analysis

1. Map structural deviations alongside regions exhibiting differences in disorder, flexibility, and binding capacity across mutants. Access the entire workflow and code at our github site, which describes the entire process with snapshots of tools used and example visualizations (*see* **Note 4**).
2. Independently analyze each mutant by inspecting atomic-level differences in Chimera. Zoom into identified regions of interest for detailed structural assessment.
3. Investigate recruitment patterns among mutants. Identify shared structural deviations or flexibility profiles to reveal potential evolutionary trends.
4. Classify mutants based on their structural and biochemical properties. Group mutants with similar characteristics. Alternatively, construct a distance tree to visualize relationships between mutants.

4 Notes

1. Mutation data for each haplotype and VOC are sourced from Tomaszewski et al [1] and processed into individual CSV files using Python.
2. Point mutations are incorporated into the original Wuhan sequence (EPI_ISL_402124) from GISAID [12] using Python to generate FASTA files for structural modeling.
3. AS2S was used to assess the similarity between the AlphaFold Wuhan reference structure and the experimentally determined structures, PDB entries 6VYB (open conformation) and 6VXX (closed conformation) [21].

4. A GitHub repository (https://github.com/gcalab/files/tree/master/Methods%20in%20Molecular%20Biology/Viruses) hosts example scripts and explanations of the code used.

Acknowledgments

This research was funded by grants from the National Institute of Food and Agriculture (ILLU-802-909 and ILLU-483-625), the Office of Research and Office of International Programs, College of Agricultural, Consumer and Environmental Sciences of the University of Illinois at Urbana-Champaign, and the New Frontiers Initiative of the National Center for Supercomputing Applications (NCSA), all to G.C-A.

References

1. Tomaszewski T, Ali MA, Caetano-Anollés K, Caetano-Anollés G (2023) Seasonal effects decouple SARS-CoV-2 haplotypes worldwide. F1000Research 12:267. https://doi.org/10.12688/f1000research.131522.1
2. Ali MA, Caetano-Anollés G (2024) AlphaFold2 reveals structural patterns of seasonal haplotype diversification in SARS-CoV-2 spike protein variants. Biology 13:134. https://doi.org/10.3390/biology13030134
3. Ali MA, Caetano-Anollés G (2024) AlphaFold2 reveals structural patterns of seasonal haplotype diversification in SARS-CoV-2 nucleocapsid protein variants. Viruses 16:1358. https://doi.org/10.3390/v16091358
4. Tomaszewski T, DeVries RS, Dong M et al (2020) New pathways of mutational change in SARS-CoV-2 proteomes involve regions of intrinsic disorder important for virus replication and release. Evol Bioinform Online 16:1176934320965149. https://doi.org/10.1177/1176934320965149
5. Jumper J, Evans R, Pritzel A et al (2021) Highly accurate protein structure prediction with AlphaFold. Nature 596:583–589. https://doi.org/10.1038/s41586-021-03819-2
6. Pettersen EF, Goddard TD, Huang CC et al (2004) UCSF Chimera—a visualization system for exploratory research and analysis. J Comput Chem 25:1605–1612. https://doi.org/10.1002/jcc.20084
7. Zhang C, Shine M, Pyle AM, Zhang Y (2022) US-align: universal structure alignments of proteins, nucleic acids, and macromolecular complexes. Nat Methods 19:1109–1115. https://doi.org/10.1038/s41592-022-01585-1
8. Ali MA (2024) AlphaFold2 reveals structural patterns of seasonal haplotype diversification in SARS-CoV-2 structural protein variants. MS Thesis, University of Illinois at Urbana-Champaign. https://www.ideals.illinois.edu/items/131533/
9. Ali MA, Caetano-Anolles G (2023) Seasonal effects decouple SARS-CoV-2 haplotypes worldwide [Model Archive]. https://www.modelarchive.org/doi/10.5452/ma-gca-mprot. Released 9 Feb 2023
10. Ali MA, Caetano-Anolles G (2024) AlphaFold2 reveals structural patterns of seasonal haplotype diversification in SARS-CoV-2 spike protein variants [Model Archive]. https://www.modelarchive.org/doi/10.5452/ma-gca-sprot. Released 15 Jan 2024
11. Ali MA, C-A (2024) AlphaFold2 reveals structural patterns of seasonal haplotype diversification in SARS-CoV-2 nucleocapsid protein variants [Model Archive]. https://www.modelarchive.org/doi/10.5452/ma-gca-nprot. Released 16 Jul 2024
12. Shu Y, McCauley J (2017) GISAID: Global initiative on sharing all influenza data—from vision to reality. Eurosurveillance 22:30494. https://doi.org/10.2807/1560-7917.ES.2017.22.13.30494
13. Mirdita M, Schütze K, Moriwaki Y et al (2022) ColabFold: making protein folding accessible to all. Nat Methods 19:679–682. https://doi.org/10.1038/s41592-022-01488-1
14. Zemla A (2003) LGA: a method for finding 3D similarities in protein structures. Nucleic Acids Res 31:3370–3374. https://doi.org/10.1093/nar/gkg571

15. Mészáros B, Erdős G, Dosztányi Z (2018) IUPred2A: context-dependent prediction of protein disorder as a function of redox state and protein binding. Nucleic Acids Res 46:W329–W337. https://doi.org/10.1093/nar/gky384
16. Grant BJ, Skjaerven L, Yao X-Q (2021) The Bio3D packages for structural bioinformatics. Protein Sci 30:20–30. https://doi.org/10.1002/pro.3923
17. Zemla A, Zhou CE, Slezak T et al (2005) AS2TS system for protein structure modeling and analysis. Nucleic Acids Res 33:W111–115. https://doi.org/10.1093/nar/gki457
18. LGA - Protein Structure Comparison Facility. http://proteinmodel.org/AS2TS/LGA/lga.html. Accessed 2 Jul 2024
19. Cock PJA, Antao T, Chang JT et al (2009) Biopython: freely available Python tools for computational molecular biology and bioinformatics. Bioinformatics 25:1422–1423. https://doi.org/10.1093/bioinformatics/btp163
20. Grant BJ, Rodrigues APC, ElSawy KM et al (2006) Bio3d: an R package for the comparative analysis of protein structures. Bioinformatics 22:2695–2696. https://doi.org/10.1093/bioinformatics/btl461
21. Wrapp D, Wang N, Corbett KS et al (2020) Cryo-EM structure of the 2019-nCoV spike in the prefusion conformation. Science 367:1260–1263. https://doi.org/10.1126/science.abb2507

Chapter 11

Phylogenomic Tree Reconstruction from Bacterial Genomes

Kyung Mo Kim and Kyuin Hwang

Abstract

The recent, rapid expansion of available prokaryotic genomes has fueled significant advancements in phylogenomics within microbial ecology and evolution. However, comprehensive protocols for reconstructing phylogenomic trees from bacterial genomes remain scarce. To address this gap, we present a series of essential bioinformatics steps. This protocol begins with setting up a Docker environment to ensure consistent implementation across different operating systems. We demonstrate phylogenomic tree reconstruction using the Genome Taxonomy Database and its toolkits with five test genomes, which were retrieved from the NCBI database, quality-checked with CheckM, aligned using GTDB-Tk, and analyzed for tree reconstruction with RAxML. Genome-level phylogenetic analyses are often complicated by gene duplication, horizontal gene transfer, and computational demands that increase with the number of taxa. We discuss practical strategies for addressing these issues and highlight the value of orthologous gene concatenation in generating accurate phylogenomic trees.

Keywords Maximum-likelihood, Sequence alignment, Substitution model, 16S ribosomal RNA, Phylogenetic analysis

1 Introduction

The term "phylogeny" was first introduced in 1866 by Ernst Haeckel in *Generelle Morphologie der Organismen* [1], where he developed a framework for describing the evolutionary history of organisms. However, a primitive representation of a phylogenetic tree, illustrating evolutionary divergence from a common ancestor, was already presented in 1859 as the sole figure in Charles Darwin's *On the Origin of Species* [2]. Around the same time, the concept of homology emerged from comparative anatomical studies of various invertebrates [3]. Building on these nineteenth-century foundations, the field of cladistics—a framework for modern phylogenetic theory—was developed by Dr. Willi Hennig and his colleagues in the mid-twentieth century [4]. In the early stages of phylogenetic research, DNA sequencing was not yet available, and studies relied largely on morphological traits. These phenotypic characteristics, often influenced by convergent

Gustavo Caetano-Anollés (ed.), *Evolutionary Genomics*, Methods in Molecular Biology, vol. 2981,
https://doi.org/10.1007/978-1-0716-4836-0_11,

evolution, presented challenges due to their varied quantification among scientists. However, these limitations were significantly mitigated with the introduction of the Sanger DNA sequencing method in the 1970s. Along with the development of PCR, these advancements had underlain modern molecular phylogenetics.

Since then, phylogenetic studies of animal and plant taxa have flourished through the analysis of mitochondrial and chloroplast genes [5, 6]. In contrast, Carl Woese focused on ribosomal RNA (rRNA) genes for reconstructing prokaryotic phylogenies [7]. The rRNA genes are considered universal phylogenetic markers for several reasons: (1) they are present across all forms of life, (2) their genetic divergence correlates with species divergence, (3) PCR primers can be easily designed due to conserved regions interspersed with hypervariable ones, and (4) the varying levels of sequence conservation within these genes are effective for resolving both closely related and evolutionarily distant lineages. Woese's pioneering work not only identified the domain Archaea [7] but also demonstrated its evolutionary proximity to Eukaryotes compared to Bacteria [8]. For decades, rRNA genes have served as the gold standard for prokaryotic phylogenetics.

However, relying on a single marker gene like rRNA often leads to incomplete lineage resolution and polytomies, particularly in large-scale phylogenetic reconstructions, such as the Tree of Life [9]. This limitation has been addressed by the rapid accumulation of prokaryotic genomic data, which offers vastly more information than single genes. Recently, the Genome Taxonomy Database (GTDB) and its associated toolkit (GTDB-Tk) have been established as standard frameworks for prokaryotic phylogenomics [10, 11]. At the time of writing, the GTDB includes approximately 600,000 prokaryotic reference genomes. GTDB-Tk allows users to align their genomes with existing multiple sequence alignments created from ubiquitous single-copy proteins of these reference genomes. In the case of bacteria, the resulting alignment consists of ~5000 amino acid sites derived from the concatenation of 120 universal single-copy proteins, providing more robust phylogenetic information than traditional 16S rRNA alignments. Consequently, the integration of GTDB and GTDB-Tk enhances resolution in prokaryotic phylogeny compared to classic single gene-based methods. This protocol demonstrates methods for 1) quality-checking genomes, 2) aligning genomes, and 3) reconstructing a phylogenomic tree using GTDB-Tk.

2 Materials

2.1 Building a Docker Environment

The bioinformatics project for reconstructing phylogenomic trees using bacterial genomes requires several programs to be installed. However, computing environments—including operating systems, software, and libraries—vary among users, often leading to compatibility issues during installation. To ensure that this protocol is compatible across different computing environments, both software installation and bioinformatics analysis will be performed on Ubuntu within a Docker container, a virtual environment. First, install Docker Desktop from https://www.docker.com/products/docker-desktop/. After installation, launch Docker Desktop (version 4.28.0 in my case) and allocate CPU cores and memory limits using the Resources tab of the Docker menu settings. Once Docker Desktop is set up, you need to install the Ubuntu operating system. To do this, search for "ubuntu" on the Docker Hub website (https://hub.docker.com/) to find the appropriate docker pull command. Open a terminal window on your computer and execute the command (e.g., *$docker pull ubuntu:latest* in my case) to download the latest Ubuntu Docker image. If a terminal program is not available on your operating system, you can use another program like *PowerShell* as an alternative. You can check if the image has been successfully downloaded by running *$docker image list*. Next, create a new Docker container by executing the command *$docker run -v ~/Desktop/Docker/:/mnt/host -it ubuntu:latest /bin/bash*. In this command, the *-v* option connects the file systems between the host directory (*~/Desktop/Docker/*) and the folder (*/mnt/host*) of the newly created Docker container. The *-it* option and */bin/bash* allow you to use a bash shell on Ubuntu within the Docker container. If you need to exit the Docker container, enter the command *exit* or the shortcut Ctrl+D in the terminal. You can start and stop the created Docker container using *$docker start –a –i* <container_id> and *$docker stop* <container_id>, respectively. To find the container ID, use *$docker ps –a*. Once you run *$docker start*, you will enter the Ubuntu environment in your Docker container. In the terminal, create a working home directory for this project with *$mkdir /home/Bacterial_Phylogenomics* and enter the directory with the command *cd*. Most bioinformatics tools are available at anaconda.org and can be easily installed via conda. To install conda on the Ubuntu Docker container, run the following commands: (1) *$apt update* for updating *apt*, (2) *$apt install wget* for installing *wget*, and (3) *$wget* https://repo.anaconda.com/miniconda/Miniconda3-latest-Linux-x86_64.sh for downloading the latest version of Miniconda. Then run *$bash Miniconda3-latest-Linux-x86_64.sh* to install it. Answer "yes" to the question "*conda init --reverse $SHELL?*" during program installation. Then, type *$source /root/.bashrc* to activate Miniconda.

2.2 Preparing Bacterial Genomes

We will reconstruct a phylogenomic tree for the genus *Planktophila*, which belongs to the *Nanopelagicales* order. This genus is notable for being the smallest freshwater bacteria in terms of genome size and the most abundant in terrestrial aquatic ecosystems [12]. To avoid conflicts between libraries and programs in Ubuntu, we will first create a separate working environment using conda. In the terminal, type *$conda create -n Nanopelagicales_phylogeny python=3.12*.

This command will create a working space named *Nanopelagicales_phylogeny*. It is important to set the Python version to 3.12 to avoid errors during the installation of subsequent programs. Next, activate the conda environment by typing *$conda activate Nanopelagicales_phylogeny*. You will know that the activation is successful when the shell prompt changes from (*base*) to (*Nanopelagicales_phylogeny*). Next, we need to prepare the bacterial genomes for analysis. Instead of directly downloading genomes form the NCBI assembly database, we prefer to use a conda package named ncbi-genome-download (version 0.3.3, https://doi.org/10.5281/zenodo.8192432). You can find the installation command by searching for "ncbi-genome-download" at anaconda.org, which should look like *$conda install bioconda::ncbi-genome-download*. Run this command in your terminal to install it. For this project, we will analyze the following five genomes [12]: *Candidatus* Planktophila lacus (GCF_002288385.1), *Candidatus* Planktophila dulcis (GCF_002287965.1), *Candidatus* Planktophila sulfonica (GCF_002288065.1), *Candidatus* Planktophila versatilis (GCF_002288265.1), and *Candidatus* Planktophila limnetica (GCF_002288365.1). The values in parentheses indicate the NCBI genome RefSeq assembly accessions. In your working home directory (e.g., */home/Bacterial_Phylogenomics*), download the genomes by typing *$ncbi-genome-download bacteria -A GCF_002287965.1,GCF_002288385.1,GCF_002288065.1, GCF_002288265.1,GCF_002288365.1 -F fasta*. Here, the *-A* option specifies a comma-separated list of genome assembly accessions to download, and the *-F* option specifies the sequence format. If you need to retrieve genomes with GenBank assembly accessions that start with "GCA_," add the *-s genbank* option to the command above. The retrieved genomes in FASTA format will be automatically saved in the folder *refseq/bacteria* below your working home directory. To decompress all the FASTA sequence files, run the following command: *$gzip -d ./refseq/bacteria/GCF_00228*/GCF*.fna.gz*. Note that the asterisk is used as a wildcard character. For organizational purposes, you can consolidate all the decompressed genomes into one folder. First, create a new directory by typing *$mkdir input_genomes*. Then, copy all the genomes into this new folder with *$cp ./refseq/bacteria/GCF*/GCF*.fna ./input_genomes*.

2.3 Software

This protocol requires multiple computer programs, all of which are freely available online. URLs for downloading these programs are included in the text, where each program is introduced. Embedded programs and dependencies within the software packages are not listed separately, but their functions are described in the text.

1. *Docker Desktop* is used to create and run Docker containers on your computer.

2. *Ubuntu* is a Linux-based operating system where all the bioinformatics programs mentioned in this protocol can be installed and executed.
3. *Miniconda* is a lightweight installer for Conda that enables users to create isolated environments for managing various programs and libraries.
4. *ncbi-genome-download* is a command-line program to download genomic sequence data from the NCBI databases.
5. *CheckM2* is a command-line tool to assess the sequencing quality of microbial genomes.
6. *GTDB-Tk* is a tool used to assign taxonomic classifications to prokaryotic genomes.
7. *ModelTest-NG* is a program designed to select the optimal models for multiple sequence alignment.
8. *RAxML* is a software tool used to reconstruct phylogenetic trees using the maximum-likelihood method.
9. *Dendroscope* is a software tool with an interactive, user-friendly interface designed for visualizing phylogenetic trees.
10. *FastTree* is a tool for rapidly and approximately reconstructing maximum-likelihood phylogenetic trees.

3 Methods

3.1 Running CheckM

Bacterial genomes are often incomplete and may contain contamination. The completeness and contamination levels of genomes of interest can be assessed using CheckM2 [13]. This program can be installed using the command *$conda install -c bioconda -c conda-forge checkm2*, and the database can be downloaded using *$checkm2 database --download*. Note that before installing CheckM2, the conda environment *Nanopelagicales_phylogeny* should be activated. To run CheckM2, use *$checkm2 predict --threads <numberOfthreads> --input <folder_with_genomes> --output-directory <output_folder>*. In */home/Bacterial_Phylogenomics*, we ran like *$checkm2 predict --threads 8 --input ./input_genomes --output-directory ./CheckM_output*. Add the *--lowmem* option at the end of the command if you use a low-memory computer. The CheckM result can be found in *./CheckM_output/quality_report.tsv*. CheckM's output includes completeness and contamination for the query genomes (Fig. 1). Typically, genomes from pure cultures have near 100% completeness and near 0% contamination. Genomes from metagenomics and single-cell genomics, however, often have lower quality. To evaluate genome quality, three standards are applied [14]: (1) <50% completion and <10% contamination for low quality, (2) >50% and <10%

Name	Completeness	Contamination	Completeness_Model_Used	Translation_Table_Used	Coding_Density	Contig_N50	Average_Gene_Length	Genome_Size	GC_Content	Total_Coding_Sequences	Total_Contigs	Max_Contig_Length	Additional_Notes
GCF_002287965.1_ASM228796v1_genomic	99.99	0.1	Neural Network (Specific Model)	11	0.952	1361776	315.865062	1361776	0.48	1371	1	1361776	None
GCF_002288065.1_ASM228806v1_genomic	99.95	0.11	Neural Network (Specific Model)	11	0.954	1344614	318.243504	1344614	0.49	1347	1	1344614	None
GCF_002288265.1_ASM228826v1_genomic	100	0.26	Neural Network (Specific Model)	11	0.947	1325420	314.81982	1325420	0.48	1332	1	1325420	None
GCF_002288365.1_ASM228836v1_genomic	99.76	0.32	Neural Network (Specific Model)	11	0.955	1328793	314.60905	1328793	0.45	1348	1	1328793	None
GCF_002288385.1_ASM228838v1_genomic	99.98	0.21	Neural Network (Specific Model)	11	0.951	1410107	319.241453	1410107	0.48	1404	1	1410107	None

Fig. 1 Output file from running CheckM2. A completeness value closer to 100% and a contamination value closer to 0% indicate higher genome quality

for medium quality, and (3) >90% and <5% for high quality. Genomes that do not meet the chosen standard can be excluded from downstream analysis.

3.2 Running GTDB-Tk

For genomes selected through CheckM2, a multiple sequence alignment (MSA) is necessary before conducting phylogenomic analysis. In this protocol, GTDB-Tk v2.4.0 [11] is used to generate the MSA for the five *Planktophila* genomes. This tool can be installed using Bioconda, Docker, or Pip; here, we will describe installation via Bioconda, referring to the instructions at https://ecogenomics.github.io/GTDBTk/installing/bioconda.html. Due to frequent updates to the GTDB database and toolkit, we recommend checking for the latest versions before installation. To avoid conflicts with existing software and libraries, it is recommended to install GTDB-Tk in a dedicated conda environment. Begin by exiting the *Nanopelagicales_phylogeny* conda environment using *$conda deactivate*. Next, create a new environment named gtdbtk-2.4.0 and install GTDB-Tk v2.4.0 with the following command: *$conda create -n gtdbtk-2.4.0 -c conda-forge -c bioconda gtdbtk=2.4.0 python=3.12 numpy=1.26*. Activate this environment by typing *$conda activate gtdbtk-2.4.0*. Then, in your working home directory, run *$download-db.sh* to automatically download, extract the GTDB reference genome data, and set up the system path. Note that approximately 110 GB of disk space is required to complete this task. Now, run *$conda deactivate* followed by *$conda activate gtdbtk-2.4.0* so that GTDB-Tk is ready to use. Then, type *$gtdbtk classify_wf --skip_ani_screen --cpus 8 -x fna --genome_dir ./input_genomes --out_dir ./GTDB_out*. The classify_wf workflow includes several stages: (1) Open reading frames (ORFs) are identified using Prodigal, (2) 120 bacterial and 53 archaeal orthologous marker genes are detected using HMMER (*see* **Note 1**), (3) Marker gene sequences are aligned to their corresponding HMM models, and (4) The individual marker gene alignments are concatenated, resulting in approximately 5000 aligned amino acid sites (*see* **Note 3**). In this command, -x specifies the extension of the sequence files, and --skip_ani_screen forces GTDB-Tk to perform the ORF finding and HMM alignments. Alongside the MSA of the query genomes, GTDB-Tk generates several output files in the folder *GTDB_out* below your working home directory, including gtdbtk.*. summary files with taxonomic classifications based on ANI values or tree topology (Fig. 2). If pplacer is not runnable due to memory

user_genome	classification
GCF_002287965.1	d__Bacteria;p__Actinobacteriota;c__Actinomycetia;o__Nanopelagicales;f__Nanopelagicaceae;g__Planktophila;s__Planktophila dulcis
GCF_002288065.1	d__Bacteria;p__Actinobacteriota;c__Actinomycetia;o__Nanopelagicales;f__Nanopelagicaceae;g__Planktophila;s__Planktophila sulfonica
GCF_002288265.1	d__Bacteria;p__Actinobacteriota;c__Actinomycetia;o__Nanopelagicales;f__Nanopelagicaceae;g__Planktophila;s__Planktophila versatilis
GCF_002288365.1	d__Bacteria;p__Actinobacteriota;c__Actinomycetia;o__Nanopelagicales;f__Nanopelagicaceae;g__Planktophila;s__Planktophila limnetica
GCF_002288385.1	d__Bacteria;p__Actinobacteriota;c__Actinomycetia;o__Nanopelagicales;f__Nanopelagicaceae;g__Planktophila;s__Planktophila lacus

Fig. 2 Taxonomic assignment based on genome analysis using GTDB-Tk

limitations, and if you do not need to determine taxonomy using pplacer but only require the genome alignment, you can proceed as follows: *$gtdbtk identify --cpus 8 -x fna --genome_dir ./input_genomes --out_dir ./GTDBidentify*. Then, type *$gtdbtk align --skip_gtdb_refs --cpus 8 --identify_dir ./GTDBidentify --out_dir ./GTDB_out*. The first command handles stages (1) and (2) above, while the second command is responsible for stages (3) and (4).

3.3 Alignment and Substitution Model

In the folder *GTDB_out/classify* below your working home directory, two phylogenomic tree files are generated: *gtdbtk.backbone.bac120.classify.tree* and *gtdbtk.bac120.classify.tree.5.tree*. The former positions query genomes on a GTDB reference tree representing one genome per family (or order), while the latter provides more detailed placements, potentially down to the species level. These trees allow approximate phylogenetic placements of genomes within the GTDB reference trees for Bacteria. However, note that phylogenetic relationships based on pplacer are less accurate than those produced by standard phylogenomic methods [15]. To obtain a more precise phylogeny, a de novo tree reconstruction using maximum-likelihood methods is recommended (*see* **Note 2**). In *GTDB_out/align* below your working home directory, two multiple sequence alignment (MSA) files are provided: (1) *gtdbtk.*.msa.fasta.gz* contains both GTDB reference and query genomes, and (2) *gtdbtk.*.user_msa.fasta.gz* includes only the query genomes. The first file is useful for observing genomic placements within the bacterial reference tree, while the second file can be used to perform a de novo phylogenomic analysis. Here, decompress the alignment files by typing *$gzip -d gtdbtk.*.fasta.gz*. Before reconstructing the tree, determine the best protein substitution model for the query MSA (*gtdbtk.*.user_msa.fasta*). ModelTest-NG v0.1.7 is recommended, since it is faster than jModelTest and ProtTest with similar accuracy [16]. Start by exiting the GTDB conda environment with *$conda deactivate*, then activate our project environment with *$conda activate Nanopelagicales_phylogeny*. Next, search for "modeltest-ng" on anaconda.org and install it (e.g., *$conda install bioconda::modeltest-ng*). Then, create a directory named ModelTest below your working home directory and enter it: (1) *$mkdir ModelTest* and (2) *$cd ModelTest*. Run ModelTest-NG on the MSA file using: *$modeltest-ng -d aa -i ../GTDB_out/align/gtdbtk.bac120.user_msa.fasta -o user_msa_modeltest -p 8 -t ml*. In this command, *-d* specifies the data type (e.g., nt for nucleotide and aa for amino acid), *-o* sets the output prefix, *-p* indicates the number of CPU cores, and *-t ml* directs ModelTest-NG to use maximum-likelihood trees as the starting topology for evaluating all the substitution models based on Akaike (AIC), corrected Akaike (AICc), and Bayesian (BIC) information criteria. For large MSAs, ModelTest-NG becomes computationally intensive, so it can be practical to use substitution models widely applied to related taxonomic groups. For example, the WAG model with gamma-distributed rate heterogeneity used for GTDB bacterial trees may be suitable for broader trees, such as the Tree of Bacteria [10]. For the MSA of the five *Planktophila* genomes in *gtdbtk.*.user-msa.fasta*, ModelTest-NG selected LG+G4+F as the best model based on BIC, AIC, and AICc. LG specifies the substitution matrix, G4 represents a four-category Discrete GAMMA model for among-site rate heterogeneity, and F indicates empirical base frequencies. Details on model selection can be found in the file *user_msa_modeltest.out*.

3.4 Reconstructing Maximum-Likelihood Trees

The software RAxML is used for reconstructing trees based on the maximum-likelihood methods [17]. First, search for the conda installation command for RAxML on anaconda.org and execute the command (e.g, *$conda install bioconda::raxml*) in the terminal. To reconstruct an original tree using RAxML v8.2.13, follow these steps: (1) type *$mkdir User_MSA_Phylogeny* below your working home directory, (2) enter this new folder, and (3) run the following command: *$raxmlHPC-PTHREADS -s ../GTDB_out/align/gtdbtk.bac120.user_msa.fasta -n Nanopelagicales_original_ML -T 8 -m PROTGAMMALGF.* Note that RAxML designates the model LG+G4+F as PROTGAMMALGF. Be aware that model designations can vary between phylogenetic programs. To find the correct model designation for software you are using, refer to the "Commands" section in the output file of ModelTest-NG (Fig. 3).

The *RAxML_info.** file, one of the outputs from RAxML, provides the number of aligned sites treated as missing data. Next, bootstrapping can be used to evaluate the tree reliability (*see* **Note 3**). To reconstruct bootstrap trees, use the following command: *$raxmlHPC-PTHREADS -s ../GTDB_out/align/gtdbtk.bac120.user_msa.fasta -n Nanopelagicales_BOOT -T 8 -x 12345 -#100 -m PROTGAMMALGF -p 12345.* Here, *-T* specifies the number of CPU threads, *-s* represents the input alignment, *-n* names the output file, *-x* sets an

```
Commands:
  > phyml  -i ../GTDB_out/align/gtdbtk.bac120.user_msa.fasta -d aa -m LG -f e -v 0 -a e -c 4 -o tlr
  > raxmlHPC-SSE3 -s ../GTDB_out/align/gtdbtk.bac120.user_msa.fasta -m PROTGAMMALGF -n EXEC_NAME -p PARSIMONY_SEED
  > raxml-ng --msa ../GTDB_out/align/gtdbtk.bac120.user_msa.fasta --model LG+G4+F
  > paup -s ../GTDB_out/align/gtdbtk.bac120.user_msa.fasta
  > iqtree -s ../GTDB_out/align/gtdbtk.bac120.user_msa.fasta -m LG+G4+F
Done
```

Fig. 3 Output file named *user_msa_modeltest.out* from ModelTest-NG. The "Commands" block at the bottom of the file provides specifies the phylogenetic program-specific designation for the optimal model, which is LG+G4+F

integer number (random seed) for bootstrapping, and *-p* indicates a random number seed for the parsimony inferences. If your MSA is lengthy or contains many taxa, consider changing PROTGAMMA into PROTCAT, which is designed to accelerate computations. For more details, refer to RaxML's manual at https://cme.h-its.org/exelixis/resource/download/NewManual.pdf. Once the original tree reconstruction and bootstrapping are completed, you can map the bootstrap values onto the corresponding internal branches of the original tree with the following command: *$raxmlHPC-PTHREADS -f b -T 8 -m PROTGAMMALGF -s ../GTDB_out/align/gtdbtk.bac120.user_msa.fasta -z RAxML_bootstrap.Nanopelagicales_BOOT -t RAxML_bestTree.Nanopelagicales_original_ML -n GTDB_Original_with_Bootstrap*. This command assigns the bootstrap values from the 100 trees in *RAxML_bootstrap.Nanopelagicales_BOOT* onto the original tree *RAxML_bestTree.Nanopelagicales_original_ML*, resulting in the maximum-likelihood tree with bootstrap values named *RAxML_bipartitions.GTDB_Original_with_Bootstrap*. This way, you do not need to manually record bootstrap values from bootstrap trees onto the original tree. Next, execute *$cp RAxML_bipartitions.GTDB_Original_with_Bootstrap /mnt/host*. Files in */mnt/host* of your Ubuntu Docker container can be accessed through your host file system. In my case, the file in */mnt/host* is accessible at *~/Desktop/Docker*, which is connected to Docker's */mnt/host* through the *-v* option when executing the *docker run* command as mentioned earlier. Finally, add the .tre extension to the tree file and open it with Dendroscope [18] to visualize the phylogenetic tree (Fig. 4).

3.5 Faster Maximum-Likelihood Estimation

As the number of taxa increases, RAxML's tree reconstruction becomes significantly slower. In cases where the computational load is excessive, FastTree, which is several orders of magnitude faster [19], can be an effective alternative (*see* **Note 2**). The FastTree program

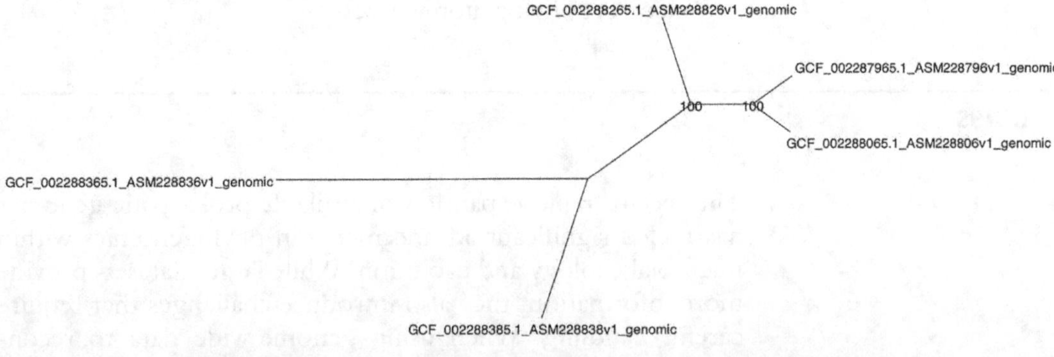

Fig. 4 Phylogenomic tree of the five *Planktophila* genomes visualized using Dendroscope. The tree is unrooted and includes bootstrap values at internal nodes. Additional modifications, such as rooting and editing taxon names, can be made using Dendroscope. The output svg file can be exported to an Illustrator tool for further graphic adjustments

is installed along with the GTDB-Tk. Let's test here if FastTree is runnable with the big alignment (>100,000 taxa) that consists of GTDB representative and the *Planktophila* genomes. For large datasets, running ModelTest-NG may be impractical; instead, use the WAG + GAMMA model, as in the GTDB Tree of Bacteria [10]. First, run *$conda activate gtdbtk-2.4.0*. Then, create a folder below your working home directory by typing *$mkdir Tree_of_Bacteria*, enter this new folder, and execute *$fasttree -wag -gamma -fastest /home/Bacterial_Phylogenomics/GTDB_out/align/gtdbtk.bac120.msa.fasta > GTDB.tre*. The *-fastest* option accelerates calculations and reduces memory usage, which is ideal for datasets over 50,000 taxa. If more speed is needed, try FastTreeMP, the multithreaded version. First, specify the number of CPU cores available for this program (e.g., *$export OMP_NUM_THREADS=8*) and run *$FastTreeMP -wag -gamma -fastest /home/Bacterial_Phylogenomics/GTDB_out/align/gtdbtk.bac120.msa.fasta > GTDB.tre*. Note that FastTree calculates node confidence using the Shimodaira-Hasegawa test [20] instead of traditional bootstrapping [21], so no bootstrap trees are needed to be reconstructed. Additionally, it is notable that GTDB-Tk offers a workflow (de_novo_wf) for reconstructing phylogenomic trees. If you are in any other conda environment, run *$conda deactivate*. Next, run the following commands: (1) *$conda activate gtdbtk-2.4.0*, and (2) In your working home directory (*/home/Bacterial_Phylogenomics*), *$gtdbtk de_novo_wf --genome_dir input_genomes/ --taxa_filter g__Planktophila --outgroup_taxon g__Nanopelagicus --out_dir de_novo_GTDB_out/ --cpus 8 --bacteria*. Here, *--genome_dir* specifies the folder containing genomes of interest (ingroup taxa) in FASTA format, *--bacteria* notifies that the user genomes are taxonomically included in Bacteria, and *--taxa_filter* and *--outgroup_taxon* designate sister and outgroup taxa, respectively. Note that while *de_novo_wf* is convenient in tree reconstruction, it approximates RAxML and bootstrapping with FastTree and the Shimodaira–Hasegawa test. Furthermore, the misdesignation of sister and outgroup taxa often leads to incorrect phylogenomic trees.

4 Notes

1. The recent, rapid expansion of available prokaryotic genomes has fueled significant advancements in phylogenomics within microbial ecology and evolution. While larger datasets provide more information, they also introduce challenges that require careful handling. When using genome-wide data to reconstruct phylogenomic trees, it is essential to ensure compatibility between genes, as some genes may introduce "phylogenetic noise" due to differing evolutionary histories. For example, if genes have undergone duplication, they may include paralogs,

which should be filtered out using tools like Mestortho [22], leaving only orthologs for downstream analysis. Alternatively, researchers often select single-copy genes (e.g., *rpoB*, *recA*, and *gyrB*) as phylogenomic markers, since they are less likely to contain paralogs. The GTDB database, a standardized framework for prokaryotic phylogenomics, also mitigates duplication effects by using only universal single-copy genes [10]. Horizontal gene transfer (HGT) introduces another source of phylogenetic noise. Even single-copy genes can be affected by HGT, which complicates phylogenetic relationships. Tree reconciliation methods, along with sequence composition metrices like G+C content, oligonucleotide frequency, and codon usage bias, can help identify horizontally transferred genes [23]. In addition, informational genes involved in the central dogma of cellular processes like DNA replication, transcription, and translation are typically less impacted by HGT than operational genes related to metabolisms [24], making them suitable markers to reduce HGT effects.

2. Multiple sequence alignment (MSA) is foundational for reconstructing the phylogenomic trees, since it produces aligned evolutionary homologs at each position in sequences. Tools like CLUSTAL [25] and MUSCLE [26] are widely used for MSA, balancing accuracy, and computational efficiency. As the number of sequences grows, obtaining a global maximum alignment becomes computationally impractical. Instead, MSA programs often find local maxima, reflecting a tradeoff between accuracy and speed. This balance also applies to phylogenomic tree reconstruction. For example, the standard maximum-likelihood methods like RAxML [17] search more exhaustively for the best tree candidates, requiring significant computational resources. In large-scale analyses, approximate methods like FastTree can provide faster results, though it is slightly less accurate [19].
3. As the number of taxa increases, most analyses face challenges due to a lack of phylogenetically informative characters [27]. This often leads to unresolved relationships among taxa in phylogenetic trees. One approach to address this issue is orthologous gene concatenation [28], which creates a long multiple sequence alignment (MSA) by combining individual alignments of orthologous genes. This process involves identifying orthologous proteins across the target proteomes using tools like Proteinortho [29], aligning the orthologous proteins individually (e.g., using MUSCLE [26]), concatenating the multiple sequence alignments, editing the concatenated alignment (e.g., with Gblocks [30]), and finally reconstructing a phylogenomic tree (e.g., using RAxML). The resulting concatenated alignment is often longer and more phylogenetically

informative than the GTDB alignment (~5000 amino acid sites), allowing for greater phylogenetic resolution. However, as the number of genomes analyzed increases, the number of universal orthologs tends to decrease. In this case, the GTDB-based approach may be preferable to orthologous gene concatenation. Once a phylogenetic tree is reconstructed, its reliability can be assessed statistically, with methods like jackknife resampling, Bayesian posterior probabilities, and bootstrapping [21, 31]. Bootstrapping, the most commonly used technique, resamples the MSA to estimate branch support. Using 1000 bootstrap replicates provides robust support, but only 100 replicates may be computationally feasible for larger datasets despite less stable bootstrap values.

Acknowledgments

This work was supported by Korea Polar Research Institute (grant number PE25130).

References

1. Haeckel E (1866) Generelle Morphologie der Organismen: Bd. Allgemeine Anatomie der Organismen, vol 1, G. Reimer
2. Darwin C (1859) On the origin of species by means of natural selection. J. Murray, London
3. Owen R (1843) Lectures on comparative anatomy and physiology of the invertebrate animals, delivered at the Royal College of Surgeons, in 1843. Hunterian Lectures 1:1–392
4. Hennig W (1966) Phylogenetic systematics. University of Illinois Press, Urbana
5. Ingman M, Kaessmann H, Pääbo S, Gyllensten U (2000) Mitochondrial genome variation and the origin of modern humans. Nature 408(6813):708–713
6. Soltis D, Soltis P, Endress P, Chase M, Manchester S, Judd W, Majure L, Mavrodiev E (2018) Phylogeny and evolution of the angiosperm. University of Chicago Press
7. Woese CR, Fox GE (1977) Phylogenetic structure of the prokaryotic domain: the primary kingdoms. Proc Natl Acad Sci 74(11):5088–5090
8. Woese CR, Kandler O, Wheelis ML (1990) Towards a natural system of organisms: proposal for the domains Archaea, Bacteria, and Eucarya. Proc Natl Acad Sci 87(12): 4576–4579
9. Pace NR (2009) Mapping the tree of life: progress and prospects. Microbiol Mol Biol Rev 73(4):565–576
10. Parks DH, Chuvochina M, Waite DW, Rinke C, Skarshewski A, Chaumeil P-A, Hugenholtz P (2018) A standardized bacterial taxonomy based on genome phylogeny substantially revises the tree of life. Nat Biotechnol 36(10):996–1004
11. Chaumeil P-A, Mussig AJ, Hugenholt P, Parks DH (2020) GTDB-Tk: a toolkit to classify genomes with the Genome Taxonomy Database. Oxford University Press
12. Neuenschwander SM, Ghai R, Pernthaler J, Salcher MM (2018) Microdiversification in genome-streamlined ubiquitous freshwater Actinobacteria. ISME J 12(1):185–198
13. Chklovski A, Parks DH, Woodcroft BJ, Tyson GW (2023) CheckM2: a rapid, scalable and accurate tool for assessing microbial genome quality using machine learning. Nat Methods 20(8):1203–1212
14. Bowers RM, Kyrpides NC, Stepanauskas R, Harmon-Smith M, Doud D, Reddy T, Schulz F, Jarett J, Rivers AR, Eloe-Fadrosh EA (2017) Minimum information about a single amplified genome (MISAG) and a metagenome-assembled genome (MIMAG) of bacteria and archaea. Nat Biotechnol 35(8):725–731
15. Matsen FA, Kodner RB, Armbrust EV (2010) pplacer: linear time maximum-likelihood and Bayesian phylogenetic placement of sequences onto a fixed reference tree. BMC Bioinf 11:1–16

16. Darriba D, Posada D, Kozlov AM, Stamatakis A, Morel B, Flouri T (2020) ModelTest-NG: a new and scalable tool for the selection of DNA and protein evolutionary models. Mol Biol Evol 37(1):291–294
17. Stamatakis A (2014) RAxML version 8: a tool for phylogenetic analysis and post-analysis of large phylogenies. Bioinformatics 30(9):1312–1313
18. Huson DH, Scornavacca C (2012) Dendroscope 3: an interactive tool for rooted phylogenetic trees and networks. Syst Biol 61(6):1061–1067
19. Price MN, Dehal PS, Arkin AP (2009) FastTree: computing large minimum evolution trees with profiles instead of a distance matrix. Mol Biol Evol 26(7):1641–1650
20. Shimodaira H, Hasegawa M (1999) Multiple comparisons of log-likelihoods with applications to phylogenetic inference. Mol Biol Evol 16(8):1114
21. Felsenstein J (1985) Confidence limits on phylogenies: an approach using the bootstrap. Evolution 39(4):783–791
22. Kim KM, Sung S, Caetano-Anolles G, Han JY, Kim H (2008) An approach of orthology detection from homologous sequences under minimum evolution. Nucleic Acids Res 36(17):e110–e110
23. Jeong H, Arif B, Caetano-Anollés G, Kim KM, Nasir A (2019) Horizontal gene transfer in human-associated microorganisms inferred by phylogenetic reconstruction and reconciliation. Sci Rep 9(1):1–18
24. Kim KM, Caetano-Anollés G (2011) The proteomic complexity and rise of the primordial ancestor of diversified life. BMC Evol Biol 11(1):140
25. Thompson JD, Higgins DG, Gibson TJ (1994) CLUSTAL W: improving the sensitivity of progressive multiple sequence alignment through sequence weighting, position-specific gap penalties and weight matrix choice. Nucleic Acids Res 22(22):4673–4680
26. Edgar RC (2004) MUSCLE: multiple sequence alignment with high accuracy and high throughput. Nucleic Acids Res 32(5):1792–1797
27. Brown JR, Douady CJ, Italia MJ, Marshall WE, Stanhope MJ (2001) Universal trees based on large combined protein sequence data sets. Nat Genet 28(3):281–285
28. Ciccarelli FD, Doerks T, Von Mering C, Creevey CJ, Snel B, Bork P (2006) Toward automatic reconstruction of a highly resolved tree of life. Science 311(5765):1283–1287
29. Lechner M, Findeiß S, Steiner L, Marz M, Stadler PF, Prohaska SJ (2011) Proteinortho: detection of (co-) orthologs in large-scale analysis. BMC Bioinf 12(1):124
30. Castresana J (2000) Selection of conserved blocks from multiple alignments for their use in phylogenetic analysis. Mol Biol Evol 17(4):540–552
31. Huelsenbeck JP, Ronquist F, Nielsen R, Bollback JP (2001) Bayesian inference of phylogeny and its impact on evolutionary biology. Science 294(5550):2310–2314

Check for updates

Chapter 12

Feature Frequency Profile (FFP) Method: A Language Model for Whole-Genome ("Organism") Phylogeny

JaeJin Choi, Byung-Ju Kim and Sung-Hou Kim

Abstract

For almost half a century, **molecular phylogenetic methods** (based on genomic sequence variations among a small number of genes/proteins that are highly homologous among the study population) have dominated in inferring the evolution of diverse organisms despite many important intrinsic limitations. Strictly speaking, these molecular trees represent the inferred evolution of those **molecules** selected, neither of **organisms**, nor of **whole genomes** (as simplified yet quantifiable proxy for the organism) selected. This is primarily because of two reasons: (a) the technology of economic sequencing and assembling the whole genome of an organism did not emerge until recently, especially for eukaryotic whole genomes, and (b) computational methods of comparing whole-genome sequences efficiently did not emerge until recently due to the sparsity of available whole-genome sequences. An example of a whole-genome Tree of Life is presented.

Keywords Molecular tree, Whole-genome tree, Organism tree, Pan-genomic Tree-of-Life (ToL), Rooting of ToL, Feature frequency profile (FFP), Jensen–Shannon divergence

1 Introduction

The broad universality of the inherited genetic language (genomic alphabets) and of the shared biochemistry used by *all extant* living organisms strongly suggests that they had one last common ancestry, a small ancestral population. Thus, the variation of the *whole-genome sequences* of the organisms may provide information about the degree of diversity of all extant *organisms*. Furthermore, since the whole genome sequence of an organism is, at present, the best available and quantitative representation of the organism, a "good" quantitative estimate of all pair-wise whole-genome differences (divergence) among representative members can be used to draw an *inference* on the evolutionary relationship among the members of the extant organisms.

Pan-Genomic Organism Tree vs. Homologous Molecular Tree: However, for almost half a century, *molecular phylogenetic methods* have

Gustavo Caetano-Anollés (ed.), *Evolutionary Genomics*, Methods in Molecular Biology, vol. 2981,
https://doi.org/10.1007/978-1-0716-4836-0_12,

dominated the field [1, 2] despite many important intrinsic limitations [3]. This is primarily because of two reasons: (a) the technology of sequencing and assembling the whole genome of an organism did not emerge until recently, especially for eukaryotic whole genomes, and (b) the computational methods of comparing whole genome sequences efficiently also did not emerge until recently. Thus, all molecular phylogenetic methods have been limited to using only the *highly conserved* sequence segments of a few genes/proteins that are *functionally similar among all diverse organisms of a given study*, which account for a very small fraction of the whole genome/proteome. Such limitations of the types of selected sequences may produce systematic biases for the selected molecules. Therefore, strictly speaking, such a tree represents a tree of those molecules selected (*molecular tree*), not of the organisms (*organism tree*) [3]. We present our experiences with the FFP method of building a *whole-genome tree* [4, 5], as a proxy of an organism tree, using a variation of the linguistic language model in Information Theory [6, 7] to represent the genomic language of the whole genome of organisms (*see* **Note 1**).

2 Materials

As a practice for constructing an organism Tree-of-Life (ToL) using the FFP method, we sampled a small data set consisting of about 100 organisms from each of six groups (Archaea, Bacteria, Fungi, Land Plants, Animals, and "Protists") (*see* **Note 2**). In this study, the protist group consists of all organisms *other than* those in the first five groups, *plus* all algae. Thus, we expanded the protist group to include algae, rather than they being a part of a broadly defined plant group containing both land and aquatic plants, because most algae are aquatic organisms and lack many of the distinct cell and tissue types found in land plants.

3 Methods and a Tutorial

3.1 Feature Frequency Profile (FFP) Method

Both linguistic and genomic languages are represented in *linearly arranged* alphabets. However, the rules of arranging their alphabet are different. For example, the English language consists of a linearly arranged "words" separated by "space", and the "Word Frequency Profile" for a given text is used to represent the text in the language analysis. However, the genomic language contains no "words" or "spaces", thus, each chromosome contains one long string of uninterrupted alphabets. Thus, we define a "feature" ("genomic word") as the ordered alphabet within a sliding window of a given length, as shown in Fig. 1 [8], and a Feature Frequency Profile (FFP) represents the whole genomic sequence of the chromosome for a given "feature length". This difference brings in an important point between

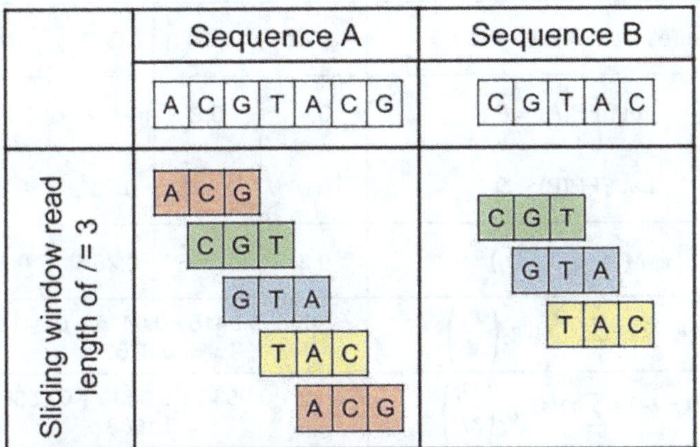

Feature type	Sequence A Count	Sequence A Frequency (probability)	Sequence B Count	Sequence B Frequency (probability)
ACG	2	2 / 5	0	0
CGT	1	1 / 5	1	1 / 3
GTA	1	1 / 5	1	1 / 3
TAC	1	1 / 5	1	1 / 3
n = 4		FFP(A)		FFP(B)

Modified from Zielezinski et al. **(8)**

Fig. 1 Converting sequence to Feature Frequency Profile (FFP)

"words" and "features": each "word" has no information about its neighbors, but each "feature" has information about its neighbors, i.e., it is in context. Therefore, in contrast to the "Word Frequency Profile," which has insufficient information to reconstruct the original text, FFP contains all the information needed to reconstruct the original sequence.

Figure 2 shows stepwise description of calculating the information divergence between two FFPs using Jensen–Shannon Divergence (D_{JS}; [9]), which is a symmetrized variation of more general Kullback–Leibler Divergence (D_{KL}; [10, 11]). The collection of D_{JS}'s of all pairwise FFPs provides a "distance" matrix, where "distance" is represented by the Information divergence of D_{JS}.

The same procedures apply whether one uses the FFP of the 4-alphabet nucleotide code or that of the 20-alphabet amino acid code.

3.2 Protocol for Constructing a Sample Pan-Genomic Organism Tree of Life (ToL) Using FFP Method

Since our earlier experiences [4, 5] showed that 4-letter alphabets of nucleotide sequences are insufficient for obtaining a topologically stable whole-genome ToL (*see* Fig. 3), we used 20 amino acid alphabets of the proteome sequences for each organism in this practice. All the steps below–using the same 600 proteome dataset–will be provided as a tutorial in the link: https://github.com/jaejinchoi/FFP/

Feature index (i) from 1 to n = 4	ACG	CGT	GTA	TAC
Let FFP(A) = P	0.4	0.2	0.2	0.2
Let FFP(B) = Q	0.0	$0.33\dot{3}$	$0.33\dot{3}$	$0.33\dot{3}$
$Let\ M_i = \frac{1}{2}(P_i + Q_i)$	0.2	$0.26\dot{6}$	$0.26\dot{6}$	$0.26\dot{6}$
$D_{KL}(P, M) = \sum_{i=1}^{n} P_i \times \log_2\left(\frac{P_i}{M_i}\right)$	0.2 - 0.0415 - 0.0415 – 0.0415 = 0.0755			
$D_{KL}(Q, M) = \sum_{i=1}^{n} Q_i \times \log_2\left(\frac{Q_i}{M_i}\right)$	0 + 0.0536 + 0.0536 + 0.0536 = 0.1608			
$D_{JS}(P, Q) = \frac{1}{2}\left(D_{KL}(P, M) + D_{KL}(Q, M)\right) = 0.1182$				

D_{KL} : Kullback-Leibler divergence (**10, 11**)
D_{JS} : Jensen-Shannon divergence (**9**)
The dot above a number indicates repeating the number indefinitely.

Fig. 2 Calculating Jensen–Shannon Divergence (D_{JS}) between two FFPs. D_{KL} : Kullback–Leibler divergence [10, 11]. D_{JS} : Jensen–Shannon divergence [9]

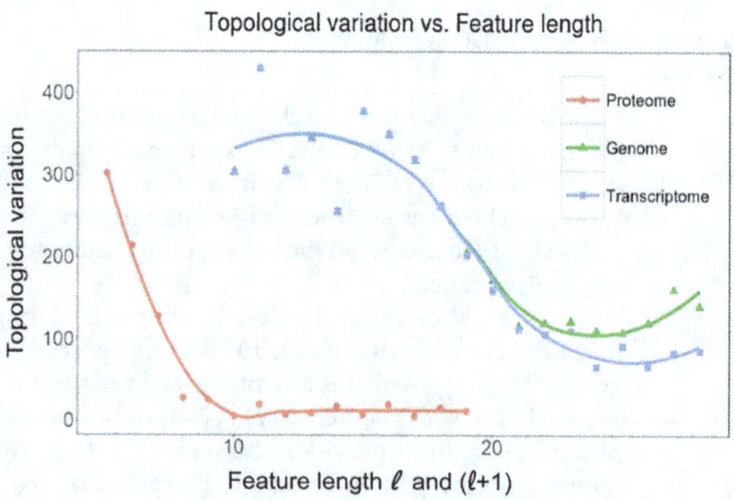

Fig. 3 Trend of Robinson–Foulds metrices [12] in three different languages: proteome, genome and transcriptome alphabets

tree/master/example. The guide will provide a wrapper (Python) script to automate steps.

1. Collect all the predicted proteome sequences: the protein sequences coded by CDSs (coding DNA sequences, which

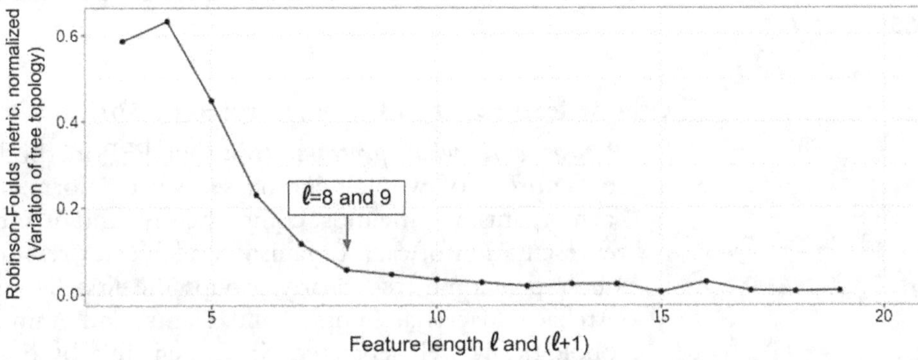

Fig. 4 Robinson–Foulds metric [12] between a pair of trees, one built using D_{JS} for feature length *l* and the other for (*l*+1) for the test data

include multiple transcripts from a single gene) for each of all members of the study population are saved as individual FASTA-formatted files. Any organelle-derived (e.g., mitochondrion or chloroplast) sequences are not considered in this test because they may have been endocytosed at an early evolutionary period.

2. Construct a "synthetic" outgroup member by shuffling the amino acid sequence of the proteome of an organism with the smallest proteome size and including it in the proteome data (*see* Step 4 below for a simpler way of introducing an outgroup member later into the proteome data).
3. Construct FFP Feature for each organism plus the outgroup member for each of the feature length (*l*) from 4 to 12 (*see* **Note 3** "Selection of optimum feature length"). Use the FF-Profiling program (*see* https://github.com/jaejinchoi/FFP/tree/master/versions/2v.4.x), which is developed to use in a command-line environment, e.g., Linux, or Cygwin for Windows, to construct a set of features and their frequencies for given feature length *l*.
4. Construct D_{JS} distance matrices for all study organisms plus the outgroup member for each *l* within the range 4–12.
5. Construct the neighbor-joining tree for each *l* using, for example, NEWICK or NEXUS tree format.
6. Plot Robinson–Foulds metric (RF; [12]) vs. two trees with feature length *l* and *l*+1 (Fig. 4) to find the optimal *l* at the "nadir" point, the beginning of the low plateau (region of *topologically stable* trees), as the optimal *l*.
7. Use the tree with the optimal *l* as the final pan-genomic whole-proteome ToL for detailed analysis inferred by the tree.

4 Notes

1. **Unique features of FFP whole-proteome ToL**

   (a) *Pan-genomic whole-proteome tree*: Our FFP method takes the entirety of whole-genome sequence information of each organism as the most comprehensive and quantifiable representation of each organism available at present. This whole-proteome tree clearly groups the first five groups (Archaea, Bacteria, Fungi, Land Plants, and Animals) as topologically well-separated clades definable by organism phenotypes, and the sixth group, an “expanded” Protist group, into several subgroups (*see* Fig. 5a).

   (b) *Deep “burst” of the founders of all six groups*: As observed in our earlier result [5] the founders of all six groups of the descendants of the last common ancestry emerged “in a burst” near the origin of the tree root within about 1/100 of the genomic divergence scale (*see* Fig. 5b, c).

   (c) *“Out-group” for Tree of Life (ToL) and the “rooting” of ToL*: The “outgroup” of a study population can be constructed to have given characteristics of a user’s choice: one by shuffling the sequence of an organism with the smallest proteome size, the other by constructing an "hypothetical taxon” with D_{JS} values to all in-group members assigned to be 1.0, and added to the in-group distance matrix.

   (d) *Branch lengths estimation*: The branch length between two organisms in FFP is estimated by genomic-sequence divergence caused by *all* mechanisms, known and unknown, of genomic alterations, while, in multiple sequence-alignment methods, they are estimated by exclusively one mechanism, point substitution rates.

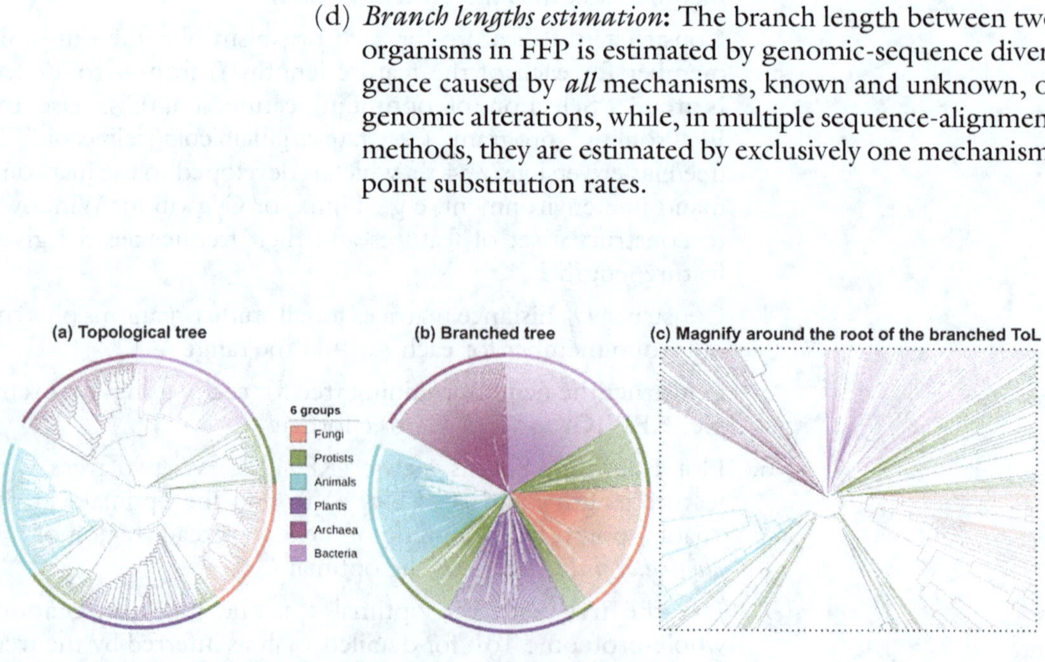

Fig. 5 BIONJ trees, visualized using ITOL [13]. (a) Topological whole-proteome ToL. (b) Branched whole-proteome ToL. (c) Expanded portion around the root of the branched ToL

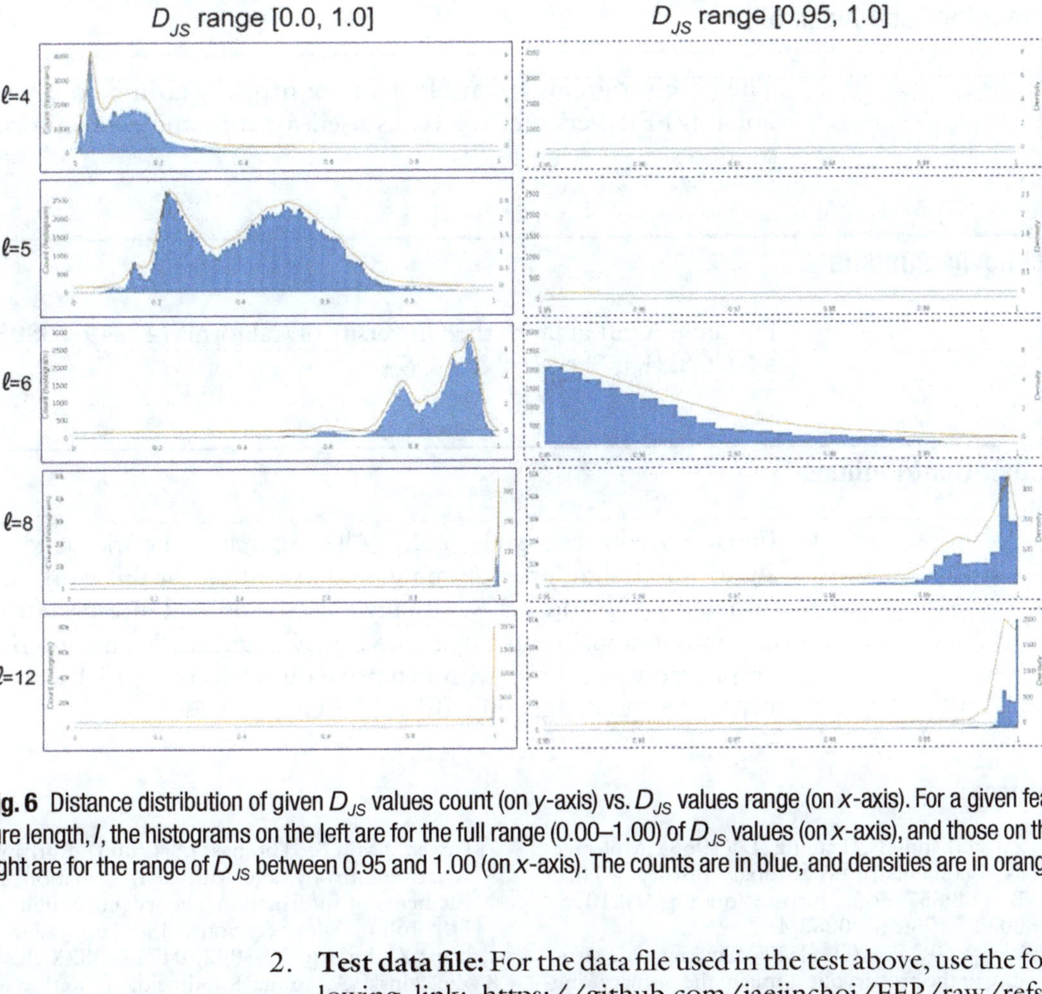

Fig. 6 Distance distribution of given D_{JS} values count (on *y*-axis) vs. D_{JS} values range (on *x*-axis). For a given feature length, *l*, the histograms on the left are for the full range (0.00–1.00) of D_{JS} values (on *x*-axis), and those on the right are for the range of D_{JS} between 0.95 and 1.00 (on *x*-axis). The counts are in blue, and densities are in orange

2. **Test data file:** For the data file used in the test above, use the following link: https://github.com/jaejinchoi/FFP/raw/refs/heads/master/example/600_tutorial_proteome/600-taxons_NCBI-proteome-source.xlsx.

3. **Selection of "optimum" feature length**

   The selection process consists of two considerations: (a) Examination of Fig. 4 indicates that the tree topology remains unstable until $l = 8$, when the tree topology with $l = 8$ and that of $l = 9$ is almost identical (i.e., Robinson–Foulds (RF) index close to zero) and remain so after $l = 8$. (b) However, as l increases more and more features become "unique" and more D_{JS}s in the matrix elements of any in-group members reach the value of 1.0, which is the maximum possible value by definition. With the given 600 taxon dataset, the overwhelming portion of all D_{JS} elements with $l = 12$ have values of greater than 0.99, as shown in Fig. 6, suggesting that most of the 600 organisms are evolutionarily unrelated, which is against our starting point that all extant living organisms have a common ancestry. Thus, we choose the beginning point of the lowest plateau as the "optimal" feature length for the 600 taxon data, which is $l = 8$.

Computer Code Availability

The core programs are available here: https://github.com/jaejinchoi/FFP. Version 2v.4.1 was used for this study and tutorial purposes.

Acknowledgments

Funding: A gift grant to the University of California (41349-10805-44-CCSHK), Berkeley, CA, USA.

Author Contributions

Initial conceptual design by S.H.K. All computational work including algorithm design, programming, and execution for this work was done by J.J.C. and B.J.K. equally. Discussion and interpretation of computational results by S.H.K., J.J.C., and B.J.K.; manuscript preparation by S.H.K. with extensive discussions with J.J.C. and B.J.K.; figures designed by B.J.K., J.J.C., and S.H.K.

References

1. Zuckerkandl E, Pauling L (1965) Molecules as documents of evolutionary history. J Theor Biol 8:357–366. https://doi.org/10.1016/0022-5193(65)90083-4
2. Woese CR, Fox GE (1977) Phylogenetic structure of the prokaryotic domain: the primary kingdoms. Proc Natl Acad Sci USA 74:5088–5090. https://doi.org/10.1073/pnas.74.11.5088
3. Pace NR (2009) Mapping the tree of life: progress and prospects. Microbiol Mol Biol Rev MMBR 73:565–576. https://doi.org/10.1128/MMBR.00033-09
4. Sims GE, Jun S, Wu GA, Kim S (2009) Alignment-free genome comparison with feature frequency profiles (FFP) and optimal resolutions. Proc Natl Acad Sci USA 106:2677–2682. https://doi.org/10.1073/pnas.0813249106
5. Choi J, Kim S-H (2019) Whole-proteome tree of life suggests a deep burst of organism diversity. Proc Natl Acad Sci USA 117:756155. https://doi.org/10.1101/756155
6. Claude S (1948) A mathematical theory of communication. Bell Syst Tech J 27:379–423
7. Cover TM., & Thomas JA (2005) Entropy, Relative Entropy, and Mutual Information. in Elements of Information Theory (ed. Schilling, D.) John Wiley & Sons, Ltd. pp. 13–55. https://doi.org/10.1002/047174882X.ch2
8. Zielezinski A, Vinga S, Almeida J, Karlowski WM (2017) Alignment-free sequence comparison: benefits, applications, and tools. Genome Biol 18:1–17. https://doi.org/10.1186/s13059-017-1319-7
9. Lin J (1991) Divergence measures based on the Shannon entropy. IEEE Trans Inf Theory 37:145–151. https://doi.org/10.1109/18.61115
10. Kullback S, RA L (1951) On information and sufficiency. Ann Math Stat 22:79–86
11. Bishop CM (2006) Pattern recognition and machine learning. Springer, New York
12. Robinson DF, Foulds LR (1981) Comparison of phylogenetic trees. Math Biosci 53:131–147
13. Letunic I, Bork P (2021) Interactive tree of life (iTOL) v5: an online tool for phylogenetic tree display and annotation. Nucleic Acids Res 49:W293–W296. https://doi.org/10.1093/nar/gkab301

Chapter 13

Phylogeny Building for Structural Phylogenomic Research

Arshan Nasir and Gustavo Caetano-Anollés

Abstract

Over the last two decades, we have routinely published phylogenies describing the evolution of organisms and viruses using protein structure information. The structure-based trees offer several advantages over traditional approaches, including improved resolution of basal branches of the trees, a more realistic representation of evolutionary events such as gene duplication, loss, gain, and horizontal gene transfer, better handling of fast-evolving (micro)-organisms and organisms with parasitic tendencies (viruses), and reduced susceptibility to artifacts arising from sequence alignment and reconstruction in complex genomic datasets. Here, we present a generic protocol for phylogenomic analysis of molecular structure, which is timely considering the recent revolution in AI-driven models of protein structure prediction. While the protocol is illustrated with viral protein structures, the procedure is generic in nature and can be easily adapted for other structures (e.g., RNA) and molecular characteristics (e.g., molecular functions, pathways), thereby enriching the phylogenetic toolkit available to molecular biologists.

Keywords Protein structure, Structural domain, Phylogenetics, Tree of Life, SCOP, SUPERFAMILY, HMM, Python

1 Introduction

The traditional approach in molecular phylogenetics is to use genetic sequences of conserved proteins and nucleic acids, including ribosomal proteins and RNA, to build phylogenies that describe the evolutionary relationships of sampled genes, and by inference, of organisms [1]. While such trees have many useful applications, especially in pathogen genomic research, they tend to perform poorly when the datasets include a mix of stable and fast-evolving or distantly related organisms [2] or when the objective is to resolve evolutionary splits that likely happened billions of years ago [3]. The major limitation comes from the very fact that nucleotide and/or amino acid sequences accumulate mutations over time [4] and at an uneven rate in different lineages (e.g., RNA viruses evolve much faster than DNA viruses and eukaryotic organisms [5]). These

Gustavo Caetano-Anollés (ed.), *Evolutionary Genomics*, Methods in Molecular Biology, vol. 2981,
https://doi.org/10.1007/978-1-0716-4836-0_13,

underlying features make it challenging to identify homologs of conserved genes/proteins in distantly related and fast-evolving organisms [6] and then to reliably align them for phylogenetic studies [7].

Fortunately, protein structures provide an alternative solution. Protein structure directly or indirectly dictates protein function. Even minor changes in the overall three-dimensional (3D) backbone of a protein may cause loss of function that can directly impact the life of an organism. Protein structures are thus (relatively) better conserved in evolution [8]. For these reasons, we have previously advocated the use of protein (and RNA) structures in phylogenomic investigations [9], especially when the datasets are complex and diverse [6]. For example, we have reconstructed structure-based trees of life to investigate the origin and evolution of viruses [10–12], to place newly discovered "candidate phyla radiation" lineages into bacterial phylogenies [13], to describe the reductive tendencies in modern proteomes [14, 15] and to reconstruct the proteomic makeup of the last universal common ancestor [16]. We have also reconstructed phylogenies describing the evolution of *structural domains*—the structural, functional, and evolutionary building blocks of the protein world—to study the origin and evolutionary history of proteins [17, 18], ribosomes [19] and metabolic enzymes [20, 21], to trace the birth and intrinsic disorder of domains in protein evolution [22], and to calibrate a molecular clock of protein folds [23], among a wide diversity of applications [24]. Figure 1 shows examples of phylogenies reconstructed in one study [18].

Phylogenomic studies have benefitted from the parallel development of public databases describing experimentally-determined protein structures and associated annotations such as the PDB [25], SUPERFAMILY [26], CATH [27], SCOP [28, 29] and its extended backward-compatible version SCOPe [30]. Each of these popular databases has been around for decades, but still protein structures remain underutilized in phylogenomic investigations. Here, we provide instructions to produce a data matrix for structural phylogenomic research, which follows a widely used experimental scheme [11]. The data matrix can be imported into a phylogenetic analysis software such as PAUP* (ver. 4.0) [31] to reconstruct a phylogenomic tree describing the evolution of taxa (either proteomes representing organisms or structural domains) using structural domains grouped into SCOP families as phylogenetic characters. The character states are the normalized counts of each SCOP family detected in a proteome. We demonstrate the function of our pipeline on four (04) proteomes of "giant" viruses and provide instructions on how run this matrix in PAUP*. Giant viruses are intriguing as their genomes and particle sizes overlap with the genomes and physical sizes of numerous cellular parasites and they encode several unique proteins never encountered previously in viruses [32, 33] (refer to [11, 34]

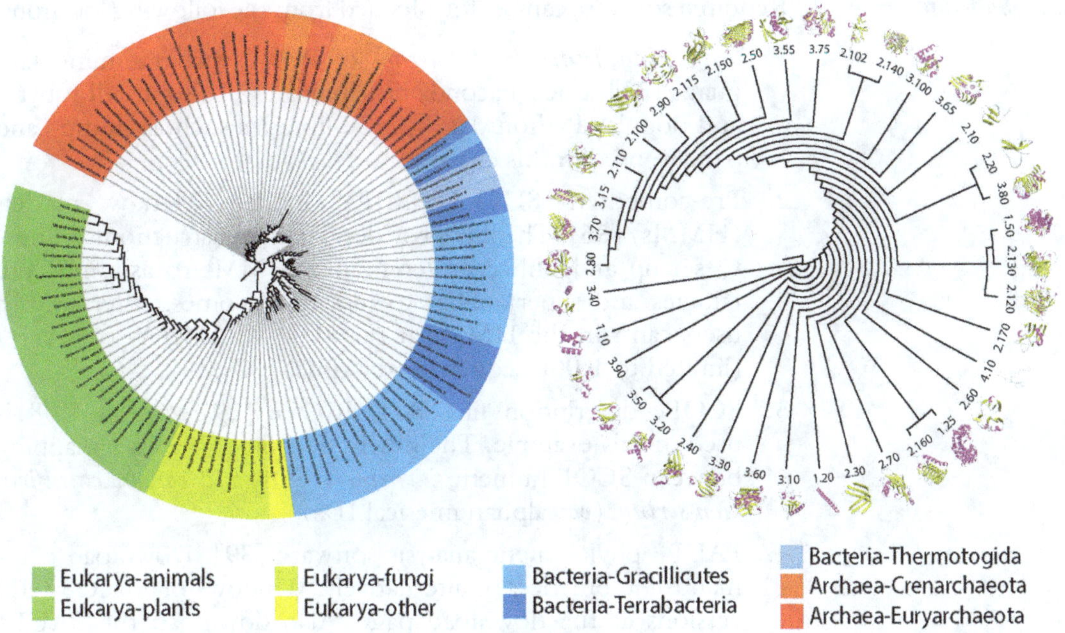

Fig. 1 A tree of proteomes (ToP) and a tree of domains (ToD) in circular tree format generated from domains defined at the homologous superfamily (H) and architecture (A) levels of CATH classification, respectively [18]. Note that both trees are rooted

for a detailed structure-based investigation of giant virus origin, emergence, and evolution).

Throughout this protocol, we follow some typographical conventions: (1) menu items or those contained in dialog boxes are given in bold (e.g., **File > Open** means click "File" in main menu and then select "Open" from the menu list); (2) text in command-line prompts or in dialog boxes are spelled in single quotation marks and in lower case and plain fixed-width font (e.g., the instruction to type "cstatus" means that you are typing a command that will summarize the data in PAUP*).

2 Materials

We assume the user has some experience working with shell commands, the Python programming language, and has administrative privileges. We also assume the user is familiar with SCOP and SUPERFAMILY databases and knows how to download proteomic data for organisms of interest and how to assign SCOP classified families and superfamilies to downloaded proteomes using public tools. We recommend using a computer with a 64-bit operating system possessing at least 8 GB RAM.

2.1 Software

Required software can be downloaded from the following locations:

1. *Anaconda Individual Edition* of Python programming language [35]. The Anaconda distribution will also install Jupyter and popular Python data science libraries such as numpy and pandas, used in this chapter.
2. Pre-computed SUPERFAMILY hidden Markov models (HMMs) [36]. The link provides detailed instructions on how to set up and utilize pre-computed HMMs to assign SCOP families and superfamilies to input proteomes. Alternatively, users can run the HMMs at the SUPERFAMILY web server (limited to 1,000 sequences at a time) [37].
3. SCOPe description file [38]. SCOPe [29] version 2.08 is used in this example. The description file provides a mapping between SCOP numeric *identifiers* (IDs) and *concise classification strings* (*ccs*, alphanumerical IDs).
4. PAUP* phylogenetic analysis software [39]. Download command-line binaries, or alternatively, Windows or MacOS GUI versions at the download page. Also download the PAUP* manual, which contains descriptions of all commands and provides useful examples and tutorials.
5. A graphical viewer of phylogenetic trees, such as FigTree [40] or iTol [41]. Note that the GUI versions of PAUP* have publication-quality tree drawing capabilities but cannot manage large trees. iTol, for example, can visualize trees with up to 50,000 taxa.

2.2 Other Materials

1. Data and relevant files and materials are available on GitHub [42].

3 Methods

3.1 Building Data Matrices

The steps below will process pre-computed domain assignment files (".txt" extension) outputted by the SUPERFAMILY HMMs [37] and will produce raw and normalized data matrices listing abundance counts for each SCOP family detected in every input proteome. In the example below, we have pre-computed assignment files for four proteomes corresponding to giant viruses: *Acanthamoeba polyphaga mimivirus* (GenBank Accession NC_014649; 979 proteins), *Acanthamoeba polyphaga moumouvirus* (NC_020104; 894 proteins), *Megavirus chiliensis* (NC_016072; 1,120 proteins), and Powai lake megavirus (NC_075034; 996 proteins). Note that the unit of classification in SCOP is the *structural domain*, and not the whole protein. Since proteins may contain one or more domains, it is possible to have multiple SCOP family assignments to a single protein. Moreover, structure coverage in public databases such as the PDB remains far from complete, and viruses often encode virus-specific proteins not detected in any other organism [34, 42]. Hence, several proteins may not map to any SCOP family (indicated by "-" in the HMM assignment file in the command-line version).

1. The commands below list files in the present directory, including the input FASTA files and output ".txt" files from the SUPERFAMILY webserver (Fig. 2). For example, there are four virus datasets (FASTA files) and four output families with domain assignments. The number of domain assignments can be greater or smaller than the number of proteins, depending on how the user runs the HMMs (e.g., command-line vs. webserver) and the biological complexity of the genome, since some proteins can encode multiple domains.

   Regardless of how the user assigns domains to input proteins, the output is a delimited file, which includes columns about the SUPERFAMILY model ID, *E*-value, sequence accession number, and SCOP family ID, among other information. In Fig. 2, the SCOP domain IDs are listed in the last column (e.g., 48404). Once the assignment files are ready, you can start generating the abundance matrices. Before doing so, check the output file (*see* **Note 1**).

2. Use the command "import" to import the required Python libraries to be used in the pipeline (Fig. 3). Import Python pandas and numpy libraries, along with os and glob libraries, which allow interacting with the shell.

```
In [1]: # count and print number of proteins in each FASTA dataset in the present directory
        !grep -c '>' *.fasta

        NC_014649_mimivirus.fasta:979
        NC_016072_Megavirus_chiliensis.fasta:1120
        NC_020104_moumovirus.fasta:894
        NC_075034_Powai_lake_megavirus_isolate1.fasta:996

In [2]: # count and print pre-computed domain assignments for each virus (identified by a '.txt' extension)
        !wc -l *.txt

             843 NC_014649_mimivirus.txt
             970 NC_016072_Megavirus_chiliensis.txt
             762 NC_020104_moumovirus.txt
             645 NC_075034_Powai_lake_megavirus_isolate1.txt
            3220 total

In [3]: # print an example domain assignment
        !head -1 NC_014649_mimivirus.txt

        YP_003987447.1 0048259 143-464 7.1e-85 3 AVRWASGNGHLEVVKYLVSQGANIRADNDH---AIGLASYYGYLEVVKYLVSQGADIRSDND---YAVRMASRN
        GHIEVVEYLVSQGANIRSDND---YAVRLASQNGHLEVVKYLVSQGADIKSDND---YAVRLASQNGHLEVVEYLVTQGTNIRVNNNY---AVEWASKNGNLEVVKYLISQGADI
        IADNN---FAVRWASRNGHLEVVKYLVSLGADIKSDNDY---AVRWASGNGHLEVVKYLVSQGSDIRVENDY---AVRWASRNGHFDVIKYLVSQGADIRSDNDY---AVKWASE
        NGHLEVVKFLVSLGADIKAEDDY---AVRWASEKGHLEVVEYLVSQGA 0.0000565 79778 48404
```

Fig. 2 Example input and output files used in this workflow

```
In [4]: # import required libraries
        import pandas as pd
        import numpy as np
        import os, glob
```

Fig. 3 Import required libraries

3. Combine the four individual assignment files into a single dataset (Fig. 4). The resulting data frame has three columns: (1) Sequence ID, which is the truncated FASTA header for each protein in the FASTA files and includes the sequence (protein) accession number, (2) SCOP ID which is the unique SCOP family identifier, and (3) Genome name to which the assignment belongs to. This name was extracted from the input FASTA file name. You may use any other code or numeric ID in the FASTA file names to separate or identify different taxa. Note that SCOP family 48404 was detected at least three times in the mimivirus genome when printing only the top five rows of the concatenated dataset. This redundancy suggests domain duplication or gain via horizontal gene transfer, which needs to be accounted for in the phylogenomics analysis for accurate reconstruction of evolutionary histories.
4. Convert the dataset into an abundance matrix, where column headers indicate SCOP FF IDs, column values indicate the number of times that FF was detected, and rows indicate genomes under analysis (see Fig. 5 for an example).
5. Transform the data frame into a normalized and rescaled data matrix compatible with pathogen biology and PAUP* (Fig. 6). The transformation step accounts for unequal genome sizes and unequal variances expected in the real and more complex datasets [10, 11] input to phylogenetics software.

Out[5]:

	scop_id	genome
seq_id		
YP_003987447.1	48404	NC_014649_mimivirus
YP_003987447.1	48404	NC_014649_mimivirus
YP_003987446.1	48404	NC_014649_mimivirus
YP_003987444.1	54696	NC_014649_mimivirus
YP_003987440.1	53475	NC_014649_mimivirus

Fig. 4 Combine assignment files into a Python-compatible data frame

Out[6]:

scop_id	46566	46580	46590	46610	46690	46768	46775	46925	46943	46962	...	161061	161066	161071	161094	1612
genome																
NC_014649_mimivirus	3	2	0	0	1	1	1	1	1	1	...	0	0	1	1	
NC_016072_Megavirus_chiliensis	4	0	0	1	2	1	0	1	1	0	...	1	1	0	1	
NC_020104_moumovirus	4	3	1	0	1	0	0	1	1	1	...	0	0	0	1	
NC_075034_Powai_lake_megavirus_isolate1	4	0	0	0	1	1	0	0	0	0	...	0	0	0	0	

4 rows × 596 columns

Fig. 5 Raw abundance matrix for FF count data in every proteome

Out[7]:

scop_id	46566	46580	46590	46610	46690	46768	46775	46925	46943	46962	...	161061	161066	161071	161094	1612
genome																
NC_014649_mimivirus	6	5	0	0	3	3	3	3	3	3	...	0	0	3	3	
NC_016072_Megavirus_chiliensis	7	0	0	3	5	3	0	3	3	0	...	3	3	0	3	
NC_020104_moumovirus	7	6	3	0	3	0	0	3	3	3	...	0	0	0	3	
NC_075034_Powai_lake_megavirus_isolate1	7	0	0	0	3	3	0	0	0	0	...	0	0	0	0	

4 rows × 596 columns

Fig. 6 Log-transformation of the raw data frame into a normalized and rescaled abundance data matrix.

Out[10]:

scop_id	a.2.3.1	a.2.5.1	a.2.7.1	a.2.11.1	a.4.1.1	a.4.2.1	a.4.3.1	a.4.11.1	a.5.4.1	a.6.1.3	...	f.52.1.1	f.53.1.1	f.54.1.1	f.57.·
genome															
NC_014649_mimivirus	6	5	0	0	3	3	3	3	3	3	...	0	0	3	
NC_016072_Megavirus_chiliensis	7	0	0	3	5	3	0	3	3	0	...	3	3	0	
NC_020104_moumovirus	7	6	3	0	3	0	0	3	3	3	...	0	0	0	
NC_075034_Powai_lake_megavirus_isolate1	7	0	0	0	3	3	0	0	0	0	...	0	0	0	

4 rows × 596 columns

Fig. 7 The final data matrix that can be exported for phylogenetic analysis

This procedure applies a logarithmic transformation to each column value and normalizes it by dividing by the logarithm of the maximum value in the entire data matrix. The resulting value is then scaled by a factor of 23 (or another value depending on the desired resolution) and converted into an alphanumeric format. The result, in this case, is a data matrix with four taxa and 596 SCOP family characters. The column headers represent SCOP family identifiers, and the column values correspond to the normalized abundance counts of each SCOP family detected in the respective proteome.

6. **OPTIONAL:** You may replace the SCOP numeric IDs by their *ccs*, which are alpha-numeric IDs, and may improve readability. Download the SCOP mapping file from the SCOPe database [43] and then create a mapping between SCOP IDs and *ccs*. The resulting data frame maps each SCOP numeric ID to its corresponding *ccs* (Fig. 7). The column headers now represent *ccs* rather than numeric IDs. The *ccs* may be more meaningful as strings starting from "a" indicate all-alpha secondary structure and strings starting from "b" indicate all-beta secondary structures. Please refer to SCOP documentation about different codes and what they mean.

3.2 Building Phylogenies

Phylogenetic reconstruction maximizes the fit of phylogenetically informative data to character changes in the branches of the trees. Computationally, this optimization is done by traveling through the enormous taxon-delimited tree solution space, generally via branch-swapping operations and retaining trees that are most optimal. In this protocol, we use maximum parsimony as the optimality criterion in PAUP* to illustrate retrodiction (*see* **Note 2**). PAUP* reconstructs phylogenies using maximum parsimony, maximum likelihood, quartet, and distance methods. The software reads data files and outputs tree files in *nexus* data format [46]. A *nexus* file is composed of a number of blocks describing aspects of the phylogenetic analysis (e.g., data, assumptions, trees, paup). Blocks hold commands that PAUP* will execute, starting with "begin" and concluding with "end" and including a "matrix" command that contains the phylogenetic data matrix. All commands and associated options are followed by semicolons. The modular design of *nexus* is expandable (additional software-specific modules can be added), inclusive (embeds all information), portable (independent of the operating system), and process-enhancing (reads useful and skips undesired information).

Building phylogenies involves five general directives: (1) Construct a data file that can be read by an appropriate phylogenetic reconstruction software; (2) Manage data, assumptions and optimality criteria; (3) Run the optimization protocol with a specific tree search strategy; (4) Root the optimal trees that were retained; (5) Assess confidence using bootstrap analysis; and (5) Visualize and describe the optimal tree(s). Below, we walk through each of these five steps.

3.2.1 Construct the Data File

1. Convert the text file with the exported data matrix into a *nexus* file with a *nex* or *nxs* extension. Use a text editor or code to build the file, which must begin with the *nexus* file declaration #NEXUS and must contain at least one block of commands containing the matrix. Figure 8 shows an example *nexus* file, which we will use to build a phylogenomic tree of the proteomes of giant viruses. Note that all text typed between brackets is invisible to the program and is generally used to index relevant information.
2. Construct the different blocks of the *nexus* file. The example file described in Fig. 8 contains three blocks: data, assumptions, and paup.

The first block is the data block defined by the command "begin data." The block establishes the number of taxa and characters with the "dimensions" subcommands "ntax" and "nchar," respectively. In our example, the data matrix has four taxa and 596 characters. The "format" command specifies the format of the data file, with the "symbols" subcommand defining permissible symbols used to designate character states (whitespace does not count). In our example, symbols describing the 24-character states describing domain abundance in proteomes are "0123456789ABCDEFGHIJKLMN", which are abbreviated to "0~9 A~N". The "matrix" command contains a sequence of taxon names followed by character state

```
#NEXUS
[Nasir, Kim and Caetano-Anollés, Meth Mol Biol, 2025]

begin data;
   dimensions ntax=4 nchar=596;
   format symbols="0~9 A~N";
matrix
NC_014649_mimivirus 6 5 0 0 3 3 3 3 3 3 5 3 0 3 3 0 7 3 0 3 3 5 3 3 0 3 5 0 0 3 5 3 0 7 5 3 3 0 7 3 0 0 3 3 5 3
5 3 0 0 0 0 N 0 0 3 6 3 0 3 0 0 5 3 3 3 0 3 0 0 3 3 0 5 0 3 3 0 0 3 5 3 3 3 3 3 0 0 0 5 3 B 6 B 0 0 3 3 0 0 3 0
0 5 0 3 3 7 9 0 3 0 5 5 5 0 0 0 0 0 3 3 0 0 0 3 3 7 3 0 7 0 6 3 5 0 9 6 3 7 5 3 3 3 0 3 6 3 7 3 7 0 3 3 3 0 0 5
5 3 0 0 0 3 8 5 0 5 3 0 3 3 0 3 3 0 0 3 0 0 0 9 0 3 5 3 9 3 3 5 0 3 5 0 0 0 8 0 0 5 9 5 F B 6 3 3 6 3 3 5 3 0 3
3 3 3 5 0 3 0 3 0 5 0 0 5 5 0 0 5 3 3 3 3 3 0 3 5 9 3 0 3 5 3 5 7 5 0 0 0 0 0 3 7 3 0 0 0 3 3 3 0 3 3 6 0 0 0 0
0 0 0 0 0 5 5 3 0 3 3 5 0 3 5 9 3 5 3 0 5 0 3 3 5 0 0 0 3 3 0 6 0 0 0 0 3 0 3 3 3 3 3 3 6 5 6 3 3 3 5 0 3 6 5 3
6 5 7 5 0 3 3 0 3 3 0 5 0 3 3 0 6 3 0 0 0 3 3 5 3 5 E B 5 3 3 3 0 E 6 3 3 3 5 5 3 6 5 3 3 A 3 3 B 3 0 7 0 6 3 3
3 5 3 0 5 3 0 3 5 3 6 0 0 3 E 3 3 3 0 0 3 0 0 5 3 3 0 6 3 3 0 0 3 5 0 8 3 0 0 3 0 0 5 0 0 6 3 3 0 3 5 0 0 0 5 0
6 0 3 0 0 0 3 5 3 3 3 0 5 0 0 0 0 3 3 0 0 0 0 0 3 3 5 3 0 3 3 0 0 0 3 3 3 0 0 3 3 3 0 0 0 3 0 0 0 3 3 3 3 0 3 0
3 0 0 0 3 0 0 0 3 0 0 3 3 0 B 3 3 3 0 0 0 0 7 3 3 0 5 3 3 0 0 3 0 0 0 0 0 0 0 0 3 3 0 5 0 0 3 3 3 0 3 3 3 0 0 3
0 0 3 5 3 0 0 0 0 5 0 0 0 0 0 3 0 3 0 0 5 0 3 0 3 0 0 3 0 0 0 3 3 0 0 3 0 0 3 3 0 0 3 3 3 3
NC_016072_Megavirus_chiliensis 7 0 0 3 5 3 0 3 3 0 9 3 0 0 3 5 8 0 0 0 3 3 0 0 3 3 7 3 3 3 5 3 0 8 5 0 0 3 7 0
0 0 0 5 5 0 3 3 3 3 3 0 N 0 3 5 0 0 0 3 3 3 3 8 0 3 0 5 3 5 0 3 3 5 3 0 3 3 3 3 5 3 0 3 3 3 0 0 3 3 3 9 5 6 3 5
3 3 3 3 3 3 5 5 0 0 3 7 7 3 3 0 6 0 B 7 0 0 0 3 3 3 0 3 3 0 3 9 3 3 6 3 8 3 5 3 7 3 5 7 3 3 0 3 0 0 0 3 7 5 7 3
3 3 3 3 0 5 5 3 0 3 3 0 8 3 0 3 5 3 3 0 0 3 0 0 0 3 0 3 3 7 3 3 5 3 A 3 3 5 3 3 5 3 0 3 7 3 3 5 9 0 D B 6 3 3 5
0 3 3 0 3 0 3 3 3 6 3 0 3 0 3 3 3 3 6 5 3 3 3 0 3 3 0 3 3 3 5 7 3 0 3 5 3 5 8 5 3 3 0 3 0 3 6 0 3 0 0 0 3 0 3 3
3 5 3 0 5 3 3 3 3 3 6 0 3 0 3 3 3 5 0 3 3 A 3 5 0 3 6 3 3 3 0 5 3 0 0 3 0 6 0 3 3 0 6 0 3 0 0 0 5 3 8 6 6 0 3 0
3 0 0 5 5 0 B 0 G 5 0 3 0 3 0 5 3 0 3 3 0 3 A 3 0 3 3 0 3 0 0 0 F C 3 0 0 3 3 H 0 0 3 3 5 5 3 3 0 0 0 B 0 0 B 7
5 7 0 3 6 0 0 0 3 3 5 9 0 3 5 0 5 3 3 3 E 3 0 0 3 3 5 0 3 6 3 0 3 0 0 3 0 6 3 5 0 7 0 3 3 3 3 3 0 0 3 6 0 3 3 0
5 3 0 0 3 3 5 0 0 0 3 0 3 5 3 3 3 3 5 3 3 3 6 0 0 0 3 5 3 3 3 0 0 0 0 0 0 3 3 0 0 0 3 0 0 3 0 0 3 3 0 3 3 3 3 0
0 3 3 6 3 3 3 3 0 3 3 3 3 0 0 3 0 0 0 3 E 0 0 0 3 3 3 0 7 0 0 0 0 3 0 0 3 0 0 3 3 3 0 3 0 3 6 0 3 0 3 3 5 0 3 3
3 0 3 0 3 3 0 0 0 6 0 3 0 0 3 0 0 3 3 0 3 0 3 0 0 3 6 0 3 0 3 3 3 0 0 3 0 0 3 3 3 3 3 3 0 3 0 3 3 3 0 0
NC_020104_moumovirus 7 6 3 0 3 0 0 3 3 3 5 0 3 0 5 3 5 0 3 0 3 0 0 3 0 3 3 5 3 3 3 0 3 A 5 0 0 0 8 0 3 5 0 3 5
0 5 3 0 0 3 3 M 3 0 3 0 0 3 0 0 0 3 3 0 3 3 5 0 3 0 0 0 5 0 0 3 5 0 3 5 3 0 0 0 3 3 3 3 0 3 8 6 6 0 0 3 3 0 3 6
0 3 5 3 0 3 8 7 3 3 3 0 5 9 5 0 0 3 0 5 3 3 0 0 0 3 9 5 0 5 3 6 3 3 5 8 5 5 8 3 3 0 3 3 0 5 3 5 3 7 3 3 0 3 0 3
5 6 0 3 0 0 0 0 3 3 0 3 0 3 3 0 3 0 3 3 3 3 0 0 6 0 3 5 3 A 3 3 6 0 3 5 0 3 0 7 0 0 5 6 0 B A 6 3 3 3 0 3 3 0 3
3 3 3 3 6 0 0 3 0 0 0 7 3 6 5 0 0 5 0 3 3 0 5 0 3 0 6 3 3 3 3 3 3 5 5 0 0 3 3 3 3 3 0 0 3 3 0 3 0 0 5 3 5 3 3 0
0 5 0 0 0 0 0 3 0 3 3 0 5 0 0 3 9 3 5 0 0 5 0 3 3 0 0 0 3 3 3 3 5 3 0 0 3 5 3 0 0 0 0 3 3 7 7 5 0 0 0 6 3 0 5 5
3 6 0 A 5 0 0 0 0 0 3 0 3 0 0 0 0 8 3 3 0 0 0 3 3 0 0 E B 5 0 0 5 0 B 6 0 3 3 5 5 3 5 0 0 0 5 0 3 A 5 0 7 3 0 6
0 0 3 3 3 5 5 3 3 5 0 6 0 0 3 D 0 0 0 0 0 3 3 3 6 0 0 0 3 3 3 3 0 6 3 3 3 0 0 0 3 0 3 5 3 0 6 0 0 0 0 5 0 3 3 3
0 5 3 0 3 0 3 6 5 3 3 3 0 6 0 0 3 5 0 5 3 0 3 3 0 3 0 0 0 3 0 0 0 0 3 3 0 5 3 3 0 0 0 0 3 3 0 0 0 0 3 3 3 0 3 0
0 5 0 3 0 0 0 0 3 0 3 3 0 0 0 F 0 0 0 0 0 3 3 9 0 0 3 0 3 0 3 0 0 3 0 0 0 5 0 3 0 6 0 0 0 0 0 3 0 3 0 3 3 3 3 3
3 3 3 3 5 0 0 3 3 0 0 3 0 3 3 3 0 0 0 3 3 6 3 3 3 3 0 0 0 3 0 3 0 3 0 3 5 0 0 0 3 3 0 0 5 0 0
NC_075034_Powai_lake_megavirus_isolate1 7 0 0 0 3 3 0 0 0 0 0 0 0 0 3 0 7 0 0 0 3 0 0 0 0 3 0 3 3 3 3 0 0 8 5 0
0 0 7 0 0 0 0 3 5 0 3 3 0 0 3 0 M 0 0 0 0 0 0 0 0 0 0 3 0 3 0 0 0 3 0 3 0 3 0 0 0 3 3 0 3 3 0 3 3 3 0 0 3 3 3 8
5 5 0 0 3 3 0 0 5 0 5 5 0 0 0 7 7 5 3 0 0 3 E 5 3 3 0 0 3 3 0 3 0 0 3 8 3 3 7 3 7 0 3 0 8 3 5 6 5 0 0 3 0 0 0 3
7 5 7 3 3 0 3 0 0 5 5 3 0 5 0 0 7 3 0 0 3 0 3 3 3 3 0 0 0 3 0 3 0 5 0 3 5 3 A 0 3 3 0 3 5 0 0 0 7 0 0 0 5 0 B B
6 3 0 3 0 3 3 0 0 0 3 3 3 5 0 0 3 0 0 0 3 3 6 5 0 0 3 0 3 3 0 3 0 3 5 6 3 0 3 3 3 3 8 5 0 3 0 3 0 3 6 0 0 0 0 0
3 0 0 3 3 3 3 0 5 0 0 0 0 0 0 0 6 0 6 3 0 5 3 0 0 6 0 3 0 0 3 0 0 0 0 0 0 0 0 0 0 5 0 0 0 0 5 3 3 0 0 0 5 3 7 5
5 0 0 0 3 0 0 5 5 0 8 0 F 5 3 0 0 0 0 0 0 0 0 0 0 0 0 3 0 0 0 0 3 0 0 0 F C 3 0 0 3 0 A 0 0 3 3 5 5 0 0 0 0 0 B
0 0 B 0 3 5 0 3 6 0 0 0 3 3 5 5 0 3 0 0 0 0 3 3 D 0 0 0 0 0 0 0 3 6 0 0 0 0 0 3 0 3 0 5 0 3 0 0 3 3 0 0 0 0 0 6
0 0 0 0 5 0 0 0 0 0 6 0 0 0 0 0 3 5 3 3 3 0 5 0 0 0 6 0 0 0 0 0 3 0 3 0 0 0 0 0 0 0 0 0 0 3 0 0 3 0 0 0 0 0 0 0
0 0 0 0 0 3 3 0 0 0 3 0 0 0 3 0 0 0 0 0 0 0 0 0 C 0 0 0 0 0 0 0 6 0 0 0 0 0 0 3 0 0 0 0 0 0 0 0 0 0 6 0 0 0 0 0
0 0 3 0 3 0 0 0 0 3 0 0 0 5 0 0 0 0 0 0 0 0 0 0 0 0 3 0 0 3 5 0 3 0 3 0 0 0 0 0 0 0 0 0 0 0 0 0 0 0 0 0 0 0 0 0
;
end;

begin assumptions;
   typeset *organism = ord:all;
   ancstates  *ancestor = 0:all;
end;

[The following block initiates phylogenetic analysis; it can be activated by eliminating brackets]

[begin paup;
log start file=tree_of_proteomes.log;
exclude uninf;
set increase=auto;
set autoinc=100;
set taxlabels=full;
hsearch addseq=random nreps=100 nchuck=1 chuckscore=2;
roottrees rootmethod=lundberg;
describetrees /plot=phylogram diag=yes;
savetrees brlen=yes file=tree_of_proteomes.tre format=nexus;
bootstrap search=faststep nreps=1000 treefile=tree_of_proteomes_boot.tre format=nexus;
randtrees nreps=1000;
log stop;
end;]
```

Fig. 8 A *nexus* file describing main headers, data block, and assumptions block with minimal information about character transformation

information for that taxon; in the example *nexus* file, the matrix was derived from the "data_matrix_family_abundance.txt" file.

The second block in the file holds assumptions about the data, i.e., the model that will drive the reconstruction exercise. The block is defined by the command "begin assumptions." The "typeset" command specifies a set of character-type assignments for each character, either predefined (e.g., unordered, ordered, Dollo) or defined by a "usertype" command (not used in this protocol). The typeset command in the example file of Fig. 8 (named "organism") applies the ordered character-type "ord" (Wagner characters) to all 596 characters following a transformation series of increasingly derived character states (*see* **Note 3**). The "ancstates" command defines ancestral states for every character, which, according to the model specified in the file, are the minimum possible states (0) for all 596 characters.

The third block holds PAUP* commands necessary to reconstruct the phylogeny. While a total of 98 commands is available for this task, we will only demonstrate a few. In Fig. 10, commands of this block are in brackets. Removing brackets will prompt the software to execute all commands of the block. We recommend executing individual PAUP* commands to learn what they do and their associated options before entering into a batch processing mode. Follow the instructions of the next section while examining the commands issued in Fig. 8.

3.2.2 Manage Data, Assumptions, and Optimality Criteria

1. Request processing of your file by using the command "execute" in the command-line interface of the portable version or **File>execute** of your file in the drop-down menu of the GUI version of the PAUP* program. Commands will be processed until the end of the file is reached or a "quit" command is issued. Request that analyses be logged out to a file with the "log start" and "log stop" commands. Commands and outputs stored in the display buffer will be saved to a file with a log suffix.
2. Explore and include/exclude characters and taxa. Command "cstatus" or the menu equivalent **Data>Show character status** will prompt a description of the current optimality criterion and of characters that are available, including their type, weight, and informativeness. Command "tstatus" or **Data>Show taxon status** will prompt a description of taxa. If uninformative characters are present, exclude them with the "exclude uninf" command. In our example, 381 characters were uninformative and were excluded. Deleting or restoring taxa is available through the "delete" or "undelete" commands or **Data>Delete-Restore taxa**.
3. Define the optimality criterion used in phylogeny reconstruction with the "set optimality" command. Note that "set optimality=parsimony" or **Analysis>Parsimony** is the default. The example file performs a maximum parsimony search.

3.2.3 Run the Optimization Protocol with a Specific Tree Search Strategy

1. Search for optimal trees with one of three strategies: (1) An exhaustive search of tree solution space with the "alltrees" command or the menu equivalent **Analysis>Exhaustive Search**; (2) A branch-and-bound search with the "bandb" command

or **Analysis>Branch-and-Bound Search**; or (3) A heuristic search with the "hsearch" command or **Analysis>Heuristic Search**. Note that each one of these commands contains useful subcommands you need to use to define search details and output (see next step). In the example exercise we use the heuristic approach for illustration purposes. However, the analysis should be performed with exhaustive or branch-and-bound searches given that only four taxa are being analyzed (well below the ~20-taxon limit of exact methods).

2. Heuristic searches are fast and effective ways to find optimal trees. However, results are approximations that depend on the way searches are conducted (*see* **Note 4**). For that reason, spend time exploring the options that are available and experiment with strategies that will better navigate the tree solution space. You have three levels of control: starting trees, stepwise addition of taxa, and branch swapping. The default option of starting trees is to get them by stepwise addition. However, you can also obtain them by neighbor joining, or using a tree you may have in memory. The addition sequence of taxa can be "simple", "closest", "asis," or "random", each one increasing the computational effort involved. Finally, branch swapping can be selected using the NNI, SPR, and TBR approaches (TBR is the default). In our example, we use a random addition sequence with 100 replicates using subcommands "addseq= random nreps=100". The "nchuck" and "chuckscore" commands allow for a "channeled" search in which "nchuck" specifies the number of trees that can be saved in a replicate if they have scores greater than or equal to a score specified by "chuckscore". Together with commands "chucklen" and "keep", the approach allows for navigating the world of trees more effectively.
3. The output of the search is provided in Fig. 9. A single most-parsimonious tree was retained with a length of 924 steps (changes). A single island was identified. An identical tree of the same length was retained with an exhaustive search, showing that the heuristic strategy was effective.

3.2.4 Root the Optimal Tree(s) That Were Retained

1. Computed trees are generally unrooted for the simple computational reason that the solution space of unrooted trees is lower in complexity than that of rooted trees. Once most parsimonious unrooted trees are retained, they can be rooted using the "roottrees" command or the menu equivalent **Trees>Root Trees**. Rooting is necessary to unfold history, provides a direction to the phylogeny, and fully explains the underlying evolutionary process materializing from a single evolutionary origin of taxa to each leaf of the crown [44]. A number of alternative rooting options are available (*see* **Note 5**). We use Lundberg

```
Heuristic search settings:
  Optimality criterion = parsimony
    Character-status summary:
      381 characters are excluded
      Of the remaining 215 included characters:
        All characters are of type 'ord' (Wagner)
        All characters have equal weight
        All characters are parsimony-informative
  Starting tree(s) obtained via stepwise addition
    Addition sequence: random
    Number of replicates = 100
    Starting seed = generated automatically
    Number of trees held at each step = 1
  Branch-swapping algorithm: tree-bisection-reconnection (TBR) with reconnection limit = 8
    Steepest descent option not in effect
    No more than 1 tree of score (length) greater than or equal to 2 will be saved in each
replicate
  Initial 'Maxtrees' setting = 100 (will be auto-increased by 100)
  Branches collapsed (creating polytomies) if maximum branch length is zero
  'MulTrees' option in effect
  No topological constraints in effect
  Trees are unrooted

Heuristic search completed
  Total number of rearrangements tried = 2
  Score of best tree(s) found = 924
  Number of trees retained = 1
  Time used = 0.00 sec (CPU time = 0.00 sec)

Tree-island profile:
                      First       Last                              First    Times
Island      Size       tree       tree              Score       replicate      hit
-----------------------------------------------------------------------------------
     1         1          1          1                924               1     100*

Note(s):
     * Islands of size 1 (= limit on trees saved per replicate) found in different replicates may
in fact belong to the same island
```

Fig. 9 Output with description of the heuristic search

rooting [47], which can be invoked with the "roottrees rootmethod=Lundberg" command. Use the "describetrees" command or its menu equivalent, **Trees>Describe Tree,** to display the trees and associated statistics. Figure 10 describes the single unrooted tree obtained following our heuristic search and trees rooted with Lundberg optimization using the "ancestor" and "standard" implementations, which were isomorphic. Goodness-of-fit statistics are provided for optimal trees. Note that the "ancestor" implementation of Lundberg uses the ancestor we previously defined in Step 2 of **Sect. 3.2.1**, and that rooting increases the length of the rooted trees from 924 steps to 1545 steps. Tree length increases in the rooted tree to account for the basal branch of the root leading to the ingroup taxa.

3.2.5 Assess Confidence Using Bootstrap Analysis

1. Perform a bootstrap analysis with the "bootstrap" command and the "search=faststep nreps=1000" subcommands, following the methodology of Felsenstein [48]. The method constructs a series of bootstrap replicates by sampling the data set with replacement, keeping always the size of the original dataset. While taxa are held constant, characters are sampled with replacement. Each replicate is then subjected to tree reconstruction, using either heuristic or branch-and-bound approach, depending on the number of taxa. In our example, we use a fast heuristic search. Finally, a majority-rule consensus

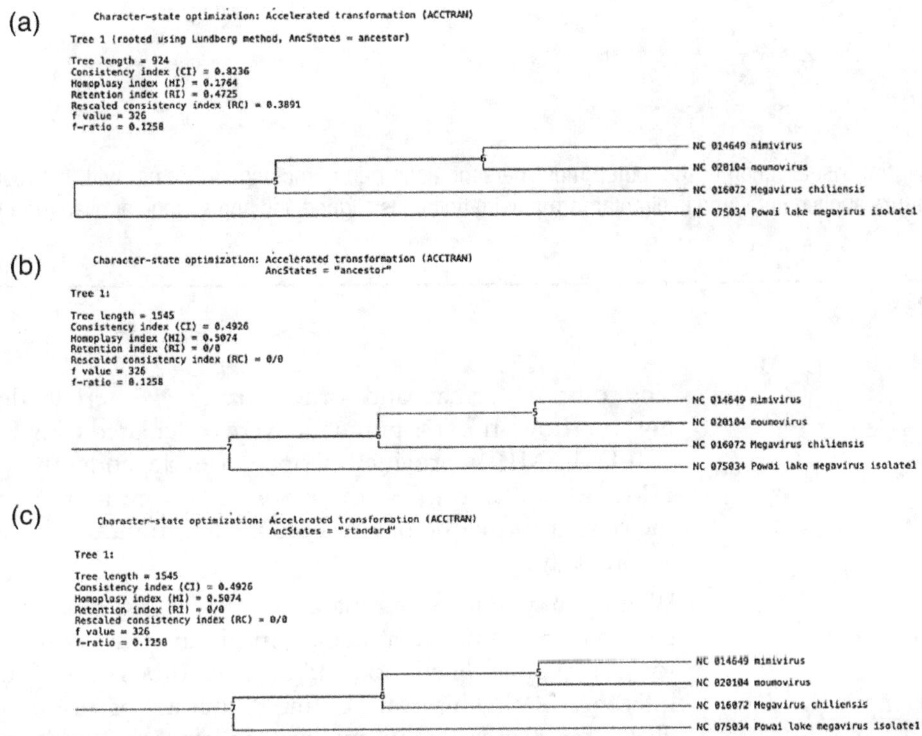

Fig. 10 The single unrooted most-parsimonious tree (**a**) obtained following exhaustive or heuristic searches was rooted using Lundberg with the "ancestor" (**b**) or the "standard" implementation (**c**). Goodness-of-fit statistics are provided for each reconstruction

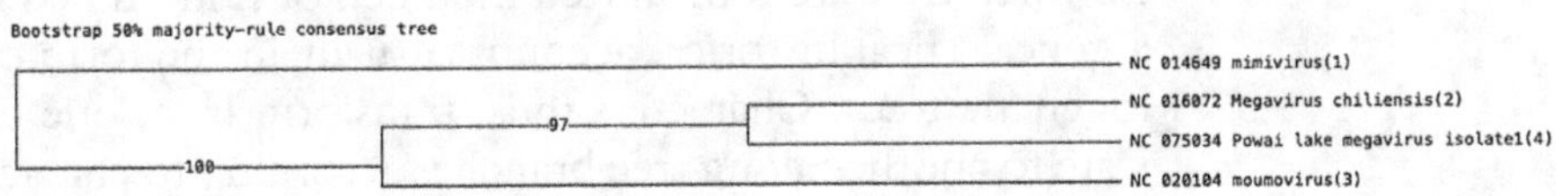

Fig. 11 Bootstrap 50% majority-rule consensus tree

is constructed that summarizes confidence in each split (node) of the tree as a percentage of trees containing the split. Figure 11 shows the consensus tree derived from the 1000 bootstrap replications.

3.2.6 Visualize and Describe the Optimal Tree(s)

1. Most parsimonious trees are saved to a *nexus* file (default), which can be the input to appropriate tree visualization software. Trees can also be saved in formats suitable for other programs, including FreqPars, Newick (for RAxML, PhyML, PAML, and MEGA), Hennig86, and Phylip. Figure 12 shows the rooted tree saved by PAUP* visualized using the program FigTree.

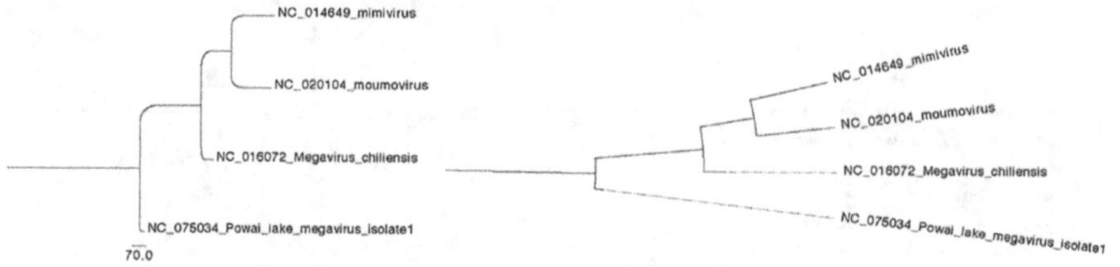

Fig. 12 Figtree visualization of the rooted most-parsimonious tree in rectangular format with full root length and curvature applied (left), and in circular format with tip labels aligned and angle range applied (right)

4 Notes

1. The output format and order may be different depending on how the assignments were calculated (e.g., using SUPERFAMILY provided scripts or independently via the HMMER software). We therefore encourage users to review the output format before proceeding and to make adjustments accordingly.
2. When using multi-state characters describing serial homologies, such as domain abundance, maximum parsimony outperforms maximum likelihood and other methods of phylogenetic inference [49]. This is, for example, the case of morphological characters that are commonly used in systematic biology.
3. Characters are the fundamental elements used to formalize hypotheses of homology for any phylogenetic analysis [45]. They are often attributes of a living entity that are capable of distinguishing it from another. Phylogenetic characters are useful when they are transmitted modified or unmodified through a genealogical historical account (usually in the form of a phylogenetic tree). Characters must transform from one character state to another along tree branches according to character state transformation models (transformation series). These models are usually defined with character state graphs (CSGs) that connect vertices representing character states with edges representing allowed transformations. One commonly used CSGs (of the "linear" type) defines "ordered" characters, which are also known as Wagner characters. These characters are widely used in the analysis of serial homologies. For example, the CSG of an ordered character with four character states is: 0–1–2–3. Transformation from character state 0 to character state 1 is less costly (one step) than transformation from character state 0 to character state 3 (three steps). Note that ordered characters are fully reversible, do not impose an evolutionary direction of change (they do not polarize change), and cannot be used to root phylogenetic trees during the search for optimal trees.

For example, a transformation from character state 0 to character state 1 costs the same as a transformation from character state 1 to character state 0. Another commonly used CSG defines "unordered" characters, which are also fully reversible. The CSG of an unordered character with four character states is a complete undirected graph in which every pair of vertices is connected by a unique edge, making a total of six possible character state transformations (each costing one step).

4. Heuristic hill-climbing methods offer an alternative to exact methods, as these cannot be used with large datasets. However, heuristic searches do not guarantee optimality in their exploration of the tree solution space. An initial tree is selected to start the exploration process of rearranging the tree in an attempt to increase its optimality (i.e., decrease its length in a maximum parsimony search). In the absence of further improvement, the process is stopped. The problem is that there are no guarantees that optimality has reached a global optimum and not a local one. PAUP* heuristics, however, are quite effective thanks to the management of how to find initial trees by *stepwise addition* and how to subject them to rearrangements by *branch swapping*. The former maximizes coverage of tree space, and the latter maximizes finding optimal solutions.

   Stepwise addition involves adding taxa to a growing tree until all of the taxa are connected. In each step of this process, a taxon is generally added to all possible branches of the growing trees, and only the tree that is most optimal is saved for the next step. However, there are many ways to conduct stepwise addition. The "*as is*" option adds taxa in the order presented in the data matrix. In the computationally more demanding "*closest*" option, every taxon must be connected to the tree at every possible branch, and all possible three-taxon tree statements are evaluated at the start. The "*simple*" option uses the "simple algorithm" of Farris [50] to add taxa in an order that increases an "advancement index" that measures a distance from a reference taxon (which can have all character states of an ancestor). Finally, the "*random*" option uses a pseudorandom number generator to permute taxa to be used in the addition sequence. By using different starting points for branch swapping in the search, the procedure maximizes the chances of finding global instead of local optima. The ultimate goal is to explore all possible "islands" of trees. Islands (also known as families) are the modules of the tree solution space; trees of the same island are more closely related than those from another island. Swapping trees from different starting trees increases the chances of "island hoping," especially if starting trees are trees of the random addition sequences.

Branch swapping offsets the greediness of the stepwise addition process. The procedure produces sets of predefined tree rearrangements that increase the chances of finding better trees. PAUP* uses three branch-swapping algorithms. The *nearest neighbor interchanges* (NNI) algorithm interchanges subtrees connected to an internal branch. The four possible subtrees are neighbors, and each subtree swap becomes an interchange. *The subtree pruning and regrafting* (SPR) algorithm "prunes" a subtree and reattaches it to a different location. The grafting procedure explores all possible removals and reattachments of subtrees. Finally, the *tree bisection and reconnection* (TBR) algorithm bisects a tree into two subtrees and then reconnects the subtrees along a different pair of branches. All possible bisections and reconnections are explored.

Stepwise addition and branch swapping must still be complemented with other strategies to avoid getting trapped in local optima during swapping. These include keeping equally parsimonious trees with the MulPars option to navigate plateaus that may exist in tree solution space, perform iterative branch-swapping rounds to find all trees in islands, and use the "steepest descent" option to avoid stopping branch-swapping when an optimal tree is found.

5. Rooting identifies and "pulls down" a branch that will hold the ancestor of the ingroup (i.e., the taxa of interest). A number of rooting options are available under the "roottrees" command, which converts trees in memory from unrooted to rooted representations. These include: Outgroup, Lundberg, or Midpoint, which are activated with the "method" subcommand. Because trees are generally reconstructed unrooted, one can reorient the unrooted tree for better interpretation without changing its length. However, if irreversible characters, asymmetric stepmatrix characters or a hypothetical ancestor are present during reconstruction, their presence results in a priori character polarization and the length of trees depends on the position of the root. The number of assumptions about polarity in these rooting approaches can be eliminated/minimized by either including outgroups in the analysis or attaching ancestors (outgroups) to optimal trees a posteriori [45]. Outgroups root the tree where the outgroup connects to the ingroup, but involve characters that separate the outgroup from the ingroup. Ancestors root an unrooted tree for ingroup taxa a posteriori but require that the character states of the ancestor be specified with the "ancstates" command. Lundberg attaches and ancestor (either "standard" or user-defined) using an optimization strategy (MinF) that is independent of the position of the root and requires that characters be either ordered or unordered.

Acknowledgments

Research was supported by grants from the National Science Foundation (MCB-0749836 and OISE-1132791) and the United States Department of Agriculture (ILLU-802-909 and ILLU-483-625) and supported by Blue Waters supercomputer allocations from the National Center for Supercomputing Applications (NCSA) to G.C.A.

Conflict of Interest A.N. is presently an employee of Moderna and may hold stock/stock options in the company. The work described in this chapter was conducted prior to A.N.'s employment at Moderna, and Moderna had no role in the study design, data collection, analysis, interpretation, or publication of this work.

References

1. Kapli P, Yang Z, Telford MJ (2020) Phylogenetic tree building in the genomic age. Nat Rev Genet 21:428–444
2. Da Cunha V, Gaia M, Gadelle D, Nasir A, Forterre P (2017) Lokiarchaea are close relatives of Euryarchaeota, not bridging the gap between prokaryotes and eukaryotes. PLoS Genet 13:e1006810
3. Gouy R, Baurain D, Philippe H (2015) Rooting the tree of life: the phylogenetic jury is still out. Philos Trans R Soc Lond B Biol Sci 370:20140329
4. Illergård K, Ardell DH, Elofsson A (2009) Structure is three to ten times more conserved than sequence—a study of structural response in protein cores. Proteins 77:499–508
5. Peck KM, Lauring AS (2018) Complexities of viral mutation rates. J Virol 92:e01031-17
6. Nasir A, Mughal F, Caetano-Anollés G (2021) The tree of life describes a tripartite cellular world. BioEssays 43:2000343
7. Holmes EC, Duchêne S (2019) Can sequence phylogenies safely infer the origin of the global virome?. MBio 10:e00289–19
8. Caetano-Anollés G, Wang M, Caetano-Anollés D, Mittenthal JE (2009) The origin, evolution and structure of the protein world. Biochem J 417:621–37
9. Caetano-Anollés G, Nasir A (2012) Benefits of using molecular structure and abundance in phylogenomic analysis. Front Genet 3:172
10. Nasir A, Kim KM, Caetano-Anolles G (2012) Giant viruses coexisted with the cellular ancestors and represent a distinct supergroup along with superkingdoms Archaea, Bacteria and Eukarya. BMC Evol Biol 12:156
11. Nasir A, Caetano-Anollés G (2015) A phylogenomic data-driven exploration of viral origins and evolution. Sci Adv 1:e1500527
12. Mughal F, Nasir A, Caetano-Anollés G (2020) The origin and evolution of viruses inferred from fold family structure. Arch Virol 165:2177–2191
13. Bokhari RH, Amirjan N, Jeong H, Kim KM, Caetano-Anollés G, Nasir A (2020) Bacterial origin and reductive evolution of the CPR Group. Genome Biol Evol 12:103–121
14. Wang M, Yafremava LS, Caetano-Anollés D, Mittenthal JE, Caetano-Anollés G (2007) Reductive evolution of architectural repertoires in proteomes and the birth of the tripartite world. Genome Res 17:1572–85
15. Wang M, Kurland CG, Caetano-Anollés G (2011) Reductive evolution of proteomes and protein structures. Proc Natl Acad Sci USA 108:11954–11958
16. Kim KM, Caetano-Anollés G (2011) The proteomic complexity and rise of the primordial ancestor of diversified life. BMC Evol Biol 11:140
17. Caetano-Anollés G, Caetano-Anollés D (2003) An evolutionarily structured universe of protein architecture. Genome Res 13(7):1563–1571
18. Bukhari A, Caetano-Anollés G (2013) Origin and evolution of protein fold designs inferred from phylogenomic analysis of CATH domain structures in proteomes. PLoS Comput Biol 9(3):e1003009
19. Harish A, Caetano-Anollés G (2012) Ribosomal history reveals origins of modern protein synthesis. PLoS One 7:e32776
20. Caetano-Anollés G, Kim HS, Mittenthal JE (2007) The origin of modern metabolic networks inferred from phylogenomic analysis of protein architecture. Proc Natl Acad Sci USA 104(22):9358–9363
21. Mughal F, Caetano-Anollés G (2019) MANET 3.0: hierarchy and modularity in evolving metabolic networks. PLoS One 14:e0224201

22. Caetano-Anollés G, Mughal F, Aziz MF, Caetano-Anollés K (2024) Tracing the birth and intrinsic disorder of loops and domains in protein evolution. Biophys Rev. https://doi.org/10.1007/s12551-024-01251-0
23. Wang M, Jiang Y-Y, Kim KM, Qu G, Ji H-F, Mittenthal JE, Zhang H-Y, Caetano-Anollés G (2011) A universal molecular clock of protein folds and its power in tracing the early history of aerobic metabolism and planet oxygenation. Mol Biol Evol 28:567–582
24. Caetano-Anollés G, Aziz MF, Mughal F, Caetano-Anollés D (2021) Tracing protein and proteome history with chronologies and networks: folding recapitulates evolution. Exp Rev Proteom 18(10):863–880
25. Burley SK, Berman HM, Bhikadiya C et al (2019) RCSB Protein Data Bank: Biological macromolecular structures enabling research and education in fundamental biology, biomedicine, biotechnology and energy. Nucleic Acids Res 47:D464–474
26. Gough J, Chothia C (2002) SUPERFAMILY: HMMs representing all proteins of known structure. SCOP sequence searches, alignments and genome assignments. Nucleic Acids Res 30:268–272
27. Orengo C, Michie A, Jones S, Jones D, Swindells M, Thornton J (1997) CATH—a hierarchic classification of protein domain structures. Structure 5:1093–1109
28. Murzin AG, Brenner SE, Hubbard T, Chothia C (1995) SCOP: a structural classification of proteins database for the investigation of sequences and structures. J Mol Biol 247:536–540
29. Andreeva A, Kulesha E, Gough J, Murzin AG (2020) The SCOP database in 2020: expanded classification of representative family and superfamily domains of known protein structures. Nucleic Acids Res 48:D376–382
30. Fox NK, Brenner SE, Chandonia JM (2014) SCOPe: structural classification of proteins–extended, integrating SCOP and ASTRAL data and classification of new structures. Nucleic Acids Res 42:D304–9
31. Swofford, DL, 2003. PAUP*: Phylogenetic analysis using parsimony (*and other methods), Version 4.0b10. Sinauer Associates, Sunderland, MA, USA
32. La Scola B, Audic S, Robert C et al (2003) A giant virus in amoebae. Science 299:2033
33. Philippe N, Legendre M, Doutre G et al (2013) Pandoraviruses: amoeba viruses with genomes up to 2.5 Mb reaching that of parasitic eukaryotes. Science 341:281–286
34. Nasir A, Romero-Severson E, Claverie JM (2020) Investigating the concept and origin of viruses. Trends Microbiol 28:959–967
35. Anaconda Individual Edition. Python programming language. https://www.anaconda.com/products/individual. Accessed 1 Jan 2025
36. Superfamily HMM library and genome assignments server. Version. 1.75. https://supfam.mrc-lmb.cam.ac.uk/SUPERFAMILY/downloads.html. Accessed 1 Jan 2025
37. Superfamily Database. https://supfam.org. Accessed 1 Jan 2025
38. SCOPe, downloads. https://scop.berkeley.edu/downloads/. Accessed 1 Jan 2025
39. PAUP* download page (2024) https://paup.phylosolutions.com. Accessed 1 Jan 2025
40. FigTree, version 1.4.4. http://tree.bio.ed.ac.uk/software/figtree/. Accessed 1 Jan 2025
41. iTol, version 7. https://itol.embl.de. Accessed 1 Jan 2025
42. Download materials at GitHub. https://github.com/gcalab/files/. Accessed 1 Jan 2025
43. Legendre M, Alempic J, Philippe N et al (2019) *Pandoravirus celtis* illustrates the microevolution processes at work in the giant *Pandoraviridae* genomes. Front Microbiol 10:430
44. SCOP Mapping File. https://scop.berkeley.edu/downloads/parse/dir.des.scope.2.08-stable.txt. Accessed 1 Jan 2025
45. Caetano-Anollés G, Nasir A, Kim KM, Caetano-Anollés D (2018) Rooting phylogenies and the tree of life while minimizing ad hoc and auxiliary assumptions. Evol Bioinf 14:1176934318805101
46. Maddison DR, Swofford DL, Maddison WP (1997) Nexus: an extensible file format for systematic information. Syst Biol 46:590–621
47. Lundberg JG (1972) Wagner networks and ancestors. Syst Zool 21:398–413
48. Felsenstein J (1985) Confidence limits on phylogenies: an approach using the bootstrap. Evolution 39:783–791
49. Goloboff PA, Torres A, Arias JS (2018) Weighted parsimony out-performs other methods of phylogenetic inference under models appropriate for morphology. Cladistics 34:407–437
50. Farris JS (1970) Methods for computing Wagner trees. Syst Zool 19:83–92

Check for updates

Chapter 14

A Pipeline for Generating Datasets of Three-Dimensional Tertiary Interaction Characters for Model-Based Structural Phylogenetics

Nicholas J. Matzke, Caroline Puente Lelievre and Matthew A. B. Baker

Abstract

Tertiary-interaction characters, termed "3Di" characters, were invented for deep structural homology search with the FoldSeek program. However, they may also be used as phylogenetic characters in model-based phylogenetic inference, just as in the Maximum Likelihood program IQ-TREE. However, to conduct such an analysis, every input amino acid (AA) sequence must be linked to a protein structure prediction file, which then must be converted to 3Di characters, and then both AA and 3Di characters need to be assembled, aligned, and joined in a partitioned data file for phylogenetic analysis. To make this process practical for medium to large datasets, we constructed a pipeline with R scripts to automate these steps. Here we present example scripts and explain the R functions used to construct a joint AA+3Di alignment file.

Keywords Structural phylogeny, Protein structure, Twilight zone, Evolution, FoldSeek

1 Introduction

Protein structure is typically more conserved than amino acid sequence [1]. This suggests that structural data might be helpful in testing phylogenetic hypotheses, especially for very deep nodes where amino acid sequence similarity has decayed into or beyond the "Twilight Zone" of 20–35% identity [2]. However, the complexity of three-dimensional protein structure has been difficult to efficiently integrate into standard probabilistic, model-based Maximum Likelihood or Bayesian phylogenetics approaches, which come with advantages such as statistical comparison of possible evolutionary models, and quantification of uncertainty. For example, several papers use measures of pairwise distance between protein structures to construct trees using neighbor-joining (NJ) [3–6]. Malik et al. [7] devised a method to characterize uncertainty by varying structures using Molecular Dynamics (MD) software, and then recalculating NJ trees

Gustavo Caetano-Anollés (ed.), *Evolutionary Genomics*, Methods in Molecular Biology, vol. 2981,
https://doi.org/10.1007/978-1-0716-4836-0_14,

from pairwise distances between sampled structural conformations (see also ref. [8], this volume). This approach is very computationally expensive. Another approach uses a secondary structure classifier, such as DSSP [9], to convert structures into a seven-character alphabet, and include these in phylogenetic analyses [10]. However, this approach may be limited by the high repetitiveness of adjacent secondary structure characters and a significant loss of information from the full 3-D structure, which may limit their use in deep homology search and phylogenetics.

We proposed a new approach that seeks to overcome these challenges to structural phylogenetics [11] by repurposing the deep homology search program FoldSeek [12]. FoldSeek uses a 20-letter alphabet of structural characters to conduct rapid homology searches on the basis of protein structure. In FoldSeek, every amino acid position in a structure is assigned one of 20 possible tertiary-interaction characters (termed "3Di" for three-dimensional), based on the closest tertiary interaction with another amino acid in the structure backbone, and the arrangement of the backbones around the pair of amino acids. Using a large database of structures, the FoldSeek authors used machine learning to optimize their character classifier for the retrieval of remote homologs. The result is that any structure file can be converted into a sequence of 20-state characters with a length identical to the original amino acid sequence. The fact that 3Di characters have 20 states means that they can be easily input into existing algorithms originally designed for amino acids (protein BLAST, alignment, phylogeny, etc.).

While originally designed for deep homology search, 3Di characters can serve as potential phylogenetic characters. The program FoldTree used 3Di characters to construct NJ trees [13]. Our proposal [11] included 3Di characters as a data partition in a Maximum-Likelihood or Bayesian analysis. This enables standard models to be used for the amino acid (AA) data partition, and 3Di-specific models to be used for the structural data partition. All of the advantages of model-based methods then accrue. For example, while Malik et al.'s (2020) analysis of the Ferritin-like Superfamily took weeks of cluster time to run the MD simulations necessary for structural bootstrapping, running the same proteins in IQ-TREE [14] with a partitioned AA+3Di dataset was performed done on a desktop machine in under an hour, complete with statistical model selection on both partitions and 1000 ultrafast bootstrap replicates to approximate Bayesian posterior probabilities.

Puente-Lelievre et al. [11] compared 60 variant analyses of the ferritin dataset, including different alignment methods (US-align for multiple alignment using the three-dimensional structures, versus famsa3di for alignment using 3Di characters), 3Di alone versus AA alone versus AA+3Di, and structures derived from the PDB [15] versus AlphaFold 2 [16, 17]. Results showed that IQ-TREE consistently selected the 3DI substitution model for 3Di data partitions, that 3Di data recovered deeper nodes and AAs recovered shallower nodes, and that AA+3Di partitioned datasets were the closest match to the Malik et al. [7] reference tree. Alignment methods and source for structures were approximately equivalent on this dataset. We concluded that these results were proof of concept that the method was likely to improve phylogenetic accuracy and resolution on datasets where structure is conserved, but AA sequence similarity is highly degraded.

However, we also called for future work to further test the method. Simulation tests are nontrivial due to the difficulty of realistically simulating protein structure evolution at gigayear timescales, and empirical tests of the method are challenging due to the lack of known true phylogenies for ancient phylogenies covering the Twilight Zone. However, as a further test, we recently applied the method to a dataset of 56 mitochondrial orthologs (the MitoCOGs dataset of [18]) that were reconstructed to exist in the last mitochondrial common ancestor. Some of these proteins have highly-conserved sequences, some less so, and as orthologs with complete or substantial mitochondrial inheritance over their history, they are perhaps a "best case" for a group of ancient proteins likely to share the same phylogenetic history. Using the Kannan et al. phylogeny of all 56 concatenated proteins as a reference tree, results showed that the single-protein phylogenies were improved in their matching to the reference tree, with AA+3Di phylogenies having a smaller Robinson-Foulds distance to the reference tree than AA-only phylogenies. This improvement was stronger for smaller and less-conserved proteins than for highly-conserved proteins, which already closely matched the all-proteins AA phylogeny.

Further exploration of the utility of augmenting phylogenetic analyses with 3Di characters has been performed by Garg & Hochberg [19], who showed that 3Di characters confirmed traditional findings wherein ancient gene duplications root the tree of cellular life between Bacteria and Archaea+Eukaryotes, but with much shorter branchlengths which should be much less susceptible to artefacts caused by long branches and model misspecification. This paper also estimated a 3Di substitution matrix, which outperformed the matrix derived from FoldSeek's alignment cost matrix for 3DI data used in Puente-Lelievre et al. Interest in augmenting AA phylogenetic inference with 3Di structural characters appears to be growing, for example, used in a recent structural phylogeny of Flaviviridae glycoproteins [20]

Our first exploration of structural phylogenetics was constructed by hand, obtaining 3Di characters from empirical PDB structures and AlphaFold-derived structural predictions. However, the major potential of the method comes from the ubiquity of AlphaFold predictions, such that AlphaFold predictions and 3Di characters can be easily added to any AA dataset. We have created a protocol to automate the procedure with a series of R functions, in order to more easily allow deployment and testing of structure-enhanced protein phylogenetics with tertiary interaction characters.

In outline, this protocol proceeds as follows:

1. Sample amino acid sequences for the problem of interest.
2. Obtain AlphaFold structural predictions for each sequence.
3. Convert each structure to a 3Di sequence.
4. Align the 3Di and AA characters via one of several methods.
5. Conduct a data-partitioned phylogenetic analysis in IQ-TREE.

2 Materials

2.1 Computer

2.1.1 Hardware

1. These scripts were constructed and tested on Mac OS X, version Sequoia 15.4, running on a 3.2 GHz 6-Core Intel Core i7 desktop machine. However, the scripts should be able to be run on Linux or a PC, with minor modifications discussed below.

2.1.2 Access to High-Performance Computing (HPC) Is Not Necessary

1. The slowest step is phylogeny inference with IQ-TREE, but for our ferritin dataset, runs took less than an hour (single protein, 53 sequences). For larger IQ-TREE runs (hundreds of sequences), running IQ-TREE jobs on a cluster may be more practical.

2.2 Software and Custom Scripts

The workflow outlined here requires the use of several freely available software packages, plus custom scripts described here, available at [21].

2.2.1 Software

1. **ChimeraX.** The largest required installation is ChimeraX, a free, cross-platform package for visualization and analysis of protein structure, produced by the Resource for Biocomputing, Visualization, and Informatics (RBVI) at the University of California, San Francisco [22]. It can be downloaded from [23]. The installation is approximately 1 GB (*see* **Note 1**).
2. **FoldSeek.** FoldSeek [12] is a command-line program that must be installed to use the included *structureto3didescriptor* function, which converts structure files to 3Di characters. FoldSeek is available at [24].
3. **Famsa3di.** Famsa3di is the FAMSA program for AA multiple sequence alignment [25] recompiled with FoldSeek's 3DI cost matrix, replacing FAMSA's default MIQS cost matrix for amino acids. Famsa's source code is available from the authors [25]. Example modified source code with the 3DI cost matrix, as well as compiled binaries for two versions of Mac OSX, are available at [21].

4. **Additional Requirements:** The R statistical computing environment [26], with the following packages installed.

   1. **Packages available from CRAN, the Comprehensive R Archive Network.** These are installed in R with the standard command, e.g., *install.packages("ape")*.

      - ape
      - seqinr
      - rvest
      - phytools
      - queryup
      - rentrez
      - rexpokit (only as a necessary dependency of BioGeoBEARS)
      - cladoRcpp (only as a necessary dependency of BioGeoBEARS)
      - BiocManager (needed to install Bioconductor packages)
      - devtools (needed to install R packages from GitHub repositories)

   2. **Available in the Bioconductor Collection of Packages.** These are installed with, e.g., *BiocManager::install("msa")*

      - msa
      - ginmappeR

   3. **Available on GitHub.** These are installed with, e.g., *devtools::install_github(repo= "nmatzke/BioGeoBEARS")*

      - BioGeoBEARS (this package, while designed for phylogenetic biogeography [27–29], contains a number of functions useful for manipulating phylogenies, data files, and tables in R)

2.2.2 Custom Scripts (available at [21])

1. *01_get_seqs/01_get_raw_seqs_from_IDs_v1.R*: Takes a user-specified list of GenBank protein sequence IDs, and downloads the amino acid sequences to a FASTA file. Users are not required to obtain their sequences this way. Any unaligned FASTA file may be used for the ChimeraX/AlphaFold step. The example given is for 34 sequence IDs for homologs of FliG, a protein of the bacterial flagellum, located in the inner membrane and playing a role in switching between clockwise and counter-clockwise rotation during chemotaxis. Some of the sequences are from MgtE, a magnesium transport protein in the inner membrane, which is a very distant homolog of part of FliG [30]. The script outputs a FASTA file of the 34 FliG/MgtE sequences, which is used in the next script.

2. *02_get_alphafolds/02_ChimeraX_FliG_MgtE.R*: Takes a user-specified FASTA file of unaligned amino acid sequences, and for each sequence, executes ChimeraX's "alphafold match" function via R's *system* command. This will identify and download the closest match to the query sequence in the AlphaFold database [31], which contains AlphaFold predictions of all UniProt sequences. The script parses the result of the search, including the percentage identity of the match, and stores the .cif structure file in the working directory. Once structure downloads are finished, the R script executes (via the *system* command) the FoldSeek function *structureto3didescriptor* to convert each structure to a 3Di sequence. Further commands in the script assemble the AA and 3Di sequences into FASTA files. The script also contains Terminal commands (run outside of R) to illustrate how to run famsa3di to align a 3Di FASTA file, and then R functions to transfer the alignment positions to the corresponding AA sequences (*see* **Note 2**).

3 Methods

3.1 Problem Selection

Structural characters are most likely to enhance phylogenetic accuracy and resolution for parts of the tree where amino acids have diverged into or past the "twilight zone." For datasets where amino acid sequence identity is already high (>50%), adding 3Di characters is unlikely to add much signal. For further advice on selecting sequences for structural phylogenetics, we recommend the advice of Langer et al. (2025, this volume).

3.2 Structure Acquisition and Processing

As illustrated in the example script *02_get_alphafolds/02_ChimeraX_FliG_MgtE.R*, R functions are deployed to automate the process of running ChimeraX's *AlphaFold fetch* function. This function may also be run in ChimeraX by manually clicking menu items Tools → Structure Prediction → AlphaFold, then pasting in an amino acid sequence and clicking Fetch. However, this becomes a tedious exercise for more than a few sequences.

The custom R functions used in the example script will be described in the order they are used, as well as other command-line programs and the decisions involved in using them.

1. *fasta_to_alphafold_cifs*: This function reads an amino acid sequence from a FASTA file, and hands it to another R function, *aa_to_alphafold_cif*, which constructs an R system command to run a command-line version of ChimeraX. An example of such a system command would be: 'ChimeraX --cmd "alphafold

match MNKNKLLSFLREASHPAELQSVAAQIKSIHP ADLA DVLNELNSNTILSLLKLLDPHTSSTIFIYLDE DRQEILLQEMPRELIPELFKCMPSDERADLY NRMEQNVKHKLLPILAPHERDDILRMSSFPDSSTGSE ATSDFVSVSPDMTVTQALEHIRMTATDKETVYIIYVI NDRSHLCGTVSLRELVL ARDHQKISDIMRPDPIFARAD CPKEKAAELIRHYDLLALPVIDGADKMIGIVTIDDA MDIEKEQDATQLARFGGTAVGDGPDL DILSSPLSRIFQ VRVF WLALLTIFGVITSTFVASQEEVLSEVIILAAFLAP VIDMGGNTGSQSATLVI RGMALGDLTVKWKYVWLVIR RELPVV IALGITISLLELILAWFSKGIGIDVMLV VGLSM LVCTVLGGVIGVVLPFIARRIGT DPATLSSPLITSIMD LLGVFIYFGFAWFFFGDLL; log save WP_160516493.1.html executableLinks true; save WP_160516493.1.cif relModel #1; close #1; quit;". The user could construct this command by hand and run it, outside of R, at the command-line terminal. However, the main utility is the ability to automate the command and iterate over a collection of amino acid sequences in a FASTA file.

The function is robust to "no match" results, as well as timeouts (no result returned); these will be recorded as "NONE" in the results table, with 0% identity matches.

The script returns several structures per minute, which is substantially faster than the AlphaFold web interface [31] sequence search function, where BLAST searches may take 5–10 min.

2. *alphafold_html_to_df*: This R function reads the HTML file returned by ChimeraX's fetch function, which includes the percent identical match between the query sequence and the closest match AlphaFold structure, as well as the coverage percentage. Typically, in our experience, 90% plus of query sequences are 100% matches for both, but sometimes the closest match is nonidentical. The result for each query sequence is stored in an R dataframe (also saved to a text file), so that users may review the hits and decide if they wish to exclude any sequences with weak hits in the AlphaFold database, or alternatively generate AlphaFold predictions from scratch using Colabfold [32].
3. *str_to_3di*: This R function runs, via *system* command, the FoldSeek function structureto3didescriptor, which creates a .dump file containing the input structure file's amino acid sequence, the corresponding 3Di sequence, as well as 3D coordinate information, which we discard. str_to_3di also runs a series of *awk* commands, which separate the .dump file information into separate files for further processing. The function returns a list of these filenames.

4. *get_AA_FASTA_from_foldseek_dumps*: This function assembles the AA (or 3Di) sequences from the individual .dump files to assemble a standard unaligned AA (or 3Di) FASTA file.
5. Once the unaligned FASTA files have been generated, users are free to align the AAs and/or 3Di sequences using a variety of command-line tools (run outside of R) developed by other authors. There are three major options. As each is supported by substantial documentation, and substantial researcher inspection and judgment is required to decide what constitutes the most adequate alignment for a given dataset (*see* **Note 3**), we will not describe their use in detail here. Instead, we will highlight the type of input and our experience of their effectiveness.
   5.1. *US-align*—This is a pioneering multiple-structure alignment program, built upon on earlier pairwise structural aligners [33]. The input is a list of structure files, and an AA multiple sequence alignment is returned. In this case, the user would have to transfer the AA alignment to the 3Di characters. Our experience with US-align is that it was effective for a dataset where all proteins shared the same highly conserved folds in the same order (for example, in the ferritin-like superfamily dataset), but resulted in implausible alignments for more complex datasets where proteins shared a fold but had substantially different sizes and presence/absence of other folds.
   5.2. *FoldMason*—This program, developed by the FoldSeek authors, inputs a list of structure files, and returns two multiple sequence alignment files are returned, one for AAs and one for 3Di characters [34]. No transferring of alignments between AA and 3Di sequences is required, as they are jointly aligned by FoldMason. The program thus has high ease of use and can achieve some plausible alignments when some proteins have additional or missing folds. However, users must carefully visually inspect alignments and perhaps correct them manually, as we have observed apparent misalignment of blocks.
   5.3. *Famsa3di*—This program, described above, rapidly aligns the 3Di sequences. The user must then transfer the alignment to the unaligned AA sequences via the R function. The alignment is therefore based entirely on the 3Di sequences, an approach which seems to work best when AAs are very highly diverged, and/or when only portions of input proteins share the same fold. One issue that can arise is if the 3Di characters are highly repetitive (sometimes seen in the 3Di translations of long alpha-helices that extend away from the compact folds). Famsa3di will "align" these highly uniform regions, but the corresponding AAs can show clear evidence of misalignment. One

solution is to use manual editing or a MAFFT [35] script to realign the amino acids within the regions where the 3Di characters are uniform, then transfer this AA alignment structure back to the 3Di characters.

6. *align_3di_to_aa*: This function will take as input an aligned 3Di FASTA file, and a FASTA file containing the corresponding unaligned AA sequences, and substitute each AA into each corresponding 3Di position. This results in an aligned AA file with an identical structure to the aligned 3Di file. This procedure only works if the original 3Di and AA sequences have identical lengths and are stored in the same order (which is the result produced by the scripts above). The function works the same if 3Di sequences are aligned to match a preexisting AA alignment (*see* **Note 4**).
7. *hcat_fastas*: To conduct a partitioned phylogenetic analysis in IQ-TREE, the different data partitions must be in the same rows of the data matrix. This means that the user must combine an AA alignment FASTA file with a 3Di alignment FASTA file, by concatenating the AA and 3Di sequence corresponding to each original protein structure. While this can be done by hand, the process is tedious and error-prone. The R function *hcat_fastas* takes the two alignment files as inputs, and horizontally concatenates (hcat) each pair of sequences that shares identical names.
8. *hcat_fastas_grepl*: This function operates as above, but uses the R function *grepl* to identify which sequence in the second file contains the name of the sequence in the first file. This can be helpful if the first file contains just sequence IDs, but the second file has longer names.

Once the user has generated unaligned and aligned versions of FASTA files for AAs, 3Di characters, and an AA+3Di dataset containing both alignments in one file, these files may be used as inputs for phylogenetic analysis in IQ-TREE [14] or any other phylogenetics program. IQ-TREE has extensive documentation available, so we will not revisit its use here, but we give sample commands in the example script. We typically run a standard IQ-TREE analysis, with automated model selection (including the 3Di substitution matrix as an option), maximum likelihood tree search, and 1000 ultrafast bootstraps (which aim to approximate Bayes factors).

4 Notes

1. For these R functions to successfully run programs like ChimeraX and FoldSeek, those programs must be executable from the system command line from any location. This can

be tested by manually running the programs at the command prompt. In Mac OSX, the simplest way of making the programs universally accessible is to copy to */usr/local/bin/* the executable file (for *foldseek*) or to include a symbolic link to the ChimeraX executable (installed in */Applications/ChimeraX.app*; right-click and click on Show Package Contents, and navigate to */Applications/ChimeraX.app/bin*, to see the executable). Other operating systems may require slightly different solutions. Users may also have to turn on execute permissions, for example, with the *chmod* command.

2. Of the 34 FliG/MgtE proteins in the example dataset, ChimeraX returns AlphaFold structures with 100% identical matches at 100% coverage for all, but the first protein WP_160516493.1, an *E. coli* protein annotated as a magnesium transporter; its closest match in AlphaFold is only 65.9% identical, with 98.9% coverage. This can be seen in the screen output, the dataframe of results, or the saved text file of results. A user might decide to cut this protein from the analysis, or manually generate an AlphaFold structure prediction using ColabFold [32]. If using a manually-generated prediction, the resulting .cif file should be saved as WP_160516493.1.cif and used to replace the .cif file generated by the R script.

3. Although initial results are promising, we emphasize that structural phylogenetics is a young field, and users should exercise critical thinking throughout their analysis. In particular, extra care must be applied to multiple sequence alignment. We encourage running different alignment programs, and carefully noting differences in alignment results, as this can help users avoid the assumption that alignment programs will operate perfectly across such deep phylogenetic divergences.

4. Because FoldSeek produces a 3Di character for each residue position in a protein structure, it is appealing to assume that the AA alignment and 3Di alignment for a given set of structures should be identical in terms of which residue positions are aligned. After all, if the amino acid at residue position #3 of *E. coli* FliG is deemed homologous to residue position #5 of *Bacillus subtilis* FliG, then it stands to reason that the structural 3Di characters assigned to those two positions should also be deemed homologous (here, *homology* does not mean "identical," it means that those positions are hypothesized to trace back to the same position in a common ancestral protein). However, it should be recognized that this represents a researcher's decision, like every other stage of data selection, multiple sequence alignment, and phylogenetic inference. In circumstances that are probably somewhat rare but definitely imaginable, certain kinds of mutation events could de-link homology at the AA and 3Di levels. For example, imagine if

Species A and B shared a homologous protein, ProtX, where the N-terminus of a protein consisted of a 20-AA alpha helix. Then, in Species B, a 10-aa insertion occurs at position 11 of this protein, and the ten new amino acids happen to adopt an alpha helical structure due to the local 3D context, while the original amino acids 1–10 become an unordered chain while retaining identical sequence. It would then arguably be legitimate to homologize A's positions 1–20 to B's 1–10, 21–30 for amino acids, but to B's positions 11–30 for 3Dis. In such situations, running separate alignment programs for AAs and 3Dis might also be defensible. However, as the premise of structural phylogenetics is that structure is typically highly conserved, such situations are likely rare or evolutionarily ephemeral at the massive timescales typically being considered.

Acknowledgments

N.J.M., C.P.L., and M.A.A.B. were supported by HFSP grant RGY0072/21. M.A.A.B. and N.J.M. were supported by ARC DP240100462, and N.J.M. was additionally supported by New Zealand Royal Society RDF 21-UOA-040, Marsden Grant 18-UOA-034, and University of Auckland DRDF Research Fund #3727963. C.P.L. and N.J.M. were supported by the University of Auckland, Faculty of Science Research Development Fund, FoS RDF #3732317.

References

1. Illergård K, Ardell DH, Elofsson A (2009) Structure is three to ten times more conserved than sequence - a study of structural response in protein cores. Proteins 3:499–508. https://doi.org/10.1002/prot.22458
2. Rost B (1999) Twilight zone of protein sequence alignments. Protein Eng 12(2):85–94. https://doi.org/10.1093/protein/12.2.85
3. Breitling R, Laubner D, Adamski J (2001) Structure-based phylogenetic analysis of short-chain alcohol dehydrogenases and reclassification of the 17-hydroxysteroid dehydrogenase family. Mol Biol Evol 18:2154–2161. https://doi.org/10.1093/oxfordjournals.molbev.a003761
4. Bujnicki JM (2000) Phylogeny of the restriction endonuclease-like superfamily inferred from comparison of protein structures. J Mol Evol 50:39–44. https://doi.org/10.1007/s002399910005
5. Garau G, Di Guilmi AM, Hall BG (2005) Structure-based phylogeny of the metallo-beta-lactamases. Antimicrob Agents Chemother 49(7):2778–2784. https://doi.org/10.1128/AAC.49.7.2778-2784.2005
6. Lundin D, Poole AM, Sjöberg BM, Högbom M (2012) Use of structural phylogenetic networks for classification of the ferritin-like superfamily. J Biol Chem 287(24):20565–20575. https://doi.org/10.1074/jbc.M112.367458
7. Malik AJ, Poole AM, Allison JR (2020) Structural phylogenetics with confidence. Mol Biol Evol 37(9):2711–2726. https://doi.org/10.1093/molbev/msaa100
8. Langer DB, Malik AJ, Klemm P, Allison JR, Poole AM (2025) MD-Phylogeny: constructing statistically supported phylogenetic trees from protein structures using molecular dynamics. In: Caetano-Anollés G (ed) Evolutionary Genomics: Methods and Protocols (Methods in Molecular Biology)
9. Kabsch W, Sander C (1983) Dictionary of protein secondary structure: pattern recognition of hydrogen-bonded and geometrical features.

Biopolymers 22(12):2577–2637. https://doi.org/10.1002/bip.360221211
10. Lai J-S, Rost B, Kobe B, Bodén M (2020) Evolutionary model of protein secondary structure capable of revealing new biological relationships. Proteins 88:1251–1259. https://doi.org/10.1002/prot.25898
11. Puente-Lelievre C, Malik AJ, Douglas J, Ascher D, Baker M, Allison J, Poole A, Lundin D, Fullmer M, Bouckert R, Kim H, Steinegger M, Matzke N (2024) Tertiary-interaction characters enable fast, model-based structural phylogenetics beyond the twilight zone. Preprint at https://doi.org/10.1101/2023.12.12.571181
12. van Kempen MKS, Tumescheit C, Mirdita M, Lee J, Gilchrist CLM, Söding J, Steinegger M (2024) Fast and accurate protein structure search with FoldSeek. Nat Biotechnol 42:243–246. https://doi.org/10.1038/s41587-023-01773-0
13. Moi D, Bernard C, Steinegger M, Nevers Y, Langleib M, Dessimoz C (2023) Structural phylogenetics unravels the evolutionary diversification of communication systems in gram-positive bacteria and their viruses. Preprint at https://doi.org/10.1101/2023.09.19.558401
14. Minh BQ, Lanfear R, Ly-Trong N, Trifinopoulos J, Schrempf D, Schmidt HA (2022) IQ-TREE version 2.2.0: tutorials and manual. Phylogenomic software by maximum likelihood. User Manual. http://www.iqtree.org/doc/iqtree-doc.pdf.
15. Burley SK, Berman HM, Kleywegt GJ, Markley JL, Nakamura H, Velankar S (2017) Protein Data Bank (PDB): the single global macromolecular structure archive. In: Wlodawer A, Dauter Z, Jaskolski M (eds) Protein crystallography: methods and protocols. Springer New York, New York, NY, pp 627–641
16. Jumper J, Evans R, Pritzel A, Green T, Figurnov M, Ronneberger O, Tunyasuvunakool K, Bates R, Žídek A, Potapenko A, Bridgland A, Meyer C, Kohl SAA, Ballard AJ, Cowie A, Romera-Paredes B, Nikolov S, Jain R, Adler J, Back T, Petersen S, Reiman D, Clancy E, Zielinski M, Steinegger M, Pacholska M, Berghammer T, Bodenstein S, Silver D, Vinyals O, Senior AW, Kavukcuoglu K, Kohli P, Hassabis D (2021) Highly accurate protein structure prediction with AlphaFold. Nature 596(7873):583–589. https://doi.org/10.1038/s41586-021-03819-2
17. Varadi M, Bertoni D, Magana P, Paramval U, Pidruchna I, Radhakrishnan M, Tsenkov M, Nair S, Mirdita M, Yeo J, Kovalevskiy O, Tunyasuvunakool K, Laydon A, Žídek A, Tomlinson H, Hariharan D, Abrahamson J, Green T, Jumper J, Birney E, Steinegger M, Hassabis D, Velankar S (2024) AlphaFold Protein Structure Database in 2024: providing structure coverage for over 214 million protein sequences. Nucleic Acids Res 52(D1):D368–D375. https://doi.org/10.1093/nar/gkad1011
18. Kannan S, Rogozin IB, Koonin EV (2014) MitoCOGs: clusters of orthologous genes from mitochondria and implications for the evolution of eukaryotes. BMC Evolut Biol 14(1):237. https://doi.org/10.1186/s12862-014-0237-5
19. Garg SG, Hochberg GKA (2024) A general substitution matrix for structural phylogenetics. Mol Bio Evo 42(6):msaf124. https://doi.org/10.1093/molbev/msaf124
20. Mifsud JCO, Lytras S, Oliver MR, Toon K, Costa VA, Holmes EC, Grove J (2024) Mapping glycoprotein structure reveals Flaviviridae evolutionary history. Nature 633(8030):695–703. https://doi.org/10.1038/s41586-024-07899-8
21. Matzke NJ (2025) AA to 3Di pipeline, version 1.0. https://github.com/nmatzke/AA_to_3Di. Accessed 29 Apr 2025
22. Meng EC, Goddard TD, Pettersen EF, Couch GS, Pearson ZJ, Morris JH, Ferrin TE (2023) UCSF ChimeraX: tools for structure building and analysis. Protein Sci 32(11):e4792. https://doi.org/10.1002/pro.4792
23. Ferrin T, Pettersen E, Goddard T, Pearson Z, Couch G, Meng E, Morris S, Huang C (2025) ChimeraX, version 1.9. https://www.cgl.ucsf.edu/chimerax/download.html. Accessed 3 Apr 2025
24. van Kempen M, Kim S, Tumescheit C, Mirdita M, Lee J, Gilchrist CLM, Söding J, Steinegger M (2024) FoldSeek release 10.0. https://github.com/steineggerlab/foldseek. Accessed 31 Jan 2025
25. Deorowicz S, Debudaj-Grabysz A, Gudys A (2016) FAMSA: fast and accurate multiple sequence alignment of huge protein families. Sci Rep 6:33964. https://doi.org/10.1038/srep33964
26. Ihaka R, Gentleman R (1996) R: a language for data analysis and graphics. J Comput Graph Stat 5(3):299–314. https://doi.org/10.1080/10618600.1996.10474713
27. Matzke NJ (2013) Probabilistic historical biogeography: new models for founder-event speciation, imperfect detection, and fossils allow improved accuracy and model-testing. Front Biogeogr 5(4):242–248. https://doi.org/10.21425/F5FBG19694
28. Matzke NJ (2014) Model selection in historical biogeography reveals that founder-event speciation is a crucial process in island clades. Syst Biol 63(6):951–970. https://doi.org/10.1093/sysbio/syu056
29. Matzke NJ (2022) Statistical comparison of DEC and DEC+J is identical to comparison of two ClaSSE submodels, and is therefore valid.

J Biogeogr 49(10):1805–1824. https://doi.org/10.1111/jbi.14346

30. Pallen MJ, Matzke NJ (2006) From *The Origin of Species* to the origin of bacterial flagella. Nat Rev Microbiol 4(10):784–790. https://doi.org/10.1038/nrmicro1493
31. Varadi M, Anyango S, Deshpande M, Nair S, Natassia C, Yordanova G, Yuan D, Stroe O, Wood G, Laydon A, Žídek A, Green T, Tunyasuvunakool K, Petersen S, Jumper J, Clancy E, Green R, Vora A, Lutfi M, Figurnov M, Cowie A, Hobbs N, Kohli P, Kleywegt G, Birney E, Hassabis D, Velankar S (2021) AlphaFold Protein Structure Database: massively expanding the structural coverage of protein-sequence space with high-accuracy models. Nucleic Acids Research 50(D1): D439–D444. https://doi.org/10.1093/nar/gkab1061
32. Mirdita M, Schütze K, Moriwaki Y, Heo L, Ovchinnikov S, Steinegger M (2022) ColabFold: making protein folding accessible to all. Nat Methods 19(6):679–682. https://doi.org/10.1038/s41592-022-01488-1
33. Zhang C, Shine M, Pyle AM, Zhang Y (2022) US-align: universal structure alignments of proteins. Nat Methods 19(9):1109–1115. https://doi.org/10.1038/s41592-022-01585-1
34. Gilchrist CLM, Mirdita M, Steinegger M (2024) Multiple Protein Structure Alignment at Scale with FoldMason. Preprint at https://doi.org/10.1101/2024.08.01.606130
35. Katoh K, Standley DM (2013) MAFFT multiple sequence alignment software version 7: improvements in performance and usability. Mol Biol Evol 30(4):772–780. https://doi.org/10.1093/molbev/mst010

Chapter 15

MD-Phylogeny: Constructing Statistically Supported Phylogenetic Trees from Protein Structures Using Molecular Dynamics

Désirée B. Langer, Ashar J. Malik, Paul Klemm, Jane R. Allison and Anthony M. Poole

Abstract

Phylogenetic tree construction from homologous sequences is a fundamental approach for studying evolutionary relationships. However, in cases where sequence similarity is too low to generate reliable alignments, comparison of protein structure provides an alternative means for inferring deep evolutionary relationships. MD-phylogeny is a method that integrates structural comparison with molecular dynamics (MD) simulations to generate robust phylogenies for protein datasets within the twilight zone of sequence similarity. This approach uses well-established structural superposition methods for building trees from structure, with a method for generating structural variants using MD simulations, from which confidence scores analogous to bootstrapping can be derived.

This protocol provides a step-by-step guide to performing MD-phylogeny, covering dataset selection, structural comparisons, tree construction, and the generation of statistical support through MD simulations.

Keywords Molecular dynamics, Structural phylogeny, Protein structure, Twilight zone, Evolution

1 Introduction

Constructing phylogenetic trees from homologous sequences enables tracking of evolutionary relationships between species as well as charting the evolutionary history of genes and even protein families. Building trees from sequence data requires reliable alignments, either of DNA or proteins. However, there are cases where sequence similarity between protein sequences is difficult to detect, yet homology is undisputed [1–3]. Protein structures contain signatures of deep evolutionary history, and similarities between structures can be probed to detect deep evolutionary relationships in cases where sequence-based alignments do not suffice [3–8]. These deep relationships are captured in a number of well-known databases, including SCOP

Gustavo Caetano-Anollés (ed.), *Evolutionary Genomics*, Methods in Molecular Biology, vol. 2981,
https://doi.org/10.1007/978-1-0716-4836-0_15,

(Structural Classification of Proteins) [9] and CATH (Class, Architecture, Topology, Homologous superfamily) [10, 11], as well as homology search tools such as FoldSeek [12].

With the growing number of both experimentally-determined [13, 14] and computationally-predicted [15, 16] structures, there is growing interest in using structural data for phylogenetic analysis. Structure can be used to augment phylogenetic analysis in a number of ways. For example, it has been used to improve the accuracy of sequence alignments [17] or to improve confidence in sequence-based phylogenies [18], and it is well established that pairwise comparisons of protein structure can be used to build trees primarily using structural data [4–8, 19, 20]. This approach has been made more accessible through the development of Structome [20] (recently renamed Structome-Q [21]), a web server that combines rapid collection of structures with structural phylogeny. Moreover, it is now possible to generate confidence values analogous to the well-known bootstrap using molecular dynamics (MD) simulations to resample protein shape [19].

In this article, we provide a step-by-step guide for performing the most established of the new methods for structure-based phylogeny: MD-phylogeny [19]. MD-phylogeny combines the well-established methods for building trees from direct comparisons of structures [4, 6] with MD simulations [22], enabling confidence measures to be generated that are analogous to a bootstrap in standard phylogenetics. This method has been used to successfully reconstruct the phylogenetic history of the Ferritin superfamily [19], where sequence similarity is too low for reliable sequence alignments to be generated. One barrier to applying this method is that it requires expertise in both phylogenetics and MD simulations, meaning the method may be challenging to implement. We hope this guide will aid in adopting these methods.

MD-phylogeny and traditional sequence-based phylogenetics differ in a few key aspects. In the case of a sequence-based phylogeny [23], the steps are alignment of protein sequences, building a Neighbor Joining (NJ) tree from a pairwise distance matrix, and then standard bootstrapping via resampling of columns with replacement from the sequence alignment. In MD-phylogeny (Fig. 1), the distance matrix derives from pairwise superimposition of structures, allowing a phylogeny to be built from the matrix of distances (Fig. 1B). It is thus similar to the generation of pairwise distances from sequence data. However, a PDB file only captures a single conformation of a protein, providing no inherent variation from which to derive statistical support in a structural phylogeny. Therefore, to assess confidence, all structures undergo MD simulation. Frames sampled from the simulations can provide support for relationships in the phylogenetic tree (Fig. 1C). Alternatively, a majority consensus tree can be generated without specific reference to the original PDB files. This is possible because, unlike with resampling of columns in an alignment, many

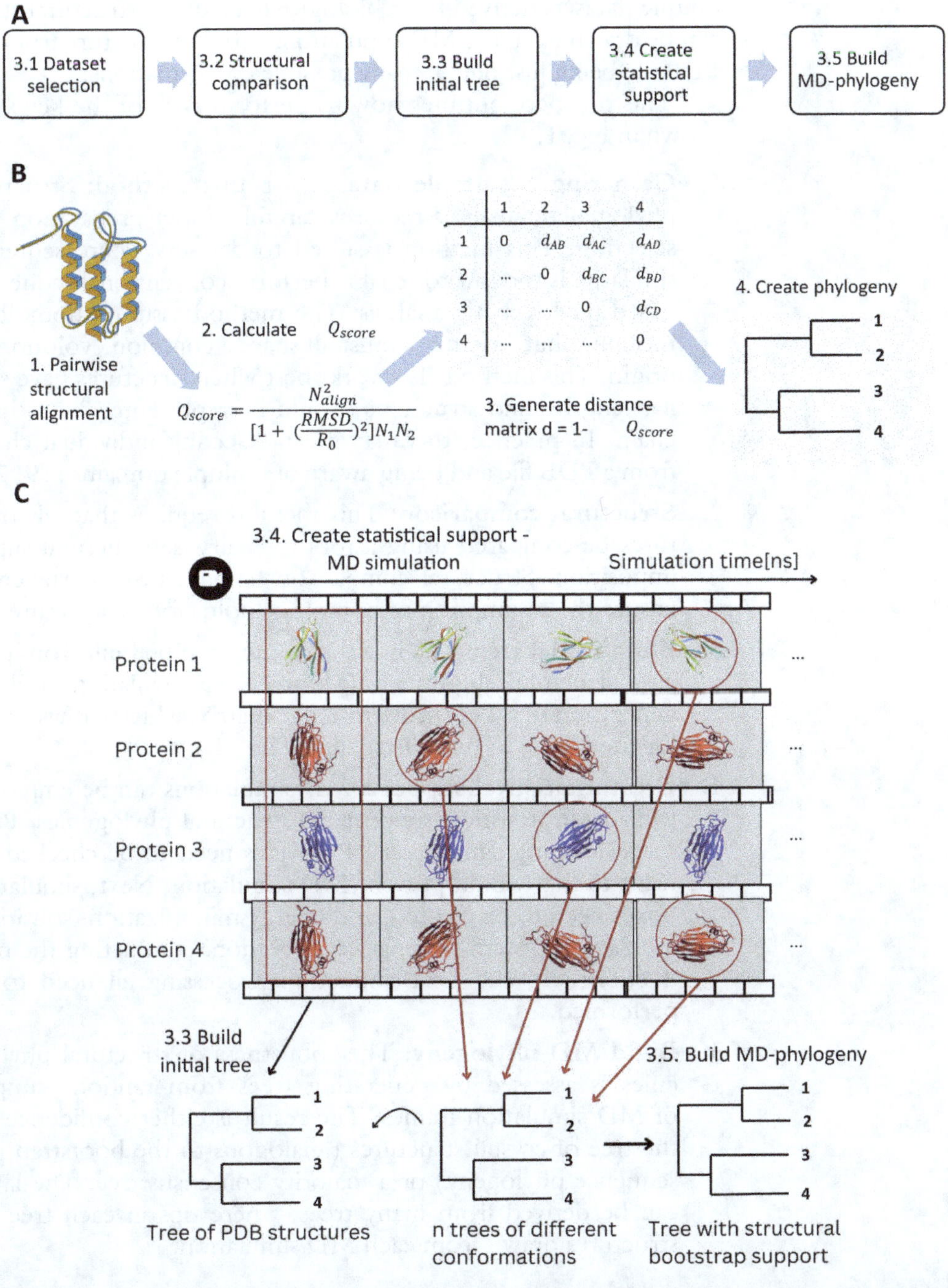

Fig. 1 An overview of the key steps in producing an MD-Phylogeny. **A**: Workflow summary of the MD-phylogeny method. Each step contains a cross-reference to the relevant subsection of the text. **B**: Detailed summary of Steps 3.2 and 3.3, showing pairwise structural alignment (1), *Q*-score calculation (2), and generation of a 1-Q distance matrix (3), which is used to build a phylogenetic tree (4). **C**: Detailed summary of Steps 3.4 and 3.5. For each protein in the dataset (1, 2, 3, 4), trajectories are randomly drawn from MD simulations of each PDB structure. These trajectories can be used to generate support for relationships in the tree. In 3.5, the example of using simulations to assess support for a tree derived from deposited PDB structures is shown. At left is an unrooted phylogeny showing the relationships between four PDB structures. For each MD-bootstrap, a structure is drawn at random from each MD simulation, and a phylogeny is generated. This is performed n times and is used to generate support for the relationships in the original PDB tree

unique datasets derived from biologically plausible structural states can be drawn from the MD simulation, with the structure from the PDB file being just one of many possible conformations.

This protocol outlines how to perform each of the key steps shown in Fig. 1.

1. **Gathering a suitable dataset for this method:** Structural phylogenetic analysis requires careful dataset preparation and selection. This method is suited to datasets where sequence similarity is too low to reliably perform conventional sequence-based phylogenetic analysis. The method assumes homology, meaning that structures must all share a common evolutionary origin. This method also works best when structures have similar lengths, and structures should be represented by a single chain. In practice, this may mean selecting individual chains from a PDB file and being aware of multiple domains [19, 20].
2. **Structural comparison:** This method requires that all structures be compared using a tool for pairwise structural superimposition. Structural comparisons should be undertaken to ensure the assumption of homology holds for all structures.
3. **Build initial tree:** The initial tree can then be built from pairwise structural alignment, *Q*-score [24] calculation, and the generation of a 1- *Q*-score distance matrix, which allows a phylogenetic tree to be constructed (Fig. 1B).
4. **Create statistical support:** MD simulations can be employed to provide statistical support for structural phylogenies. Prior to performing simulations, PDB files need to be checked for features that would prevent MD simulation. Next, simulation boxes need to be created, and energy minimization, solvation, ionization, system heating, equilibration, conducting the production run, and post-simulation processing all need to be performed.
5. **Build MD-phylogeny:** The robustness of structural phylogenies is assessed by generating trees from random samples of MD simulation frames. The result is either confidence for the tree of crystal structures (analogous to the bootstrap in a sequence phylogeny) or a majority consensus tree. The latter can be derived from many trees, where tips in each tree are structures drawn from each MD simulation.

2 Materials

2.1 Computer

2.1.1 Hardware

1. Most modern laptop or desktop computers will suffice for the preparation and analysis stages.

2. Access to high-performance computing (HPC) is strongly recommended for running MD simulations.

2.1.2 Operating System

1. Ideally, UNIX, or Mac.
2. If Windows, run Linux via a virtual machine or use Windows Subsystem for Linux (WSL/WSL2).

2.2 Software and Custom Scripts

The workflow outlined here requires the use of several online databases, software tools, and custom scripts. The latter are available from https://github.com/PooleLab/Structural-Phylogeny-with-MD-Bootstrap, along with usage information.

2.2.1 Online Databases

1. For dataset exploration and collation, we recommend FoldSeek [12] and Structome [20], along with SCOP [9], CATH [10, 11], and InterPro [25].
2. For checking sequence similarity, we recommend CLANS [26] and MMseqs2 [27].

2.2.2 Software

1. **MD engine for preparing, running, and analyzing MD simulations**: Popular options with active user communities include GROMACS [28–30], NAMD [31], CHARMM [32], and AMBER [33]. The protocols described below were developed for GROMACS version 2020 (*see* **Note 1**) and will require significant adaptation for use with other MD engines.
2. **Visualization of structures and simulations:** Visual Molecular Dynamics (VMD) [34–36] or Pymol (https://www.pymol.org/) and visualize protein structures. Visualization of simulation trajectories requires VMD [34–36].
3. **Structural superposition and calculation of Q_{score}**: Gesamt [37] (from https://www.ccp4.ac.uk/html/gesamt.html)
4. **Visualization of phylogenies**: FigTree (http://tree.bio.ed.ac.uk/software/figtree/) or iTOL [38] are recommended and are compatible with the scripts provided here.
5. **Summarization and annotation of phylogenies**: sumtrees (https://jeetsukumaran.github.io/DendroPy/programs/sumtrees.html)
6. **Additional requirements**: Python3 with the following packages:
   1. NumPy (https://2/numpy.org)
   2. dendropy (https://pypi.org/project/DendroPy)
   3. Bio Python (https://biopython.org/docs/1.75/api/Bio.html)
   4. natsort (https://pypi.org/project/natsort)

2.2.3 Custom Scripts (Available at: https://github.com/PooleLab/Structural-Phylogeny-with-MD-Bootstrap)

1. *easyproorigami.py*: Creates a topology map of a structure from a PDB file.
2. *GesamtTree.py*: Creates a structural phylogeny without support from PDB files as outlined in Sect. 3.3, or with support as outlined in Sect. 3.5.
3. *s0_sim_setup.py*: Prepares a protein for simulation locally.
4. *s0_sim_setup_slurm.py*: Prepares a protein for simulation on HPC (tailored for Slurm workload manager).
5. *s1_make_equilibration_scripts.py*: Generates Slurm batch scripts for equilibration simulations.
6. *s2_make_equilibration_execute_script.py*: Automates batch submission of equilibration simulations.
7. *s3_make_production_scripts.py*: Creates Slurm batch scripts for production simulations.
8. *s4_make_production_execute_script.py*: Automates batch submission of production simulations.
9. *BS_master_sort.py and BS_slave.tcl*: Draws protein structures at random from simulation trajectories and carries out structural bootstrapping (as described in Sect. 3.5) with *GesamtTree.py*.

3 Methods

3.1 Dataset Selection

The foundation of any structural phylogenetic analysis lies in the careful selection and preparation of an appropriate dataset. This method is aimed for sets of structures likely to share homology. Tools such as FoldSeek [12] and Structome [20], alongside established databases like SCOP [9], CATH [10, 11], and InterPro [25], are invaluable for curating a dataset.

Once a dataset is identified, it is essential to evaluate its suitability for this structural phylogenetics method. This involves ensuring the dataset lies within the "twilight zone" of sequence similarity [39], where traditional sequence-based phylogenetic methods are expected to be less reliable owing to challenges with accurate multiple sequence alignment. For sequences with clear sequence similarity, established sequence-based approaches should remain the preferred choice to build phylogenies. Assessing if the selected proteins lie within the "twilight zone" can be done by performing all-against-all pairwise sequence comparisons, which can be visualized as a network, using, e.g., CLANS [26] and MMseqs2 [27]. A well-connected network suggests conventional sequence phylogeny is suitable for the dataset. A poorly connected network where homology can nevertheless be ascertained from structural comparisons is ideal for applying MD-phylogeny.

Dataset preparation involves several critical prerequisites:

1. **Single chains**: Structures should consist of single-chain files representing the chains that have been determined to be homologous.
2. **Multiple motif or fold copies**: Identify if structures contain multiple copies of the same motif or fold within a single chain (Fig. 2a), as this could confound the alignment process and skew similarity measures.
3. **Size considerations**: Proteins or folds should be of reasonable size. Structures that are too large (e.g., ribosomes, entire nucleosomes) are often impractical for molecular dynamics (MD) simulations. Additionally, length heterogeneity across structures in the dataset may impact results [19].
4. **Source**: It should be considered whether the structure is experimentally determined (e.g., X-ray crystallography, NMR, cryo-EM) or predicted (e.g., AlphaFold [15, 16]). We recommend structures determined using X-ray crystallography, though it is important to check resolution. Predicted structures can be of varying quality, which may also impact the resulting phylogeny.

3.2 Structural Comparison

Once the dataset is assembled, the next step is to perform pairwise structural comparisons. This step generates a distance matrix, which is then required to build the phylogenetic tree. As with sequence alignment, where poor quality alignment or misalignment can distort phylogenetic reconstruction, careful analysis of pairwise structural alignments is highly recommended. This method assumes homology between the compared structures, but, as illustrated in Fig. 2, it is necessary to check what has been superimposed with what.

For instance, it is crucial to identify and separate multiple copies of the same domain within a single protein, chain, or structure. When a protein contains multiple similar domains, it can be unclear which domain aligns with a corresponding domain in another protein (domain X from Protein 2 is homologous to both domain X and X′ from Protein 1), meaning it is critical to determine which domains have been compared (Fig. 2A). To accurately trace evolutionary relationships, these domains must be clearly distinguished and, depending on the analysis, it may be useful to include pairwise superimpositions.

When selecting proteins for structural analysis, it is also essential to ensure that only relevant domains are considered. Some proteins contain additional domains that are not the focus of the analysis (Fig. 2B). For example, domain Y from Protein 2 shares homology with

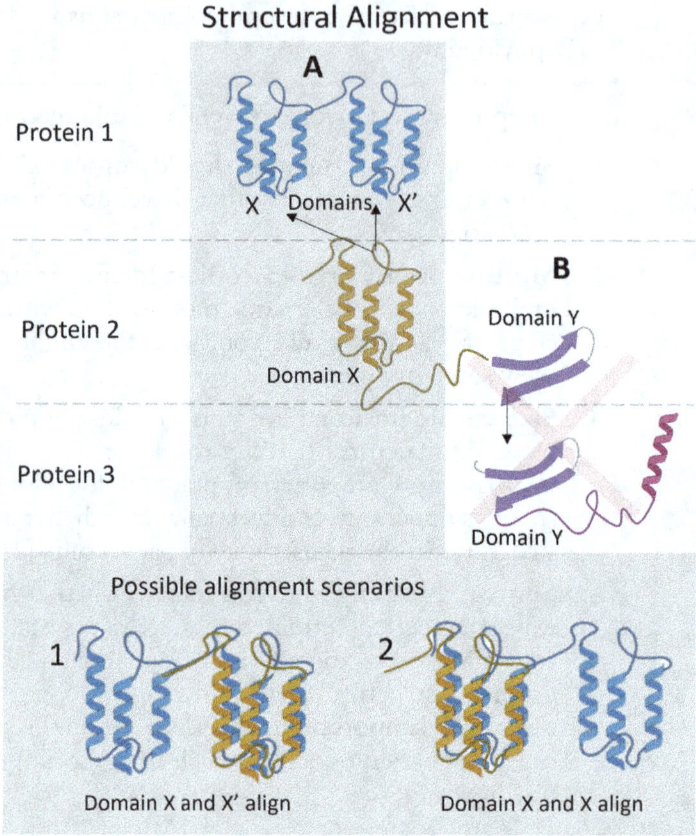

Fig. 2 Schematic representation of structural alignment in protein structure. This figure illustrates possible alignment scenarios that can impact structural alignment and influence phylogenetic analyses, highlighting potential pitfalls to avoid. (**A**) Represents the two possible ways domain X (Protein 2) could align with the two homologous domains X and X′ from Protein 1. Without checking the superimpositions (wide box at the bottom of the figure), it is not possible to establish whether a similarity score comes from domain X from Protein 2 aligning with domain X′ from Protein1 (scenario 1) or domain X from Protein 1 (scenario 2). Checking the superimpositions can resolve this. Note that in a case such as this, it might be valuable to include both scenarios (1 & 2) as domain X from Protein 2 and domain X and X′ from Protein 1 are all homologous. (**B**) In this scenario, Protein 2 has homology to Protein 1 through domain X and to Protein 3 through domain Y. Here, there is no common ancestry between domains X and Y. Including pairwise structural distances derived from both domains would therefore lead to non-homologous comparisons being included in the phylogeny. Such cases (Protein 3 in this example) must be identified and excluded to avoid generating meaningless phylogenies

domain Y from Protein 3 (Fig. 2B). A pairwise comparison between these proteins will yield a similarity score, but including this score in a distance matrix would incorporate noise since domain X (in Proteins 1 and 2) and domain Y (from Proteins 2 and 3) are not homologous. If inadvertently included during dataset selection, such non-homologous comparisons can yield high structural similarity scores despite having no evolutionary relevance to the domain of interest. To prevent this, it is necessary to verify that the similarity scores reflect evolutionary relationships of interest rather than structural similarity in unrelated regions.

One way to help establish homology is to examine secondary structure topology maps. While structures may appear similar in overall 3D shape, differences in topology could indicate that they are unrelated, and including such cases would undermine the assumption of homology. Checking topology ensures that structural comparisons are accurate for subsequent phylogenetic analysis. We have used Pro-origami [40] (https://github.com/fomightez/pro-origami-binder) and have developed a script for a user-friendly application (*easyproorigami.py*).

3.3 Build Initial Tree

The structural phylogeny method (Fig. 1B) implemented in our script *Gesamt.py* is analogous to generating a distance-based matrix from pairwise comparison of sequences, but with structural data.

1. **Pairwise structural alignment:** Perform pairwise structural alignments of the protein structures to assess their structural similarity. Various programs can be used for this step. In this protocol, we used the Gesamt [37] algorithm, which aligns the alpha-carbon atoms of the structures and calculates similarity measurements.
2. **Measure similarity:** Obtain a measure of structural similarity. Different measures, such as TM-score, *Z*-score, or RMSD, can be used to derive distance metrics. Here, we used the Q_{score} [24], as it accounts for both alignment quality and alignment length [4] (*see* **Note 2**):

$$Q_{score} = \frac{N_{align}^2}{\left[1 + \left(\frac{RMSD}{R_0}\right)^2\right] N_1 N_2}$$

$N_1, N_2 = \textit{total number of residues of two protein structures}$
$RMSD = \textit{Root mean square deviation between the aligned residues' } C-\alpha \textit{ atoms}$

$R_0 = \textit{empirical scaling factor set to } 3\text{Å}$

$N_{align} = \textit{number of aligned residue pairs}$

3. **Calculate distance:** Convert the similarity score into a distance measure. In our approach, we use $1 - Q_{score}$.
4. **Create a distance matrix:** Populate a distance matrix with these distance values. This matrix forms the basis for constructing a phylogenetic tree (*see* **Note 3**).

5. **Construct a phylogeny:** Use the distance matrix to build a phylogeny using a distance-based method. We applied Neighbor-Joining to construct the phylogeny.

3.4 Create Statistical Support

MD simulations are performed to generate statistical support for structural phylogenies. This process involves preparing the molecular system to ensure it is physically realistic and then simulating the system. It requires the installation of an MD engine (see Sect. 2); for example, we use GROMACS [28–30].

Running MD simulations can be time-intensive, so it is worth undertaking quality control steps on your dataset before starting. The preparatory steps outlined in Sects. 3.1 (dataset creation), 3.2 (structural comparison), and 3.3 (build the initial tree) are essential to prevent unnecessary computational effort and produce incorrect or unusable results. Neglecting these checks may lead to errors in the analysis. By carefully validating the dataset and confirming its suitability for analysis, you ensure that the resulting statistical support for the phylogeny is reliable.

The key steps in preparing a protein for MD simulation are outlined below. These steps can be run on a local laptop or desktop computer (*see* **Note 4**). For detailed information about each step, see [22].

3.4.1 Prepare Protein Structure for MD Simulation

1. **PDB file preparation:** GROMACS currently requires any non-standard protein residues or engineered amino acids and residues to be replaced. Examine the structure for gaps; these can be filled using, e.g., Modeler [41], but the protein can be safely simulated with gaps in some cases.
2. **Simulation box definition:** Define the size and shape of the simulation box surrounding the molecule. The protein is centered within the box, with sufficient distance from its edges to avoid artefacts caused by periodic boundary conditions. This setup prevents the protein from interacting with its periodic image (*see* **Note 5**).
3. **Energy minimization:** Resolve steric clashes and optimize geometry.
4. **Solvation:** Add water molecules to mimic a realistic environment (*see* **Note 6**).
5. **Energy minimization:** Repeat this step.
6. **Ionization:** Add ions (e.g., Na^+, Cl^-) for charge neutrality and, if desired, to mimic the natural environment of the molecule.
7. **Energy minimization:** Repeat this step.

8. **System heating:** Gradually increase the temperature to the desired simulation conditions, allowing the system to stabilize at the appropriate thermal energy level (*see* **Note 7**).
9. **Equilibration**: Equilibrate each protein and its simulation box to stabilize temperature, pressure, and density. Do an initial test with one or a few protein(s) to identify potential issues before carrying out the equilibration simulations of all proteins.

Once the system is set up and prepared, the simulation run can be performed (*see* **Note 8**). The key steps here are described below.

3.4.2 Run MD Simulations

Production: Run the data collection simulation for each protein system. Decide on a suitable simulation timeframe. This can vary depending on your biological system, timeframe and available resources. We have found a simulation time of 100 ns per protein to be a good compromise between the computational resources required and sufficient conformational sampling to generate statistical support. Again, we recommend performing an initial test on one or a few protein(s) before carrying out the production simulations of all proteins in your dataset.

3.4.3 Post-Simulation Processing

1. **Trajectory concatenation:** If each production simulation is spread across multiple trajectory files, concatenate these into a single trajectory file (*see* **Note 9**).
2. **Protein extraction:** Extract only the protein coordinates at each time point into a new trajectory file. Water, ions, and other molecules added during system preparation are not needed for the analysis and should be removed. This file will be much smaller, which will increase the speed of the remaining analysis steps.
3. **Periodic boundary correction**: Transform the concatenated trajectory to remove artefacts caused by periodic boundaries. This step additionally centers the protein in the simulation box and ensures it is visually and structurally continuous (*see* **Note 10**).
4. **Protein superposition**: Remove global rotation and translation of the protein during the simulation by fitting it to its initial coordinates. This makes it easier to analyze internal motions (*see* **Note 11**).
5. **Extract initial structure:** Extract a single structure from the start of the simulation trajectory, which is used for viewing the simulation in VMD (*see* **Note 12**).

3.5 Build MD-Phylogeny

To assess the robustness of the structural phylogeny, statistical support is generated by sampling from frames obtained during MD simulations. Sampled frames are then used to construct phylogenetic trees, which provide support for evaluating the confidence in the phylogeny (Fig. 1C). For ease of comparison to traditional phylogenetics methods, we refer to this step as MD-bootstrapping. The following steps assume the MD simulations have been completed without errors, with processed trajectory (*.*xtc* files) and initial structure (*.*gro* files) files available for each protein, as outlined above (*see* **Note 13**).

1. **Check trajectory lengths:** Ensure all trajectories have a consistent length across the simulations. Use VMD [34, 35] to load the *.xtc* and *.gro* files. This will display the simulated protein and the number of frames.
2. **Determine the number of MD-bootstraps:** Unlike sequence-based phylogenetic programs such as IQ-TREE [42], where tens of thousands of bootstraps may be feasible, structural phylogenies generated via MD-phylogeny are limited by computing resources for MD simulations. In practice, this means using a smaller number of replicates. Based on our experiments, 200 MD-bootstraps work well as a balance between computational feasibility and representativeness. You can adjust this number according to the size and complexity of your dataset and the available computational power.
3. **Generate bootstrapped samples:** This step is implemented in our script *BS_master_sort.py (and BS_slave.tcl)*
   1. For each MD-bootstrap trial, randomly select one frame from the trajectory of each protein.
   2. Convert the selected frames into PDB files, creating a resampled dataset with the same number of structures as the original dataset.
4. **Reconstruct phylogeny:** Use the workflow outlined in Sect. 3.3 for each trial to generate a phylogenetic tree (Fig. 1B). Use our script *Gesamt.py* for each of the created MD-bootstrap trials:
   1. Perform pairwise structural alignments.
   2. Calculate similarity scores (e.g., Q_{score}) and convert them to distance measures.
   3. Use calculated distances to create a distance matrix.
   4. Apply a distance-based algorithm (e.g., Neighbor-Joining) to build a phylogenetic tree.
5. **Analyze bootstrap trees:**
   1. Map MD-bootstrap results onto a reference or starting tree (e.g., tree generated from deposited PDB files) to evaluate support for specific branches (*see* **Note 14**).

2. Alternatively, generate a majority-rule consensus tree that summarizes the most frequently observed relationships across all MD-bootstrap trees (*see* **Note 15**).

4 Notes

1. The manual and the Version of GROMACS can be found in [43].
2. Our scripts only support analysis with Q_{score} at the moment. If you wish to use a different score, adjustments to the script are needed.
3. Before proceeding with the workflow, it can be useful to check whether the dataset exhibits "tree-likeness." This can be done using tools that generate a phylogenetic network rather than a bifurcating tree from the matrix of pairwise structural comparisons. To assess whether the dataset has a tree-like structure, tools like Splitstree [44, 45] using, e.g., Neigbor-Net [44, 46] can be applied. If the dataset appears star-like, it may not be meaningful to continue with phylogenetic construction and MD simulations. However, if the data suggests distinct splits, it may be reasonable to proceed. This step is a preliminary by-eye inspection of the data. See [4], for an example.
4. Use the script *sim_setup.py* to prepare a protein for simulation locally and *sim_setup_slurm.py* to prepare a protein for simulation on HPC. This script, however, is tailored for the Slurm workload manager. We have used default settings, values and adjustments here, but please check the requirements of your system and adjust the steps accordingly. For detailed information about each step, please see [22].
5. We have used a rectangular box with a distance from the protein to the box edge of 10 Å in all directions. If your protein is non-spherical, the box dimensions should be set according to the length of the longest axis of the protein.
6. There are different water models; the choice depends on the protein force field. We have used TIP3P, for which the CHARMM force fields are parameterized.
7. We have heated the system to 303.5 K, the same temperature at which we run the equilibration and production simulations. Check the requirements of your system or leave this as a default.
8. We do not recommend running the following steps on a local laptop or desktop PC. An HPC system is needed for this step.

9. We use the GROMACS *trjcat* program to concatenate multiple trajectory files (the following only works if they are numbered consecutively): gmx trjcat -f <name>.*.xtc -o <name>.all.xtc
10. We use the GROMACS *trjconv* program to extract the protein coordinates and apply periodic boundary corrections: gmx trjconv -f <name>.all.xtc -ur compact -n index.ndx -o <name>.prot.xtc -pbc mol -center -s <name>.tpr
11. We use the GROMACS *trjconv* program to fit the protein for final analysis: gmx trjconv -f <name>.prot.xtc -n index.ndx -o <name>.proc.xtc -s <name>.tpr -fit rot+trans
12. We use the GROMACS *trjconv* program to save individual gro files: gmx trjconv -f <name>.prot.xtc -b 0 -e 0 -n index.ndx -o <name>.proc.gro -s <name>.tpr
13. Make sure that only trajectories (*.xtc files) and coordinates (*.gro files) of the production run and not the equilibration are selected for this step.
14. We used summtrees to add bootstrap trees to the initial tree: sumtrees -d0 -p -o <output_name> -t <initial_tree> <path_to_all_MD-bootstraps> *.nex
15. We used summtrees to generate a majority rule consensus tree: sumtrees -s consensus -o <output_nam> -f0.6 -p -d0 Trajectories/Trials/trial_*/*.nex will produce a 60% majority rule consensus tree. Adjust parameter -f to the desired fraction.

Acknowledgments

We thank Jonah Dokman for support with testing. Désirée Langer is supported through a School of Biological Sciences PhD scholarship.

References

1. Tauer A, Benner SA (1997) The B12-dependent ribonucleotide reductase from the archaebacterium *Thermoplasma acidophila*: an evolutionary solution to the ribonucleotide reductase conundrum. Proc Natl Acad Sci USA 94:53–58. https://doi.org/10.1073/pnas.94.1.53
2. Sintchak MD, Arjara G, Kellogg BA et al (2002) The crystal structure of class II ribonucleotide reductase reveals how an allosterically regulated monomer mimics a dimer. Nat Struct Biol 9:293–300. https://doi.org/10.1038/nsb774
3. Brindefalk B, Dessailly BH, Yeats C et al (2013) Evolutionary history of the TBP-domain superfamily. Nucleic Acids Res 41:2832–2845. https://doi.org/10.1093/nar/gkt045
4. Lundin D, Poole AM, Sjöberg BM, Högbom M (2012) Use of structural phylogenetic networks for classification of the ferritin-like superfamily. J Biol Chem 287:20565–20575. https://doi.org/10.1074/jbc.M112.367458
5. Bujnicki JM (2000) Phylogeny of the restriction endonuclease-like superfamily inferred from comparison of protein structures. J Mol Evol 50:39–44. https://doi.org/10.1007/s002399910005
6. Lundin D, Berggren G, Logan DT, Sjöberg B-M (2015) The origin and evolution of

ribonucleotide reduction. Life 5:604–636. https://doi.org/10.3390/life5010604
7. Breitling R, Laubner D, Adamski J (2001) Structure-based phylogenetic analysis of short-chain alcohol dehydrogenases and reclassification of the 17beta-hydroxysteroid dehydrogenase family. Mol Biol Evol 18:2154–2161. https://doi.org/10.1093/oxfordjournals.molbev.a003761
8. Garau G, Di Guilmi AM, Hall BG (2005) Structure-based phylogeny of the metallo-beta-lactamases. Antimicrob Agents Chemother 49:2778–2784. https://doi.org/10.1128/AAC.49.7.2778-2784.2005
9. Andreeva A, Kulesha E, Gough J, Murzin AG (2020) The SCOP database in 2020: expanded classification of representative family and superfamily domains of known protein structures. Nucleic Acids Res 48:D376–D382. https://doi.org/10.1093/nar/gkz1064
10. Waman VP, Bordin N, Alcraft R et al (2024) CATH 2024: CATH-AlphaFlow doubles the number of structures in CATH and reveals nearly 200 new folds. J Mol Biol 436:168551. https://doi.org/10.1016/j.jmb.2024.168551
11. Sillitoe I, Bordin N, Dawson N et al (2021) CATH: increased structural coverage of functional space. Nucleic Acids Res 49:D266–D273. https://doi.org/10.1093/nar/gkaa1079
12. van Kempen M, Kim SS, Tumescheit C et al (2023) Fast and accurate protein structure search with Foldseek. Nat Biotechnol. https://doi.org/10.1038/s41587-023-01773-0
13. Berman HM, Westbrook J, Feng Z et al (2000) The protein data bank. Nucleic Acids Res 28:235–242. https://doi.org/10.1093/nar/28.1.235
14. Burley SK, Bhatt R, Bhikadiya C et al (2025) Updated resources for exploring experimentally-determined PDB structures and computed structure models at the RCSB protein data bank. Nucleic Acids Res 53:D564–D574. https://doi.org/10.1093/nar/gkae1091
15. Jumper J, Evans R, Pritzel A et al (2021) Highly accurate protein structure prediction with AlphaFold. Nature 596:583–589. https://doi.org/10.1038/s41586-021-03819-2
16. Varadi M, Anyango S, Deshpande M et al (2022) AlphaFold protein structure database: massively expanding the structural coverage of protein-sequence space with high-accuracy models. Nucleic Acids Res 50:D439–D444. https://doi.org/10.1093/nar/gkab1061
17. Zhang C, Shine M, Pyle AM, Zhang Y (2022) US-align: universal structure alignments of proteins, nucleic acids, and macromolecular complexes. Nat Methods 19:1109–1115. https://doi.org/10.1038/s41592-022-01585-1
18. Baltzis A, Santus L, Langer BE et al (2025) Multistrap: boosting phylogenetic analyses with structural information. Nat Commun 16: 293. https://doi.org/10.1038/s41467-024-55264-0
19. Malik AJ, Poole AM, Allison JR (2020) Structural phylogenetics with confidence. Mol Biol Evol 37:2711–2726. https://doi.org/10.1093/molbev/msaa100
20. Malik AJ, Langer D, Verma CS et al (2023) Structome: a tool for the rapid assembly of datasets for structural phylogenetics. Bioinform Adv vbad134. https://doi.org/10.1093/bioadv/vbad134
21. Malik AJ (2024) Structome-Q: https://biosig.lab.uq.edu.au/structome_q/. Accessed 15 Feb 2025
22. Collier TA, Piggot TJ, Allison JR (2020) Molecular dynamics simulation of proteins. Methods Mol Biol 2073:311–327. https://doi.org/10.1007/978-1-4939-9869-2_17
23. Baldauf SL (2003) Phylogeny for the faint of heart: a tutorial. Trends Genet 19:345–351. https://doi.org/10.1016/S0168-9525(03)00112-4
24. Krissinel E, Henrick K (2004) Secondary-structure matching (SSM), a new tool for fast protein structure alignment in three dimensions. Acta Crystallogr D Biol Crystallogr 60:2256–2268. https://doi.org/10.1107/S0907444904026460
25. Blum M, Andreeva A, Florentino LC et al (2025) InterPro: the protein sequence classification resource in 2025. Nucleic Acids Res 53:D444–D456. https://doi.org/10.1093/nar/gkae1082
26. Frickey T, Lupas A (2005) CLANS: A Java application for visualizing protein families based on pairwise similarity. Bioinformatics 20:3702–3704. https://doi.org/10.1093/bioinformatics/bth444
27. Steinegger M, Söding J (2017) MMseqs2 enables sensitive protein sequence searching for the analysis of massive data sets. Nat Biotechnol 35:1026–1028. https://doi.org/10.1038/nbt.3988
28. Abraham MJ, Murtola T, Schulz R et al (2015) GROMACS: High performance molecular simulations through multi-level parallelism from laptops to supercomputers. SoftwareX 1–2:19–25. https://doi.org/10.1016/j.softx.2015.06.001
29. Pronk S, Páll S, Schulz R et al (2013) GROMACS 4.5: a high-throughput and highly parallel open source molecular simulation toolkit. Bioinformatics 29:845–854. https://doi.org/10.1093/bioinformatics/btt055
30. Páll S, Abraham MJ, Kutzner C et al (2015) Tackling exascale software challenges

in molecular dynamics simulations with GROMACS BT. In: Markidis S, Laure E (eds). Solving software challenges for exascale. Springer International Publishing, Cham, pp 3–27
31. Phillips JC, Hardy DJ, Maia JDC et al (2020) Scalable molecular dynamics on CPU and GPU architectures with NAMD. J Chem Phys 153:44130. https://doi.org/10.1063/5.0014475
32. Brooks BR, Brooks 3rd CL, Mackerell ADJ et al (2009) CHARMM: the biomolecular simulation program. J Comput Chem 30:1545–1614. https://doi.org/10.1002/jcc.21287
33. Case DA, Aktulga HM, Belfon K et al (2023) AmberTools. J Chem Inf Model 63:6183–6191. https://doi.org/10.1021/acs.jcim.3c01153
34. Humphrey W, Dalke A, Schulten K (1996) MD—visual molecular dynamics. J Mol Graph 14:33–38
35. Stone JE, Gullingsrud J, Schulten K (2001) A system for interactive molecular dynamics simulation. Proceedings of the 2001 Symposium on Interactive 3D Graphics. Association for Computing Machinery, New York, NY, pp 191–194
36. Spivak M, Stone JE, Ribeiro J et al (2023) VMD as a platform for interactive small molecule preparation and visualization in quantum and classical simulations. J Chem Inf Model 63:4664–4678. https://doi.org/10.1021/acs.jcim.3c00658
37. Krissinel E (2012) Enhanced fold recognition using efficient short fragment clustering. J Mol Biochem 1:76–85
38. Letunic I, Bork P (2021) Interactive tree of life (iTOL) v5: an online tool for phylogenetic tree display and annotation. Nucleic Acids Res 49:W293–W296. https://doi.org/10.1093/nar/gkab301
39. Rost B (1999) Twilight zone of protein sequence alignments. Protein Eng Des Sel 12:85–94. https://doi.org/10.1093/protein/12.2.85
40. Stivala A, Wybrow M, Wirth A et al (2011) Automatic generation of protein structure cartoons with Pro-origami. Bioinformatics 27:3315–3316. https://doi.org/10.1093/bioinformatics/btr575
41. Webb B, Sali A (2016) Comparative protein structure modeling using MODELLER. Curr Protoc Bioinformatics 54:5.6.1–5.6.37. https://doi.org/10.1002/cpbi.3
42. Minh BQ, Schmidt HA, Chernomor O et al (2020) IQ-TREE 2: new models and efficient methods for phylogenetic inference in the genomic era. Mol Biol Evol 37:1530–1534. https://doi.org/10.1093/molbev/msaa015
43. Lindahl E (2020) GROMACS 2020 manual and GROMACS 2020 versions. https://doi.org/10.5281/zenodo.3562512. Accessed 15 Jan 2025
44. Huson DH, Bryant D (2006) Application of phylogenetic networks in evolutionary studies. Mol Biol Evol 23:254–267. https://doi.org/10.1093/molbev/msj030
45. Huson DH, Bryant D (2024) The SplitsTree App: interactive analysis and visualization using phylogenetic trees and networks. Nat Methods 21:1773–1774. https://doi.org/10.1038/s41592-024-02406-3
46. Bryant D, Moulton V (2004) Neighbor-Net: an agglomerative method for the construction of phylogenetic networks. Mol Biol Evol 21:255–265. https://doi.org/10.1093/molbev/msh018

Check for updates

Chapter 16

Studying the Evolution of Protein Structure One Superfamily at a Time

Adrián Cruz-González, Ricardo Hernández-Morales, José Alberto Campillo-Balderas and Rodrigo Jácome

Abstract

The field of structural biology has received a major thrust, thanks to the important advancements in cryo-electron microscopy coupled with the development of highly reliable computer-predicted structural models. Throughout evolution, the tertiary structure of proteins is more conserved than their primary structure. The superposition of protein tertiary structures has been a powerful tool in the identification of proteins with remote homology, whose evolutionary footprints are no longer detectable through their primary sequence alignment, and more recently in the reconstruction of evolutionary trees. These structure-based phylogenetic trees have been demonstrated to be quite useful in the study of deep evolutionary events and in the analysis of biological entities with extremely high mutation rates, such as RNA viruses. The construction of phylogenetic trees from tertiary structures starts by performing pairwise comparisons between a set of proteins, from which certain parameters, such as the number of spatially aligned residues and the average distance between their α-carbons, are collected. These measurements are then used to calculate a structural distance metric, which can be used as input for the construction of an evolutionary tree.

Keywords Evolution, Tertiary structure, Protein, Domain, Structural superposition

1 Introduction

1.1 Brief Historical Account

Primary structure comparisons (i.e., nucleotide or amino acid sequences) have been used for decades as hallmarks to establish evolutionary relationships between molecular markers of organisms and biological entities, including DNA and RNA viruses. However, in the last two decades, comparisons of protein tertiary structures have become increasingly used due to the fact that tertiary structures are far more conserved than primary structures during evolution [1].

The roots of the use of molecular structures to infer evolutionary relationships can be traced back to the early twentieth century with the study of comparative immunology. One of the pioneering works in this field was conducted by George H. F. Nuttall. Comparing the

Gustavo Caetano-Anollés (ed.), *Evolutionary Genomics*, Methods in Molecular Biology, vol. 2981,
https://doi.org/10.1007/978-1-0716-4836-0_16,

immunological reactions of blood sera from different animal species, Nuttall observed that the precipitation reaction—where an antibody binds to a soluble antigen, forming aggregates that precipitate—was significantly more efficient when blood sera from closely related organisms were combined [2]. His work established a foundational principle in molecular evolution in which the degree of similarity between molecular traits can indicate the evolutionary relatedness of the organisms possessing those molecules.

A few years later, Edward T. Reichert and Amos P. Brown published a work exploring the differentiation and specificity of hemoglobins in relation to biological classification and organism evolution. Comparing the hemoglobin crystals from different animals, they observed that crystal structures from closely related organisms had very similar shapes and angles, whereas those from more distantly related organisms exhibited more significant differences [3]. Thus, they recognized that structural differences in molecules may also reflect the evolutionary relationships of living beings.

Continuing the tradition of immunochemical studies, Karl Landsteiner carried out experiments on antigen–antibody interactions. His studies of the agglutination phenomena and precipitation allowed him to propose that the serological and chemical properties of proteins might be linked to the evolutionary relationships among animals, which could be represented as evolutionary trees [4, 5].

In the early 1960s, Linus Pauling and Emile Zuckerkandl applied traditional chemistry and genetic methods to investigate evolutionary relationships between organisms. By using electrophoresis and paper chromatography, they demonstrated the range of variation in the structure of hemoglobin in different animals [6]. They found that the hemoglobin degradation patterns when treated with trypsin—an enzyme that breaks down protein chains—were more similar among closely related organisms. As gene and protein sequencing methodologies were developed, more information became available for evolutionary analysis. In 1965, Zuckerkandl and Pauling proposed that sequences, either nucleotides or amino acids, could be used to study the evolutionary relationships between living beings, as they contain information that reflects their evolutionary history [7]. They conducted the first study on the differences between the hemoglobin sequences of various groups of organisms, quantitatively determining the evolutionary relationships between hemoglobins. Thus, the ideas of Zuckerkandl and Pauling bridged the fields of paleontology, evolutionary biology, and molecular biology, leading to the emergence of Molecular Evolution.

Over the next 50 years, analysis of the primary structure of proteins has been commonly used as one of the main approaches to recognize the common origin of molecular traits and infer their evolutionary relationships. However, several limitations, including high mutation rates, can hinder homology recognition or lead to misrepresentations of evolutionary relationships. These challenges led to the development of more sensitive and efficient methods to address these issues. One of these methods involves the comparison of the coordinates in space occupied by the atoms that form part of a protein tertiary structure, that is, the structural comparison of proteins. Johnson et al. [8] were the first to propose the construction of phylogenies based on a structural distance metric derived from the superpositions of protein tertiary structures. In their work, they selected six protein families with variable levels of sequence identity and built phyletic trees from the comparison of their corresponding tertiary structures, obtaining very similar results. Notably, the tertiary structure-based phyletic trees from proteins with very low levels of sequence identity were more robust than the sequence-derived ones.

1.2 The Multiple Uses of Protein Tertiary Structure Comparisons

The field of structural biology has seen major changes in the last couple of decades led by two paramount advances: the improvements associated with the use of cryo-electron microscopy, which have allowed to obtain structures of macromolecular complexes at near-atomic resolution [9], and more recently, the development of highly reliable tertiary structure predictors such as AlphaFold [10] and Robetta [11], which have significantly helped to unveil the diversity of possible protein folds in nature.

It has long been known that when two proteins are compared, there exists a direct relationship between their primary sequence identity and their tertiary structure similitude, up until the so-called twilight zone, i.e., an identity between 20% and 30% in which it is difficult to know if two proteins are evolutionarily related [12, 13]. As will be further discussed, it has been shown that, below this threshold, tertiary structure comparisons are more reliable in identifying distant homologous proteins. This, in turn, has been widely used as a good indicator of the possible function of a protein.

Apart from the latter use, tertiary structure comparisons have led to the elaboration of different classification schemes that cluster protein domains, such as SCOP [14, 15] and CATH [16–18]. These classification systems use the PDB as their main input and, as will be further described, they propose hierarchical systems that start grouping protein domains based on their common structural features. Both systems also attempt to classify protein domains into Superfamilies considering their evolutionary kinship, even between distantly related proteins in which only the tertiary structure is conserved.

An additional application of protein tertiary structure comparisons is the construction of phylogenetic trees. This methodology has proven to be quite helpful when the evolutionary relationships between the proteins can no longer be detected using their primary structure. For example, these approaches have become useful for studies of deep evolutionary events, i.e., evolutionary events that probably occurred prior to the Last Universal Common Ancestor (LUCA) or soon after its divergence into the cellular lineages. Some examples of these proteins include the ribonucleotide reductase [19], the Superfamily of right-hand polymerases [20–22], polymerases endowed with a non-canonical palm domain [23], and the cupin-like protein Superfamily [24]. Likely, the folds present in these Superfamilies were already present in the LUCA, therefore, the conservation within each of them, even at the tertiary structure level, is limited to the active site and key residues around it.

Another scenario for which tertiary structure comparisons have been extremely helpful is the evolutionary study of biological entities with high mutation rates, such as RNA viruses, whose mutation rates are 3–4 orders of magnitude higher than dsDNA viruses and cells [25]. RNA viruses have been related to the origin of life, since they have been even posited as relics from the so-called RNA World [26, 27], an ancient stage of evolution in which RNAs were both the informational and catalytic molecules [28, 29]. However, the study of the origins and evolution of RNA viruses is challenging due to their high evolutionary rates. Within a viral family, the evolutionary signal contained in their primary sequences is rapidly lost, even when proteins with relatively high levels of conservation, such as their polymerases, are compared. Conversely, the evolutionary signal imprinted in the tertiary structures of these proteins is far more preserved. Hence, the construction of evolutionary trees from RNA viral tertiary structure comparisons has been shown to be a remarkable alternative.

All RNA viruses are endowed with highly processive RNA-dependent polymerases, i.e. an RNA-dependent RNA polymerase (RdRp) or a reverse transcriptase (RT), lacking proofreading and repair capacities, which results in a very rapid but error-prone enzyme [30]. Since this is the sole conserved protein among all the known RNA viruses, several evolutionary analyses using different structural approaches have been published [31–35]. This structural approach has also allowed us to identify that RNA viruses have hijacked a DnaQ-like ExoN domain from cells at least on two occasions [36]. Another interesting hijack event identified through structural analysis involves RNA viral proteases belonging to the PA clan, all of which share a two β-barrel-like structural core. Phylogenetic analyses concluded that the RNA viral enzymes were probably taken from eukaryotic HtrA proteases [37]. The search for structural homologs also led to the identification and description of various gene duplication events in RNA viruses, which had gone unnoticed by primary structure approaches [38].

Finally, it is certain that the more frequent usage of computer-predicted models will be of great assistance towards a better understanding of viral evolution. Current examples that have used these predicted models include a proposal for the evolutionary story of RNA viral movement proteins [39], as well as the structural and evolutionary analysis from the predicted models of glycoproteins from all the known genera of the *Flaviviridae* family [40].

2 Materials

2.1 Databases Where the Protein Tertiary Structures Are Deposited

The major source for tertiary structures of biological macromolecules, i.e., proteins and nucleic acids, is the Protein Data Bank (PDB) [41]. This specialized repository was established in 1971 at the Brookhaven National Laboratory, and its earliest versions included the milestones of protein tertiary structures such as myoglobin, lysozyme, carboxypeptidase, subtilisin, chymotrypsin, papain, lactate dehydrogenase, rubredoxin, and cytochrome b5 [42]. As of 2024, over 50 years into its foundation, the PDB hosts more than 225,000 experimentally obtained structures and more than 1 million computer-generated structural models. Each of the proteins deposited in the PDB is bestowed with a unique 4-letter code. The biological and experimental information associated with the structure can be accessed directly through the PDB web page (https://www.rcsb.org/) or by downloading the corresponding PDB or CIF files.

2.2 Databases for the Structural Classification of Proteins and Domains

Protein structure databases are essential repositories in biology, designed to organize and annotate experimentally- and computer-predicted protein structures. These 3D-structure repositories are also important to facilitate research in various fields, including drug design, protein engineering, and evolutionary biology (Table 1). These databases not only provide sequence, three-dimensional coordinates, and experimental information of the unit cell and X-ray crystallography structures, but also bioinformatic tools for searching, predicting, and identifying structurally similar proteins.

2.2.1 CATH

CATH is a hierarchical classification system of protein domains available in the PDB, in which each letter stands for a different level: Class, Architecture, Topology, and Homology [16]. Active since 1997, the classification system begins by decomposing the protein into domain folds. Depending on the predominant secondary structure, each domain is placed in a Class, followed by its Architecture based on the number of secondary elements. When different folds share the same connectivity, they are assigned to a fold family at the Topology level. Finally, proteins with high structural similarity and function are placed in a Homologous Superfamily. Ultimately, each fold can be identified by a number using the following system: CATH. The web page of each domain Superfamily provides its biological distribution, its enzymatic classification, the gene ontology terms, structural superpositions, and groups the structures based on their sequence similarity into structural clusters (SC) [17]. The latest version of the CATH database has expanded this classification to include the computer-predicted structures contained in the AlphaFold DB into The Encyclopedia of Domains (TED) [43].

Table 1
Overview of some of the most important protein structure databases

Database	Primary focus	Key features	Data type	URL
Protein Data Bank—PDB	Experimentally determined 3D structures	Comprehensive archive of macromolecular structural data	X-ray, NMR, Cryo-EM structures	rcsb.org
Class, architecture, topology, homology - CATH	Hierarchical classification of protein domains	Combines automated and manual methods for protein domain classification	Domain-level structural classifications	cathdb.info
Interpro—classification of protein families	Protein classification into families and predicting protein domains	Uses predictive models for domain identification and protein classification	Sequence-based classifications and predictions	interpro-embl-ebi
Structural classification of Proteins—SCOP	Hierarchical classification of protein structures	Classifies proteins into families, superfamilies, and folds	Curated structural classifications of proteins	scop.mrc-lmb.cam.ac.uk
AlphaFold protein structure database	AI-predicted protein structures	Covers proteomes of multiple organisms	AI-predicted 3D structures	alphafold.ebi.ac.uk

2.2.2 Structural Classification of Proteins—SCOP

The Structural Classification of Proteins (SCOP) was also designed to organize the known tertiary structures deposited in the PDB in terms of their structural and evolutionary relationships [44]. Its units of hierarchical classification consist of non-redundant protein domains. Domains with evidence of being closely related are grouped into Families. More distant homologous domains, whose similarity can only be detected through structural comparisons, are clustered into Superfamilies. Depending on the global structural features of domains, Superfamilies are classified into Folds; finally, these latter are grouped into Classes depending on the main secondary structure elements. SCOP includes intrinsically unstructured protein regions as part of their main classification scheme [14, 15]. For each Family, SCOP provides a list of non-redundant representative domains and their links to other databases such as PDB or Uniprot.

2.3 Programs to Look for Structural Homologs

As previously mentioned, protein tertiary structure superposition has many purposes. For instance, it can be used to identify similarities between two proteins (pairwise alignment) by analyzing their three-dimensional conformation to know if they are evolutionarily related, i.e., to establish homology. When three or more protein structures are compared, it is thus possible to establish their evolutionary relationships, aiming to understand how recently they shared an ancestor, which can be depicted as a phylogenetic tree. There are a number of servers that allow looking for proteins with structural similarity to a domain or a protein of interest against the structures deposited in the PDB or against computer-predicted structural models deposited in databases such as the AlphaFold protein structure database [10, 45].

2.3.1 PDB Structural Similarity Search

The PDB is endowed with its own structural similarity search engine, which can be easily accessed through the web page (https://www.rcsb.org/search/advanced/structure). The search engine is based on 3D Zernike moment normalization [46]. Briefly, tertiary structures are transformed into their electron density volumes, which are then decomposed into 3D Zernike polynomials, which can be further transformed into vectors of Zernike moments [46]. This volume-based method is extremely fast and accurate in detecting similar protein assemblies to a query; however, it does not perform so well identifying individual domains, which can be found in different protein architectures [46]. The results of this search engine take no more than a few seconds.

2.3.2 Dali Server

Dali was one of the first protein structure comparison tools available on the internet [47]. Apart from the web server (http://ekhidna.biocenter.helsinki.fi/dali/), the program can be downloaded as a standalone version. Dali's algorithm looks for structural similarities of intramolecular equivalent Cα-Cα atoms between two proteins, yielding a *Z*-score, which reflects the statistical significance of the structural similarity depending on the proteins' length [48]. Dali's web server includes different search modes against the PDB: a heuristic algorithm that compares the query structure against the whole PDB, an algorithm that compares the query structure against a subset of the PDB whose maximum sequence identity is 25%, and an algorithm that compares the query structure against the AlphaFold database. Depending on the server saturation, the results of a protein search against the whole PDB database can take between a few hours and a day.

2.3.3 PDBeFold

The PDBeFold program has been available for over 20 years (https://www.ebi.ac.uk/msd-srv/ssm/cgi-bin/ssmserver). PDBeFold's algorithm, named secondary structure matching (SSM), is based on graph-matching [49]. This algorithm first identifies the secondary structure elements of a query protein and converts them into vectors with their corresponding vertices, edges, and labels, leading to a preliminary identification of correspondent amino acids (and their Cα) and protein folds. Then, all the Cα are mapped, and their interatomic distances are calculated. The program yields a *Q*-score, which weights the number of superimposed (mapped) Cα, the RMSD between them and the length of the two proteins, as well as a *Z*-score, which provides a statistical significance of the superposition. The PDBeFold server allows the user to perform the search for a given protein structure against the whole PDB, against the SCOP database, the option to undertake a pairwise comparison between a query and a target structure, and the option to compare the query structure against a list of PDB files. A search against the whole PDB using the PDBeFold server usually takes no more than a few minutes.

2.3.4 FoldSeek

As of 2025, FoldSeek is a relatively new comparison tool that is available online (https://search.foldseek.com/search). Its algorithm uses a structural alphabet to describe tertiary interactions (3Di), where the 20 states of this alphabet represent the geometric conformation of each residue in relation to its spatially closest neighbor [50]. FoldSeek operates by converting protein structures into sequences using the 3Di alphabet. Once transformed, the program uses fast sequence-based comparison methods (MMseqs2) to identify potential matches. For the best candidate matches, the program transitions back from the simplified 3Di sequences to the 3D protein structures. Lastly, those structures are aligned locally using 3Di (default) or globally with TM-align (Foldseek-TM). The FoldSeek server performs the search for a given protein against several databases such as the PDB, CATH, AlphaFold database and a new database named: The Big Fantastic Virus Database (BFVD), which, at the moment, harbors 351,242 viral predicted protein structures [51]. A search against all these databases usually takes no more than a few minutes.

2.4 Pairwise Structural Comparisons

Pairwise structural comparisons of proteins are crucial tools in structural biology, enabling the identification of similarities between protein structures and providing insights into their evolutionary relationships and functional properties. These comparisons involve superimposing the 3D structures of two proteins and quantifying their structural similarity. The comparison tools typically involve structure selection, alignment algorithms, similarity metrics, visual inspection, and normalization. Table 2 summarizes some bioinformatic tools to superimpose protein structures, which can also be used to infer evolutionary relationships, predict protein functions, and guide further experimental studies.

2.4.1 Dali's Web Server

Dali's web server includes an option to perform pairwise comparisons between a reference protein and up to 10 proteins, as well as an option to perform pairwise all-against-all comparisons with a set of up to 64 proteins. For each comparison, apart from the *Z*-score, the program yields the number of superimposed residues and the RMSD. The server also offers a program to visualize the superposition in three dimensions and a pairwise alignment of the secondary structures.

Table 2
Algorithms for pairwise structure comparisons

Algorithm	Mechanism	Key values	Speed	Scalability	Suggested applications
TM-align [54]	Pairwise alignment based on TM-score for structural similarity	Alignment score, RMSD, TM-score	Fast	Medium	Evolutionary studies of proteins
CE [52]	Fragment-based alignment of local fragments to extend to global matches	Alignment score, RMSD	Moderate	Medium	Domain comparisons, structural insights
SSM [49]	Graph-based approach using secondary structure elements	Alignment score, RMSD, *Q*-score	Moderate	Low	Protein classification, structural insights
FATCAT [53]	Fragment-based flexible alignment, allowing twisting of aligned regions	Alignment score, RMSD	Slow	High	Studying protein flexibility
DALI [47]	Distance matrix method comparing intramolecular distances	Alignment score, RMSD, *Z*-score	Moderate/ Slow	Medium	Protein family analysis

The number in parentheses corresponds to the reference.

2.4.2 PDBeFold Web Server

Similar to the Dali server, the PDBeFold has an option to perform pairwise comparisons between each of the chains of two selected proteins. For each comparison, the program displays the number of superimposed residues, the RMSD, the percentage of equivalent secondary structures, the weighted Q-score and the *Z*-score described above. Unlike Dali, PDBeFold is not endowed with an online structure visualization tool; instead, the user can download the structure-based alignment for the selected proteins.

2.4.3 CE Alignment Tool

The combinatorial extension (CE) program is a well-established tool for protein structure alignment, available as standalone software. Its algorithm identifies similarities between protein structures using aligned fragment pairs (AFPs), which are short, structurally similar fragments from each protein [52]. CE incrementally builds an alignment by extending AFPs that satisfy geometric similarity criteria, allowing for gaps within predefined limits. The program evaluates potential matches using inter-residue distances and optimizes the final alignment through iterative gap adjustment and superposition. CE can detect local and global similarities, making it suitable for detailed structural analyses.

2.4.4 FATCAT Alignment Tool

Flexible structure Alignment by Chaining AFPs with Twists (FATCAT) is a tool for protein structure alignment, available as standalone software, that incorporates structural flexibility into its algorithm. Like the CE algorithm, it identifies AFPs and links them to form an alignment while accounting for conformational changes. However, unlike rigid-body methods, FATCAT introduces twists at hinge points to connect incompatible AFPs, enabling accurate alignment of proteins with flexible regions [53]. The algorithm optimizes alignments by balancing fragment length and the number of twists required; longer AFPs and smaller root-mean-square deviations (RMSDs) are rewarded, whereas twists are penalized unless they significantly improve alignment quality. FATCAT excels in detecting both distant structural similarities and protein flexibility.

2.4.5 TM-align Tool

TM-align is a fast tool for comparing protein structures. It is available both as standalone software and online (https://zhanggroup.org/TM-align/). It aligns protein structures by maximizing the TM-score, a measure designed to assess structural similarity with a focus on global topology [54]. The TM-score is calculated by weighing residue pairs based on their distances after superimposing the structures, with closer pairs contributing more to the score. This ensures the structural alignment emphasizes overall shape rather than small local differences. TM-align starts by generating rough initial alignments using strategies like secondary structure matching or gapless residue comparisons, proceeding with the one that results in the best initial alignment. It then refines these alignments iteratively by rotating the structures and optimizing residue matches to maximize the TM-score. With its speed and accuracy, TM-align is ideal for large-scale studies, such as comparing thousands of protein structures.

2.4.6 PDB Pairwise Structure Alignment Tool

The PDB offers an integrated tool for the pairwise superposition of a reference query structure against up to ten different structures [55]. This tool is endowed with four different algorithms (jCE, jFATCAT, TM-Align, Smith-Waterman 3d), which can be selected to perform rigid structural comparisons. CE and FATCAT can also be selected to perform flexible structural comparisons. For each pairwise comparison, the tool displays the RMSD, the number of superimposed residues, the sequence identity, and the TM-score, and it also provides an integrated visualizer in which the superpositions can be inspected. Among the main advantages offered by this tool is its speed, with results available in a few seconds.

3 Methods

The reconstruction of phylogenetic trees based on tertiary structure comparisons is a relatively straightforward process. However, like any bioinformatic methodology, certain steps, such as the selection of the structures, require a thorough inspection to ensure biologically accurate results free from biases or artifacts. Figure 1 illustrates the essential steps for constructing structure-based phylogenetic trees, which will be detailed in this section.

3.1 Building the Superfamily Database

3.1.1 Searching for Homologous Structures

The first step in the reconstruction of any tertiary structure-based phylogeny is selecting an appropriate protein family or a particular structural or functional domain. As mentioned in the previous section, there are various programs and web servers that can identify tertiary structures with structural similarity to a query. These programs typically employ structural similarity scores with cut-off values, allowing for preliminary filtering.

Some biological entities are overrepresented in the PDB, including model organisms like *Escherichia coli* and *Homo sapiens*, as well as relevant human-infecting viruses such as HIV-1 or SARS-CoV-2. Therefore, it is important to eliminate redundant structures, which may clog the database, without contributing to the evolutionary analysis. Some of these can often be automated using custom scripts in languages like Perl, Python, or R.

Given the limited repertoire of secondary structures and protein folds, a visual inspection of the structural superpositions is essential, especially when the results fall near threshold values. To ensure the

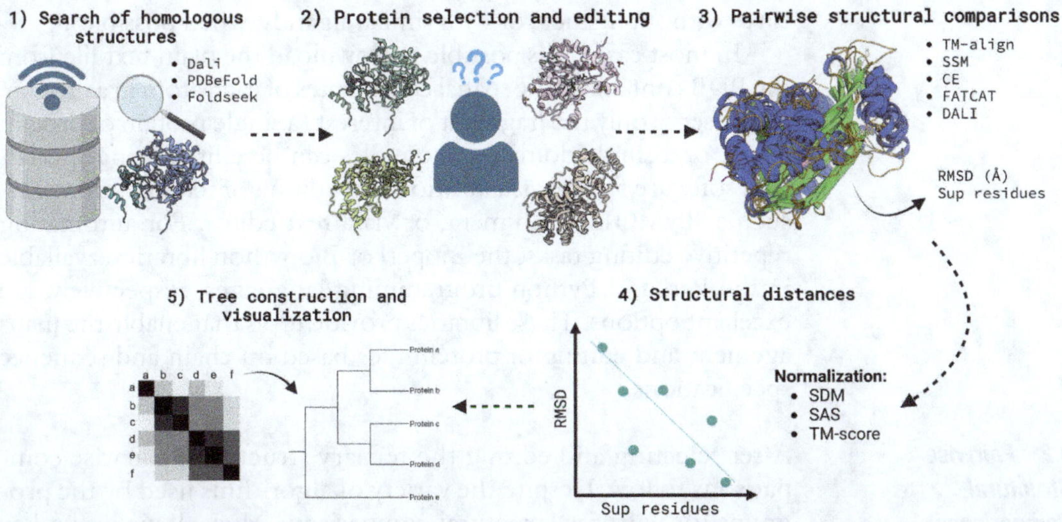

Fig. 1 Flow chart of tertiary structure-based phylogenetic trees construction. Image created with BioRender.com

broadest possible taxonomic and functional protein repertoire, it is recommended to double-check the findings with the structural databases mentioned earlier.

3.1.2 Protein Selection and Editing

Structure-based phylogenies have proven to be an effective alternative when studying the evolution of proteins with high mutation rates, such as those encoded by RNA viruses, and proteins with deep homology, whose primary structure evolutionary footprint is no longer detectable by current methodologies. Proteins typically fold into structural and/or functional fragments called domains, which can interact with other domains to perform their global functions. For instance, the HIV reverse transcriptase comprises three different domains: the right-hand polymerase domain, a connection domain, and the RNase H domain, each contributing to a specific function in viral double-stranded DNA synthesis. Proteins can consist of a single domain or assemble into large complexes in which several domains carry out different functions, such as the bacterial replisome or the nucleosome.

When undertaking a phylogenetic study, it is crucial to determine whether to focus on the evolution of a whole protein or a particular structural or functional domain within a protein. Moreover, proteins are flexible macromolecules that can adopt different conformations. Many proteins deposited in the PDB are bound to other molecules such as nucleotides, nucleic acids, ligands, or inhibitors, which can lead to conformational changes. These changes may alter the pairwise comparisons and, hence, the phylogenetic tree topology. Accordingly, one of the most critical initial steps is the selection of the structures. Based on our laboratory works, we recommend selecting proteins with (1) the best possible resolution (<4 Å), (2) the least possible number of mutations, and (3) the most similar conformation (structures without ligands, or with similar ligands, when possible).

In most cases, it is possible to download the plain text file from the PDB containing the spatial coordinates of each atom in a protein. To preserve only the fragment of interest (a single protein or protein chain, or a single domain), these files can be edited using specialized software for the visualization and editing of tertiary structures, such as PyMOL or Chimera, or via a text editor. For automating repetitive editing tasks, the Bioperl or Biopython libraries, available in the Perl and Python programming languages, respectively, are excellent options. These libraries provide tools that enable the management and editing of protein files based on chain and sequence specifications.

3.2 Pairwise Structural Comparisons

After selecting and editing the tertiary structures, pairwise comparisons follow. Despite the variety of algorithms used by the programs for pairwise structural comparisons, they all yield two key values in addition to the structural or statistical scores: (1) the number of superimposed residues or equivalent residues and the root mean square deviation, i.e., the summed average distance between the Cα of each equivalent residue. While neither of these values directly indicates homology between two proteins, a good alignment typically consists of a high number of superimposed residues and a low RMSD. A visual inspection of the superposition is often necessary to confirm that the aligned fragments are, in fact, the intended ones. For highly divergent proteins, examining the superposition around the active site and the substrate-binding regions is crucial, as these are often the only conserved fragments within a protein Superfamily. For example, the palm subdomain is the sole conserved moiety among the right-hand Superfamily of RNA and DNA polymerases [20, 22].

Most of the available programs offer integrated visualization tools for analyzing pairwise superpositions or displaying alignments of superimposed residues. These features are particularly relevant when analyzing large protein complexes with multiple functional domains. Given these considerations, a comprehensive biological and structural understanding of the protein Superfamily of interest is essential before undertaking any bioinformatic experiment.

3.3 Structural Distances

The limited number of secondary structures that amino acids can adopt within any given protein leads to a high probability of finding similar local arrangements between two random proteins during a structural superposition. The RMSD, which considers the sum of each pair of equivalent Cα atoms, can yield high values for two structurally similar proteins with flexible regions in different conformations, whereas a smaller number of superimposed residues often results in lower RMSDs [56, 57]. Hence, the RMSD alone has proven insufficient as a robust structural distance metric when comparing highly divergent proteins [58]. Therefore, additional parameters should be pondered, e.g. the number of equivalent residues or the length of the proteins, to obtain normalized structural distances between each pair of superimposed proteins [59].

Despite its simplicity, the structural alignment score (SAS) [60] has proven to be accurate, providing evolutionarily robust results across various protein Superfamilies [22–24, 33, 36, 38]. It is calculated with the following formula:

$$\text{SAS} = (\text{RMSD} * 100)/\text{number of superimposed residues}$$

Many widely used web-based programs incorporate these parameters into their formulas, such as the TM-Score [54] and the *Q*-score [49]. Table 3 shows some of the structural scores and their corresponding formulas. Once normalized structural distances are obtained, they are used to build a matrix, allowing specialized programs to calculate the relationship between proteins.

Table 3
Examples of structural distance metrics

Structural distance	Formula
Structural distance metric (SDM) [8]	$D = -100 * \log(w1 * SRMS + w2 * PFTE)$
Structural alignment score (SAS) [60]	(RMSD * 100)/Naln
Length-weighted structural alignment score (LSAS) [38]	$[100 * RMSD(L1 + L2)]/2(Naln)^2$
Structural index (SI) [68]	RMSD*min(L1, L2)/Naln
Q-score [49]	$[Naln * Naln)/[1 + (RMSD/3)^2] * L1 * L2$
TM-score [54]	$\text{Max}\,[(1 / L_{Target})\, {}^{Lali}\Sigma_i\, (1 / (1 + (d_i / (d_0 * L_{Target}))^2))]$

The number in parentheses corresponds to the reference. Naln = number of superimposed residues; L1 and L2 correspond to the length of the two compared proteins, respectively. In SDM: SRMS = 1 − RMSD/3.5; PFTE (Pairwise fractional topological equivalence) = Number of equivalent α-carbon positions/number of amino acids in the smaller structure; w1 = [(1 − SRMS) + (1 − PFTE)]/2; w2 = (PFTE + SRMS)/2; w1 + w2 = 1. In TM-Score: L_{Target} = length of the query protein; L_{ali} = number of aligned residues; d_i = distance between the ith pair of aligned residues; $d_0(L_{Target} = 1.24 * \sqrt[3]{(L_{Target} - 15)} - 1.8$.

3.4 Tree (Dendrogram) Construction and Visualization

As mentioned above, the product of the pairwise comparisons consists of a structural distance matrix, which serves as input to establish the relationships between proteins in the selected Superfamily. While several evolutionary tree-building methods are available for protein primary structures phylogenies (e.g., Maximum Parsimony, distance-based, Maximum Likelihood, Bayesian inference) [61], distance-based methods are the default option for tertiary structure-based dendrogram-building with their advantages and limitations [62]. These methods are computationally fast and include (1) the unweighted pair-group method with arithmetic means (UPGMA), (2) Neighbor-Joining [63], (3) Kitsch, and (4) Fitch. This latter is often preferred for tertiary structure-based dendograms, especially when studying distant homologous proteins, as it does not assume the existence of a molecular clock (https://phylipweb.github.io/phylip/progs.data.dist.html).

Several programs are available for tree visualization and editing, both online and as desktop software. Some of the online software include the widely used Tree of Life (iTOL—https://itol.embl.de/), Phylo.io (https://beta.phylo.io/), and EvolView (http://www.evolgenius.info/evolview.html). The desktop software encompasses programs such as FigTree (http://tree.bio.ed.ac.uk/software/figtree/), Dendroscope (http://ab.inf.uni-tuebingen.de/software/dendroscope/), and TreeViewer (https://treeviewer.org/). The aforementioned software supports various tree formats (e.g., Newick,

Nexus, or phyloXML) and offers several features like customizable tree layouts (radial, rectangular, or circular), display of rooted or unrooted trees, nodes and branches annotation with metadata, and interactive editing for better presentation, clade highlighting, trait mapping, and statistical data overlay.

4 Notes

1. Despite the accelerated growing number of tertiary structures in the PDB, there is a clear bias toward protein structures from organisms of public health, economic, or ecological interest. This may hinder building a given protein Superfamily tertiary structure-based phylogenetic tree with the widest and richest taxonomic diversity. Computer-predicted models may help overcome this limitation.
2. Related to the limitation described above, there is a significant gap between the number of genes, genomes and proteins obtained from metagenomic studies and experimentally-determined tertiary structures. For model organisms such as *Escherichia coli* or *Homo sapiens*, this is partially solved by predicted structural models in the AlphaFold database.
3. One of the key aspects in the construction of these trees is to ensure that the superpositions are biologically and biochemically robust. Manual inspection is recommended for superpositions with high RMSD, a small number of equivalent residues, or scores close to program thresholds. Conservation of the key catalytic residues, or the residues involved in the binding site of the enzyme, is a clear indication that the superimposed proteins are indeed homologous.
4. The use of distance matrices for phylogenetic tree construction limits the array of possible algorithms that can be employed for evolutionary analyses.
5. Currently, there is no standardized methodology to provide statistical or probabilistic support for tertiary structure-based phylogenetic trees, comparable to bootstrap tests in sequence-based evolutionary analyses. Posited alternatives to solve this problem include the use of hierarchical clustering [64], the construction of phylogenetic networks [65], shape fluctuations derived from molecular dynamics simulations [66], structural alphabets derived from tertiary interactions [67]. Despite the lack of statistical support, these analyses remain valuable due to the biological context surrounding the tertiary structures (biological distribution of the proteins/domains, protein's functions, and accessory domains).

6. The methodology described here has been adopted by our work group and has proven to be biologically and chemically sound. However, other groups have used different algorithms and bioinformatic programs with similar results. Readers are encouraged to explore the different options to find the best approach for their protein of interest.

References

1. Illergard K, Ardell DA, Elofsson A (2009) Structure is three to ten times more conserved than sequence—a study of structural response in protein cores. Proteins 77:499–508. https://doi.org/10.1002/prot.22458
2. Nuttall GH (1904) Blood immunity and blood relationship: a demonstration of certain blood-relationships amongst animals by means of the precipitin test for blood. Cambridge University Press, Cambridge
3. Reichert ET, Brown AP (1909) The differentiation and specificity of corresponding proteins and other vital substances in relation to biological classification and organic evolution: the crystallography of hemoglobins. Carnegie Institution of Washington, Washington, D.C.
4. Landsteiner K (1928) Cell antigens and individual specificity. J Immunol 15(6):589–600. https://doi.org/10.4049/jimmunol.15.6.589
5. Landsteiner K 1936 The specificity of serological reactions. Charles C. Thomas, Illinois, https://wellcomecollection.org/works/vanbefv9
6. Zuckerkandl E, Jones RT, Pauling L (1960) A comparison of animal hemoglobins by tryptic peptide pattern analysis. Proc Natl Acad Sci USA 46(10):1349–1360. https://doi.org/10.1073/pnas.46.10.1349
7. Zuckerkandl E, Pauling L (1965) Molecules as documents of evolutionary history. J Theor Biol 8(2):357–366. https://doi.org/10.1016/0022-5193(65)90083-4
8. Johnson MS, Sutcliffe MJ, Blundell TL (1990) Molecular anatomy: phyletic relationships derived from three-dimensional structures of proteins. J Mol Evol 30(1):43–59. https://doi.org/10.1007/BF02102452
9. Binshtein E, Ohi MD (2015) Cryo-electron microscopy and the amazing race to atomic resolution. Biochemistry 54(20):3133–3141. https://doi.org/10.1021/acs.biochem.5b00114
10. Jumper J, Evans R, Pritzel A et al (2021) Highly accurate protein structure prediction with AlphaFold. Nature 596(7873):583–589. https://doi.org/10.1038/s41586-021-03819-2
11. Baek M, DiMaio F, Anishchenko I et al (2021) Accurate prediction of protein structures and interactions using a three-track neural network. Science 373(6557):871–876. https://doi.org/10.1126/science.abj8754
12. Chothia C, Lesk AM (1986) The relation between the divergence of sequence and structure in proteins. EMBO J 5(4):823–826. https://doi.org/10.1002/j.1460-2075.1986.tb04288.x
13. Russell RB, Saqi MA, Sayle RA et al (1997) Recognition of analogous and homologous protein folds: analysis of sequence and structure conservation. J Mol Biol 269(3):423–439. https://doi.org/10.1006/jmbi.1997.1019
14. Andreeva A, Howorth D, Chothia C et al (2014) SCOP2 prototype: a new approach to protein structure mining. Nucl Ac Res 42(Database issue):D310–D314. https://doi.org/10.1093/nar/gkt1242
15. Andreeva A, Kulesha E, Gough J et al (2020) The SCOP database in 2020: expanded classification of representative family and superfamily domains of known protein structures. Nucleic Acids Res 48(D1):D376–D382. https://doi.org/10.1093/nar/gkz1064
16. Orengo CA, Michie AD, Jones S et al (1997) CATH—a hierarchic classification of protein domain structures. Structure (London, England:1993) 5(8):1093–1108. https://doi.org/10.1016/s0969-2126(97)00260-8
17. Sillitoe I, Bordin N, Dawson N et al (2021) CATH: increased structural coverage of functional space. Nucleic Acids Res 49(D1): D266–D273. https://doi.org/10.1093/nar/gkaa1079
18. Waman VP, Bordin N, Alcraft R et al (2024) CATH 2024: CATH-AlphaFlow doubles the number of structures in CATH and reveals nearly 200 new folds. J Mol Biol 436(17):168551. https://doi.org/10.1016/j.jmb.2024.168551
19. Lundin D, Berggren G, Logan DT et al (2015) The origin and evolution of ribonucleotide reduction. Life (Basel, Switzerland) 5(1):604–636. https://doi.org/10.3390/life5010604
20. Mönttinen HA, Ravantti JJ, Stuart DI et al (2014) Automated structural comparisons

clarify the phylogeny of the right-hand-shaped polymerases. Mol Biol Evol 31(10):2741–2752. https://doi.org/10.1093/molbev/msu219
21. Mönttinen HA, Ravantti JJ, Poranen MM (2016) Common structural core of three-dozen residues reveals intersuperfamily relationships. Mol Biol Evol 33(7):1697–1710. https://doi.org/10.1093/molbev/msw047
22. Jácome R, Campillo-Balderas JA, Becerra A et al (2022) Structural analysis of monomeric RNA-dependent polymerases revisited. J Mol Evol 90(3-4):283–295. https://doi.org/10.1007/s00239-022-10059-z
23. Jácome R (2024) Structural and evolutionary analysis of proteins endowed with a nucleotidyltransferase, or non-canonical palm, catalytic domain. J Mol Evol 92(6):799–814. https://doi.org/10.1007/s00239-024-10207-7
24. Cottom-Salas W, Becerra A, Lazcano A (2024) RNA or DNA? Revisiting the chemical nature of the cenancestral genome. J Mol Evol 92(5):647–658. https://doi.org/10.1007/s00239-024-10194-9
25. Duffy S, Shackelton LA, Holmes EC (2008) Rates of evolutionary change in viruses: patterns and determinants. Nat Rev Genet 9(4):267–276. https://doi.org/10.1038/nrg2323
26. Koonin EV, Senkevich TG, Dolja VV (2006) The ancient Virus World and evolution of cells. Biol Direct 1: 29. https://doi.org/10.1186/1745-6150-1-29
27. Nasir A, Kim KM, Caetano-Anollés G (2012) Viral evolution: primordial cellular origins and late adaptation to parasitism. Mob Genet Elements 2:247–252. https://doi.org/10.4161/mge.22797
28. Gilbert W (1986) Origin of life: the RNA world. Nature 319:618–618. https://doi.org/10.1038/319618a0
29. Joyce G (2002) The antiquity of RNA-based evolution. Nature 418:214–221. https://doi.org/10.1038/418214a
30. Ferron F, Sama B, Decroly E et al (2021) The enzymes for genome size increase and maintenance of large (+)RNA viruses. Trends Biochem Sci 46(11):866–877. https://doi.org/10.1016/j.tibs.2021.05.006
31. Černý J, Černá Bolfíková B, de A, Zanotto PM et al (2015) A deep phylogeny of viral and cellular right-hand polymerases. Infect Genet Evol 36:275–286. https://doi.org/10.1016/j.meegid.2015.09.026
32. Mönttinen HAM, Ravantti JJ, Poranen MM (2021) Structure unveils relationships between RNA virus polymerases. Viruses 13(2):313. https://doi.org/10.3390/v13020313
33. Jácome R, Becerra A, Ponce de León S et al (2015) Structural analysis of monomeric RNA-dependent polymerases: evolutionary and therapeutic implications. PloS ONE 10(9):e0139001. https://doi.org/10.1371/journal.pone.0139001
34. Venkataraman S, Prasad BVLS, Selvarajan R (2018) RNA dependent RNA polymerases: insights from structure, function and evolution. Viruses 10(2):76. https://doi.org/10.3390/v10020076
35. Peersen OB (2019) A comprehensive superposition of viral polymerase structures. Viruses 11(8):745. https://doi.org/10.3390/v11080745
36. Cruz-González A, Muñoz-Velasco I, Cottom-Salas W et al (2021) Structural analysis of viral ExoN domains reveals polyphyletic hijacking events. PloS ONE 16(3):e0246981. https://doi.org/10.1371/journal.pone.0246981
37. Mönttinen HAM, Ravantti JJ, Poranen MM (2019) Structural comparison strengthens the higher-order classification of proteases related to chymotrypsin. PloS ONE 14(5):e0216659. https://doi.org/10.1371/journal.pone.0216659
38. Cisneros-Martínez AM, Becerra A, Lazcano A (2021) Ancient gene duplications in RNA viruses revealed by protein tertiary structure comparisons. Virus Evol 7(1):veab019. https://doi.org/10.1093/ve/veab019
39. Butkovic A, Dolja VV, Koonin EV et al (2023) Plant virus movement proteins originated from jelly-roll capsid proteins. PLoS Biol 21(6):e3002157. https://doi.org/10.1371/journal.pbio.3002157
40. Mifsud JCO, Lytras S, Oliver MR et al (2024) Mapping glycoprotein structure reveals Flaviviridae evolutionary history. Nature 633(8030):695–703. https://doi.org/10.1038/s41586-024-07899-8
41. Berman HM, Westbrook J, Feng Z et al (2000) The Protein Data Bank. Nucl Acids Res 28(1):235–242. https://doi.org/10.1093/nar/28.1.235
42. Berman HM, Kleywegt GJ, Nakamura H et al (2012) The Protein Data Bank at 40: reflecting on the past to prepare for the future. Structure (London, England:1993) 20(3):391–396. https://doi.org/10.1016/j.str.2012.01.010
43. Lau AM, Bordin N, Kandathil SM et al (2024) Exploring structural diversity across the protein universe with The Encyclopedia of Domains. Science 386(6721):eadq4946. https://doi.org/10.1126/science.adq4946
44. Hubbard T, Ailey B, Brenner SE et al (1999) SCOP: a Structural Classification of Proteins database. Nucl Acid Res 27:254–256. https://doi.org/10.1093/nar/27.1.254
45. Varadi M, Bertoni D, Magana P et al (2024) AlphaFold Protein Structure Database in

2024: providing structure coverage for over 214 million protein sequences. Nucl Acid Res 52(D1):D368–D375. https://doi.org/10.1093/nar/gkad1011
46. Guzenko D, Burley SK, Duarte JM (2020) Real time structural search of the Protein Data Bank. PLoS Comput Biol 16(7):e1007970. https://doi.org/10.1371/journal.pcbi.1007970
47. Holm L, Sander C (1995) Dali: a network tool for protein structure comparison. Trends Biochem Sci 20(11):478–480. https://doi.org/10.1016/s0968-0004(00)89105-7
48. Holm L (2020) Using Dali for protein structure comparison. Methods in. Mol Biol (Clifton, N.J.) 2112:29–42. https://doi.org/10.1007/978-1-0716-0270-6_3
49. Krissinel E, Henrick K (2004) Secondary-structure matching (SSM), a new tool for fast protein structure alignment in three dimensions. Acta Crystallogr D Struct Biol 60(Pt 12 Pt 1):2256–2268. https://doi.org/10.1107/S0907444904026460
50. van Kempen M, Kim SS, Tumescheit C (2024) Fast and accurate protein structure search with Foldseek. Nat Biotechnol 42(2):243–246. https://doi.org/10.1038/s41587-023-01773-0
51. Kim RS, Levy Karin E, Mirdita M et al (2024) BFVD-a large repository of predicted viral protein structures. Nucl Acid Res gkae1119. Advance online publication. https://doi.org/10.1093/nar/gkae1119
52. Shindyalov IN, Bourne PE (1998) Protein structure alignment by incremental combinatorial extension (CE) of the optimal path. Protein Eng 11(9):739–747. https://doi.org/10.1093/protein/11.9.739
53. Ye Y, Godzik A (2003) Flexible structure alignment by chaining aligned fragment pairs allowing twists. Bioinformatics (Oxford, England) 19(Suppl 2):ii246–ii255. https://doi.org/10.1093/bioinformatics/btg1086
54. Zhang Y, Skolnick J (2005) TM-align: a protein structure alignment algorithm based on the TM-score. Nucleic Acid Res 33(7):2302–2309. https://doi.org/10.1093/nar/gki524
55. Bittrich S, Segura J, Duarte JM et al (2024) RCSB protein Data Bank: exploring protein 3D similarities via comprehensive structural alignments. Bioinformatics 40(6):btae370. https://doi.org/10.1093/bioinformatics/btae370
56. Yee DP, Dill KA (1993) Families and the structural relatedness among globular proteins. Protein Sci 2(6):884–899. https://doi.org/10.1002/pro.5560020603
57. Xu J, Zhang Y (2010) How significant is a protein structure similarity with TM-score = 0.5?. Bioinformatics 26(7):889–895. https://doi.org/10.1093/bioinformatics/btq066
58. Balaji S, Srinivasan N (2001) Use of a database of structural alignments and phylogenetic trees in investigating the relationship between sequence and structural variability among homologous proteins. Protein Eng 14(4):219–226. https://doi.org/10.1093/protein/14.4.219
59. Kolodny R, Koehl P, Levitt M (2005) Comprehensive evaluation of protein structure alignment methods: scoring by geometric measures. J Mol Biol 346(4):1173–1188. https://doi.org/10.1016/j.jmb.2004.12.032
60. Subbiah S, Laurents DV, Levitt M (1993) Structural similarity of DNA-binding domains of bacteriophage repressors and the globin core. Curr Biol 3:141–148. https://doi.org/10.1016/0960-9822(93)90255-M
61. Pevsner J (2015) Chapter 7 Molecular phylogeny and evolution. In Bioinformatics and functional genomics. 3rd Ed. Wiley Blackwell
62. Zou Y, Zhang Z, Zeng Y et al (2024) Common methods for phylogenetic tree construction and their implementation in R. Bioengineering 11:480. https://doi.org/10.3390/bioengineering11050480
63. Graur D, Li WH (2000) Chapter 5 Molecular phylogenetics. In Fundamentals of molecular evolution. 2nd Ed. Sinauer Associates, Inc.
64. Suzuki R, Shimodaira H (2006) Pvclust: an R package for assessing the uncertainty in hierarchical clustering. Bioinformatics 22:1540–1542. https://doi.org/10.1093/bioinformatics/btl117
65. Lundin D, Poole AM, Sjöberg BM et al (2012) Use of structural phylogenetic networks for classification of the Ferritin-like Superfamily. J Biol Chem 287:20565–20575. https://doi.org/10.1074/jbc.M112.367458
66. Malik AJ, Poole AM, Allison JR (2020) Structural phylogenetics with confidence. Mol Biol Evol 37:2711–2726. https://doi.org/10.1093/molbev/msaa100
67. Moi D, Bernard C, Steinneger M et al (2023) Structural phylogenetics unravels the evolutionary diversification of communication systems in gram-positive bacteria and their viruses. Biorxiv. 558401. https://doi.org/10.1101/2023.09.19
68. Kleywegt GJ, Jones A (1994) Crystallography 31:9–14

Chapter 17

Protein Architecture Comparisons and the Reconstruction of the Last Universal Common Ancestor (LUCA)

Ricardo Hernández-Morales, Coral C. G. Luna, Antonio Lazcano and Arturo Becerra

Abstract

The Last Universal Common Ancestor (LUCA) represents an early stage of evolution that can be studied using phylogenetic and bioinformatic methods. Research efforts focus on reconstructing LUCA's genome, proteome, metabolism, habitat, ecological context, and timeline. A novel method developed by our group identifies structural homologs by comparing protein architectures, defined as the sequential arrangement of homologous domains within a protein. In our approach, proteins are considered homologous if they share identical domain types in the same N-to-C-terminal order. Homologous proteins with the same architecture found in the primary branches of both Bacteria and Archaea are potential candidates for those present in LUCA.

Keywords Protein architecture comparison, Protein domains, Early proteins, Ancestral activities, Last Universal Common Ancestor

1 Introduction

The Last Universal Common Ancestor (LUCA) is defined as an ancestral microbial population representing one of the earliest stages of cellular evolution from which the two major prokaryotic domains diverged, which can be studied using phylogenetic and bioinformatic methods. Over recent decades, numerous studies have sought to investigate LUCA's emergence [1], reconstruct its genome and proteome [2, 3], and predict its metabolism [4–6], habitat [4], ecological context [7], timeline [8], and biological complexity [9].

Reconstructing the gene complements of distant ancestors represents a statistical approximation of the biological past, with its accuracy shaped by various factors. These include biases in genome or proteome database construction, the extent of horizontal gene transfer, significant differences in sequence substitution rates, secondary gene losses, and inherent methodological constraints [2].

Gustavo Caetano-Anollés (ed.), *Evolutionary Genomics*, Methods in Molecular Biology, vol. 2981,
https://doi.org/10.1007/978-1-0716-4836-0_17,

Despite these obstacles, some researchers have generated detailed LUCA reconstructions, encompassing its metabolic traits and gene catalogs, using diverse statistical genomic approaches. These efforts highlight the challenges and the potential for uncovering insights into life's earliest evolutionary stages [10].

However, a complete or accurate reconstruction is an unattainable goal; at best, we can make robust inferences from an unequal partial reconstruction for different ancestral states. While such reconstructions are necessarily incomplete, they provide valuable and informative biological frameworks [11]. A more comprehensive inventory of highly conserved traits would require an in-depth analysis of homologous protein sets, employing more sensitive approaches such as profile-based methods and, potentially, leveraging insights from tertiary structure databases. Despite these challenges, the available data offer valuable insights into early evolutionary stages.

Most protein domains correspond to tertiary structure elements that may fold independently and exhibit different sequence conservation levels [12]. Studying the preservation of domain ordering could help make inferences about the proteins' antiquity and evolution. Here, we report a novel protein comparison method to search for homologs. It utilizes a customized report to identify structural homologs by comparing protein architectures from the PDB database. Protein architecture, or domain arrangement, is the sequential order of domains within a protein. Hence, in our approach, two proteins are classified as homologous if they share identical types and numbers of domains, arranged in the same order from the N-terminal to the C-terminal end. Pending issues like lateral transfer and homologous proteins in the primary branches of Bacteria and Archaea are candidates dating from LUCA epochs, allowing us to infer some of their metabolic abilities and other biological traits.

2 Materials

2.1 PDB Database

Our method uses information derived from the Report of the Protein Data Bank (PDB) database [13, 14]. The PDB database provides access to 3D structure data of biomolecules. It stores atomic coordinates for protein tertiary structures is determined through various methods, including X-ray crystallography (XRC), nuclear magnetic resonance spectroscopy (NMR), and cryo-electron microscopy (cryo-EM), among others. In addition to atomic coordinates, the PDB provides detailed molecular information for each entry. This approach utilizes data from experimentally determined structures and does not include structural models derived from computational predictions.

2.2 Information Used in This Methodology

The information described below, collected from the tabular report of the PDB database for experimentally solved structures, was used to search for Pfam domains and their order in proteins, eliminate architecture redundancy, and understand the proteins' function.

- *PDB ID.* The unique identifier to each entry in the Protein Data Bank;
- *Resolution (Å).* The resolution of each tertiary structure in angstroms;
- *Entry Polymer Types.* The type of polymer, such as nucleic acids, proteins, etc;
- *Sequence.* The nucleotide or amino acid sequence;
- *Entity Polymer Type.* The specific polymeric entity, such as a protein, RNA, or DNA;
- *Sequence Length.* The length of each polymer sequence;
- *Entity Macromolecule Type.* The type of macromolecule;
- *Source Organism.* The organism from which each tertiary structure originates;
- *Taxonomy ID.* The taxonomy identifier for each entry;
- *Gene Name.* The name of the gene encoding the protein;
- *Macromolecule Name.* The name assigned to each macromolecule;
- *EC Number.* The enzymatic code to the structure, if it is an enzyme;
- *Annotation Identifier.* Identifiers associated with the tertiary structure;
- *Accession Codes.* The reference database access codes;
- *Ligand ID.* Ligands bound to each macromolecule;

2.3 Proteomes Used as a Biological Representative Sample to Infer the Phylogenetic Distribution of Protein Architectures

A sequence-based search for homologous proteins was conducted for each conserved protein architecture across reference proteomes covering the major prokaryotic phylogenetic groups. The analysis included only proteomes of Bacteria ($N = 629$) and Archaea ($N = 90$), which were downloaded from the KEGG database. Representative bacterial groups were selected based on the work of Coleman et al. [15], encompassing Proteobacteria (265), Desulfuromonadota & Desulfobacterota (23), Myxococcota (5), Bdellovibrionota (1), Firmicutes (108), Actinobacteriota (86), Cyanobacteria & Margulisbacteria (29), Chloroflexota & Dormibacterota (9), Deinococcota, Synergistota & Thermotogota (15), Planctomycetota, Verrucomicrobiota & Chlamydiota (15), Spirochaetota (13), Acidobacteriota (3), Fibrobacterota, Chlorobia & Bacteroidota (40), Fusobacteriota (5), Elusimicrobiota (1), Aquificota, Campylobacterota & Deferribacterota (10), and Nitrospirota (1). Representative archaeal groups were chosen according to the phylogenetic tree reported by Aparici-Carratalá et al. [16], and included Euryarchaeota (59), Thaumarchaeota, Aigarchaeota, Crenarchaeota & Korarchaeota (29), DPANN (1), and Asgard (1).

3 Methods

3.1 General Information on the Method Presented Here

Our method relies on a customized report from the PDB database to identify structural homologs by comparing the protein architectures. Architecture, or domain arrangement, is defined as the order of homologous domains within a protein [17]. In our approach, two proteins are considered homologous if they share the same types and numbers of domains, arranged in the same order from the N-terminal to the C-terminal (Fig. 1). Their phylogenetic distribution in the main clades is also searched (Fig. 2) to support the proposal of their presence in LUCA. An advantage of our methodology is that it allows for an easier and faster comparison of proteins compared to standard methods (*see* **Note 1**).

3.2 Raw Data of Bacteria and Archaea PDB Entries

The complete, customized, and unprocessed data is downloaded from the PDB database as plain text in a tabular format with a CSV extension. Only experimentally determined structures are considered. The information obtained from the database corresponds to Sect. 2.2. To facilitate computational manipulation of the text, the data is saved in TXT format. Two files are obtained, one for Bacteria and the

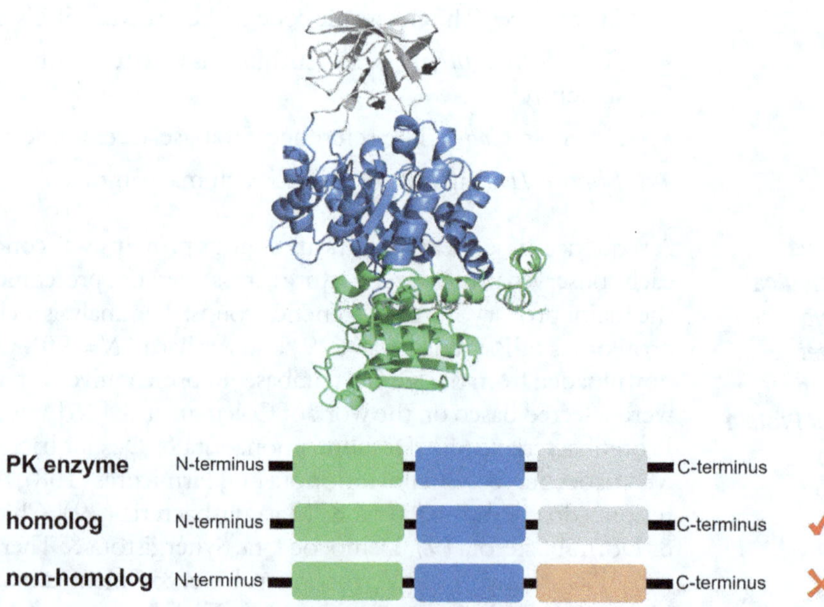

Fig. 1 Protein architecture. The tertiary structure observed at the top corresponds to pyruvate kinase (PDB 1PKN). This enzyme is involved in the last step of glycolysis. It has three distinct domains, two of which are involved in the active site (blue and green), and the last is involved in the regulation (white). The protein architecture, also known as protein domain arrangement, is represented by green, blue, and white rectangles. As described here, proteins must have the same number of domains and arrangements to be considered as homologs

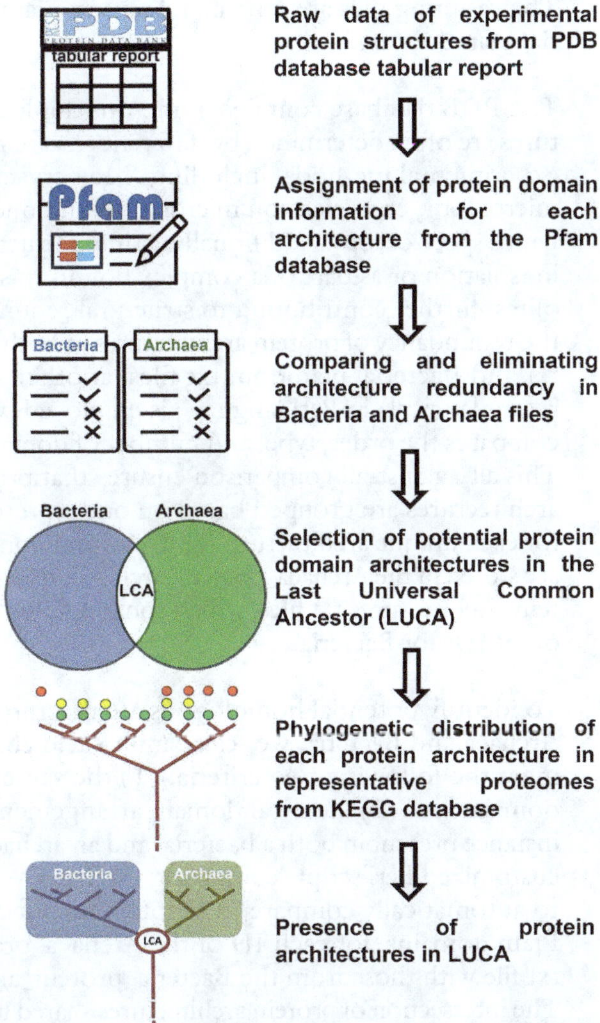

Fig. 2 Representation of the methodology to obtain the potential proteins present in LUCA. The steps followed in this method are summarized in this figure. For more information, see Sects. 3.1–3.6 in "Methods"

other for Archaea. The resulting files are named Bacteria_raw_data.txt and Archaea_raw_data.txt. The dataset includes 219,262 entries for Bacteria and 17,214 for Archaea.

3.3 Pfam Domains and Protein Architecture

Biological data on all protein domains is retrieved from the UniProt [18] and Pfam [19] databases to analyze the domain arrangements within proteins. The domain information, provided as plain text, is stored in a file called Pfam_dat. A custom Perl script (getting_pfam_domains.pl) extracts relevant details from this file, including domain names and functions. This information is stored in a new column in the files Bacteria_raw_data.txt and Archaea_raw_data.txt.

The resulting files are renamed Archaea_pfamdom.txt and Bacteria_pfamdom.txt.

3.4 Eliminating Redundancy of IDs from Bacteria and Archaea Files

The PDB database contains numerous duplicate entries, as structures are often determined by different research teams using various experimental methods, including X-ray crystallography, electron microscopy, cryo-electron microscopy, and nuclear magnetic resonance spectroscopy. Additionally, a single structure may be resolved in isolation or as part of a complex bound to substrates or metabolites, further contributing to structural redundancy. To eliminate the redundancy of protein architectures in both Archaea_pfamdom.txt and Bacteria_pfamdom.txt files, another script (eliminating_redundancy.pl) in Perl language is employed, which systematically compares the order, type, and number of domains for each protein. This all-against-all comparison ensures that proteins with identical architectures are grouped based on one representative ID selected for each unique architecture. The resulting non-redundant datasets are stored in the Archaea_protein_architectures.txt and Bacteria_protein_architectures.txt files, which contain 1,266 IDs for Archaea and 6569 IDs for Bacteria.

3.5 Homologous Protein Architectures in Bacteria and Archaea

To identify potential homologous architectures presented in both Archaea and Bacteria, we select amino acid chains of proteins that meet the following two criteria: (1) the same number of protein domains and (2) identical domain arrangements, with at least one instance present in both a bacterial and an archaeal protein. Another customized Perl script (comparing_Archaea_vs_Bacteria.pl) is used to automatically compare the protein architectures, composed of Pfam domains, for each ID of the Archaea_protein_architectures.txt file with those from the Bacteria_protein_architectures.txt files. The intersection of protein architectures shared between Archaea and Bacteria is identified and recorded in the Archaea_vs_Bacteria_homologous_architectures.txt file. A total of 787 architectures are found to be shared between Archaea and Bacteria. Some limitations can exist due to constraints on the materials and methodology, which may impact the findings of our study (*see* **Note** 2).

3.6 Phylogenetic Distribution of Homologous Architectures in the Prokaryotic Phylogenetic Tree

A comparative analysis using delta-blast [20] version 2.2.29+ is performed using the sequences of each positive hit as queries and sequences from representative cellular proteomes as subjects to determine the phylogenetic distribution of protein architectures from Archaea_vs_Bacteria_homologous_architectures.txt file and to avoid false positives on the reconstruction of LUCA (*see* **Note 3**). The parameters used are an initial BLOSUM45 scoring matrix, an -value ≤ 0.001, an identity percentage $\geq 30\%$, and query and subject coverage $\geq 70\%$. The representative proteome database is constructed based on the KEGG database [21]. Only species of

Bacteria and Archaea are considered. Endosymbiotic and parasitic species are excluded because they are subject to polyphyletic gene losses that can bias backtrack characterizations of ancestral states [22] (*see* **Note 4**). The resulting proteome dataset includes 719 proteomes from 629 Bacteria and 90 Archaea.

The architecture of each putative LUCA protein is validated based on two criteria: (1) more than 50% of the branches representing major prokaryotic phylogenetic groups contain homologous hits, and (2) these branches are scattered over the major clades. These criteria try to reduce as much as possible the number of false positives and false negatives (*see* **Notes 3 and 4**).

3.7 Functional Classification of Proteins Associated with Each Architecture

The functional information for each protein is compiled from multiple sources. Initially, data are extracted from the tabular report downloaded from the PDB database, focusing on the fields "Macromolecule Name," "EC Number," and "Accession Codes." Subsequently, the UniProt identifier provided in the "Accession Codes" field is used to retrieve functional annotations, including Gene Ontology (GO) terms [23] and the eggNOG identifier, from the UniProt database [18]. Additionally, functional information is obtained from the eggNOG database [24], which provides condensed GO Slim categories and KEGG pathway data for orthologous groups (*see* **Note 5**).

The study incorporates functional data from two distinct sources for the same protein architecture: one associated with bacterial proteins and the other with archaeal proteins. The functional annotations for the same architecture are compared to infer the potential ancestral function of the protein corresponding to the Last Universal Common Ancestor. Based on this functional information, conserved architectures are manually classified into major KEGG pathway functional categories [21], as summarized in Table 1.

4 Notes

1. Divergence time accumulates changes that erase the signal of a common origin, primarily affecting the primary structure of proteins, as selection pressure acts on function, which ultimately depends on the tertiary structure. For this reason, the analysis and comparison of tertiary structures is a highly suitable procedure for identifying proteins that may have been present in the Last Universal Common Ancestor. However, comparative analysis of tertiary structures remains computationally demanding, making it impractical for large-scale searches. On the other hand, comparing protein architectures to identify ancestral proteins is a robust and fast method that can be applied to large-scale analyses.

Table 1

Functional classification of proteins associated with each architecture. Functional classification based on the KEGG database classification scheme (https://www.genome.jp/kegg/pathway.html)

I. Genetic Information Processing	1.1 Transcription
	1.2 Translation
	1.3 Folding, sorting, and degradation
	1.4 Replication and repair
	1.5 Others
II. Metabolism	2.1 Carbohydrate metabolism
	2.2 Energy metabolism
	2.3 Lipid metabolism
	2.4 Nucleotide metabolism
	2.5 Amino acid metabolism
	2.6 Metabolism of cofactors and vitamins
	2.7 Others
III. Environmental Information Processing	3.1 Membrane transport
	3.2 Signal transduction
IV. Cellular Processes	4.1 Transport
V. Unknown	

2. Despite the advantages of the method outlined here for reconstructing LUCA's proteome, detecting false positives and negatives is a significant limitation, as with any other reconstruction approach. Unlike false positives, false negatives may significantly affect the reconstruction of ancestral traits, leading to a biased description that does not recognize the complexity of LUCA.
3. In the case of false positives, the occurrence of a specific architecture in a bacterial or archaeal representative as a result of a horizontal gene transfer (HGT) event may overrepresent the protein content in LUCA. However, searching for homologous sequences in cellular proteomes can help to minimize this possibility. The architecture's wide phylogenetic distribution supports the hypothesis of a common evolutionary origin rather than being a product of HGT.
4. In the case of false negatives, the absence of specific proteins may lead to an underrepresentation of LUCA's protein content. Several factors could contribute to this underrepresentation, including (a) secondary losses, which cannot be detected by this method, (b) biases in the protein sample available in the

Table 2

Examples of the functional information available for selected architectures. Functions associated with the same architecture were compared to infer the potential functions in LUCA. From top to bottom, we present cases with varying levels of certainty regarding the ancestral functions

Examples		Functional information		Functional inference in LUCA	
Architecture (pfam domain)	PDB	Protein from Archaea	Protein from Bacteria		
Ribosomal_L2; Ribosomal_ L2_C	8HKV 8T8C	50S ribosomal protein L2 Ribosome translation [GO:0006412] large ribosomal subunit [GO:0015934] rRNA binding [GO:0019843]; structural constituent of ribosome [GO:0003735] J_Translation, ribosomal structure and biogenesis Cytoplasmic translation Ribosome	50S ribosomal protein L2 Ribosome translation [GO:0006412] large ribosomal subunit [GO:0015934] rRNA binding [GO:0019843]; structural constituent of ribosome [GO:0003735]; transferase activity [GO:0016740] J_Translation, ribosomal structure and biogenesis Cytoplasmic translation Ribosome biogenesis Ribosome	50S ribosomal protein L2 Ribosome translation [GO:0006412] large ribosomal subunit [GO:0015934] rRNA binding [GO:0019843]; structural constituent of ribosome [GO:0003735] J_Translation, ribosomal structure and biogenesis Cytoplasmic translation Ribosome	I. Genetic Information Processing 1.2 Translation
ADH_zinc_N; ADH_N	4JBI7XY9	Alcohol dehydrogenase (Zinc) Oxidoreductase EC 1.1.1.1metal ion binding [GO:0046872]; nucleotide binding [GO:0000166]; oxidoreductase activity [GO:0016491] C_Energy production and conversion	NADP-dependent isopropanol dehydrogenase Oxidoreductase EC 1.1.1.80 isopropanol dehydrogenase (NADP+) activity [GO:0050009]; zinc ion binding [GO:0008270]	Alcohol dehydrogenase Oxidoreductase EC 1.1.1.- metal ion binding [GO:0046872]	II. Metabolism

(continued)

Table 2
continued

Examples		Functional information		Functional inference in LUCA	
Architecture (pfam domain)	PDB	Protein from Archaea	Protein from Bacteria		
GAF_2; GAF	4MN76K2H	Putative uncharacterized protein Oxidoreductase T_Signal transduction mechanisms signaling protein modification process DNA-templated transcription Metabolic pathways Cysteine and methionine metabolism	NreA Signaling protein T_Signal transduction mechanisms, signaling protein modification process, DNA-templated transcription, Two-component system	T_Signal transduction mechanisms signaling protein modification process DNA-templated transcription	I. Genetic Information Processing 1.3 Folding, sorting, and degradation
CBS domain	7MSU3K2V	Hemolysin, contains CBS domains, Membrane protein, membrane [GO:0016020], and flavin adenine dinucleotide binding [GO:0050660]	Putative D-arabinose 5-phosphate isomerase Isomerase EC 5.3.1.13 lipopolysaccharide biosynthetic process [GO:0009103] arabinose-5-phosphate isomerase activity [GO:0019146]; carbohydrate derivative binding [GO:0097367]; metal ion binding [GO:0046872]	Unknown	V. Unknown
TatD_DNase	3GUW6L25	Uncharacterized protein Structural genomics, unknown function hydrolase activity, acting on ester bonds [GO:0016788]; metal ion binding [GO:0046872] S_Function unknown	Deoxyribonuclease YcfH Metal binding protein EC 3.1.21.-DNA nuclease activity [GO:0004536]; exonuclease activity [GO:0004527]; metal ion binding [GO:0046872]	Unknown	V. Unknown

Table 2
continued

SCP2	3CNU6H6P	Predicted coding region Structural genomics, unknown function I_Lipid transport and metabolism	Ubiquinone biosynthesis protein UbiJ Lipid binding protein ubiquinone biosynthetic process [GO:0006744]; ubiquinone biosynthetic process from chorismate [GO:0032150] cytoplasm [GO:0005737]; ubiquinone biosynthesis complex [GO:0110142] identical protein binding [GO:0042802] S_Function unknown, carbohydrate metabolic process signaling	Unknown	II. Metabolism
DUF190	2DCL1O51	Hypothetical UPF0166 protein, Structural genomics, unknown S_Function unknown	Hypothetical protein TM0021, Signaling protein, T_Signal transduction mechanisms transmembrane transport	Unknown	V. Unknown

PDB database, diminishing the number of gene/protein content in LUCA, (c) instances of domain permutations, where rearrangements of domains within proteins change and are undetected due to the limitations of our operational definition, and (d) other intrinsic issues within the method itself.

5. Assigning potential functions to LUCA's proteins can be challenging, as functional annotations in the PDB database may differ even among proteins with identical architectures. Our method is particularly susceptible to being less specific regarding the function attributed to the ancestral protein. Therefore, complementary information from other databases (e.g., Uniprot, eggNOG, KEGG) should be used to support the data, and caution must be exercised when making inferences about reconstructions.

Some examples of different functions associated with the same protein architecture were identified (Table 2). We inferred the ancestral function for most architectures because the same function is reported in both Bacteria and Archaea protein architectures. An example is the Ribosomal_L2, Ribosomal_L2_C architecture that is present in bacterial and archaeal ribosomal protein L2, which binds to rRNA of the large subunit in the peptidyl-transfer center vicinity and is thought as one of the most ancient ribosomal proteins; or proteins with ADH_zinc_N, ADH_N architecture, which have similar functions, and are both involved in an oxidoreductase reaction. But we have also identified a small number of cases where the tasks reported were different (e.g., CBS domain in a hemolysin protein and a putative D-arabinose 5-phosphate isomerase from Bacteria and Archaea, respectively) or where the functional information available (e.g., TatD_DNase, SCP2, DUF190), do not allow the inference of the potential ancestral function in LUCA. Enormous efforts must be made to properly annotate and correct the information for each entry stored in the databases. Better-quality annotations will avoid problems like those described above.

Acknowledgments

This manuscript is part of the requirements for obtaining a Ph.D. degree from the Posgrado en Ciencias Biológicas, UNAM of CCGL. This work was supported by a fellowship of Consejo Nacional de Humanidades, Ciencias y Tecnologías (CONAHCyT), México (CCGL with CVU 697387). We are indebted to Alejandro Álvarez-Lugo and José Alberto Campillo-Balderas for their technical assistance and enriching discussions.

References

1. Betts HC, Puttick MN, Clark JW, Williams TA, Donoghue PCJ, Pisani D (2018) Integrated genomic and fossil evidence illuminates life's early evolution and eukaryote origin. Nat Ecol Evol 2(10):1556–1562. https://doi.org/10.1038/s41559-018-0644-x
2. Becerra A, Delaye L, Islas S, Lazcano A (2007) The very early stages of biological evolution and the nature of the last common ancestor of the three major cell domains. Annu Rev Ecol Evol Syst 38:361–379. http://www.jstor.org/stable/30033864
3. Goldman AD, Bernhard TM, Dolzhenko E, Landweber LF (2013) LUCApedia: a database for the study of ancient life. Nucleic Acids Res 41(D1):D1079–D1082. https://doi.org/10.1093/nar/gks1217
4. Weiss MC, Sousa FL, Mrnjavac N, Neukirchen S, Roettger M, Nelson-Sathi S, Martin WF (2016) The physiology and habitat of the last universal common ancestor. Nat Microbiol 1(9):16116. https://doi.org/10.1038/nmicrobiol.2016.116
5. Goldman AD, Weber JM, LaRowe DE, Barge LM (2023) Electron transport chains as a window into the earliest stages of evolution. Proc Natl Acad Sci USA 120(34):e2210924120. https://doi.org/10.1073/pnas.2210924120
6. Ledford SM, Meredith LK (2024) Volatile organic compound metabolism on early Earth. J Mol Evol 92(5):605–617. https://doi.org/10.1007/s00239-024-10184-x
7. Krupovic M, Dolja VV, Koonin EV (2020) The LUCA and its complex virome. Nat Rev Microbiol 18(11):661–670. https://doi.org/10.1038/s41579-020-0408-x
8. Moody ERR, Álvarez-Carretero S, Mahendrarajah TA et al (2024) The nature of the last universal common ancestor and its impact on the early Earth system. Nat Ecol Evol 8(9):1654–1666. https://doi.org/10.1038/s41559-024-02461-1
9. Lazcano A, Fox G, Oro J (1992) Life before DNA: the origin and evolution of early archaean cells. In: Mortlock RP (ed) The evolution of metabolic function. CRC Press. Boca Raton, Florida. pp 237–295
10. Goldman AD, Becerra A (2024) A new view of the last universal common ancestor. J Mol Evol 92(5):659–661. https://doi.org/10.1007/s00239-024-10193-w
11. Estrada A, Suárez-Díaz E, Becerra A (2022) Reconstructing the last common ancestor: epistemological and empirical challenges. Acta Biotheor 70(2):15. https://doi.org/10.1007/s10441-022-09439-1
12. Forslund SK, Kaduk M, Sonnhammer ELL (2019) Evolution of protein domain architectures. In: Anisimova M (ed) Evolutionary genomics. Methods in molecular biology. Humana, New York, NY. pp 469-504
13. Berman HM, Westbrook J, Feng Z, Gilliland G, Bhat TN, Weissig H, Shindyalov IN, Bourne PE (2000) The protein data bank. Nucleic Acids Res 28(1):235–242. https://doi.org/10.1093/nar/28.1.235
14. Berman H, Henrick K, Nakamura H (2003) Announcing the worldwide protein data bank. Nat Struct Biol 10(12):980. https://doi.org/10.1038/nsb1203-980
15. Coleman GA, Davín AA, Mahendrarajah TA, Szánthó LL, Spang A, Hugenholtz P, Szöllősi GJ, Williams TA (2021) A rooted phylogeny resolves early bacterial evolution. Science 372(6542):eabe0511. https://doi.org/10.1126/science.abe0511
16. Aparici-Carratalá D, Esclapez J, Bautista V, Bonete MJ, Camacho M (2023) Archaea: current and potential biotechnological applications. Res Microbiol 174(7):104080. https://doi.org/10.1016/j.resmic.2023.104080
17. Mulder NJ (2010) Protein domain architectures. Meth Mol Biol 609:83–95. https://doi.org/10.1007/978-1-60327-241-4_5
18. UniProt Consortium (2025) UniProt: the universal protein knowledgebase in 2025. Nucleic Acids Res 53(D1):D609–D617. https://doi.org/10.1093/nar/gkae1010
19. El-Gebali S, Mistry J, Bateman A, Eddy SR, Luciani A, Potter SC et al (2019) The Pfam protein families database in 2019. Nucleic Acids Res 47(D1):D427–D432. https://doi.org/10.1093/nar/gky995
20. Boratyn GM, Schäffer AA, Agarwala R, Altschul SF, Lipman DJ, Madden TL (2012) Domain enhanced lookup time accelerated BLAST. Biol Direct 7:12. https://doi.org/10.1186/1745-6150-7-12
21. Kanehisa M, Sato Y, Kawashima M, Furumichi M, Tanabe M (2016) KEGG as a reference resource for gene and protein annotation. Nucleic Acids Res 44(D1):D457–D462. https://doi.org/10.1093/nar/gkv1070
22. Becerra A, Islas S, Leguina JI, Silva E, Lazcano A (1997) Polyphyletic gene losses can bias backtrack characterizations of the cenancestor. J Mol Evol 45(2):115–118. https://doi.org/10.1007/pl00006209
23. Mi H, Muruganujan A, Ebert D, Huang X, Thoma PD (2019) PANTHER version 14: more genomes, a new PANTHER GO-slim and improvements in enrichment analysis tools. Nucleic Acids Res 47(D1):D419–D426. https://doi.org/10.1093/nar/gky1038

24. Hernández-Plaza A, Szklarczyk D, Botas J, Cantalapiedra CP, Giner-Lamia J, Mende DR, Kirsch R, Rattei T, Letunic I, Jensen LJ, Bork P, von Mering C, Huerta-Cepas J (2023) eggNOG 6.0: enabling comparative genomics across 12,535 organisms. Nucleic Acids Res 51(D1):D389–D394. https://doi.org/10.1093/nar/gkac1022

Check for updates

Chapter 18

Tracing the Origin and Evolution of the Secondary Structure of RNA

Fengjie Sun and Gustavo Caetano-Anollés

Abstract

The root of the tree of life remains a controversial issue in the study of the origins and evolution of early life on Earth, and so is the origin and evolution of early biochemistry. Here, we describe the methodological procedures of reconstructing global phylogenies of early life based on RNA secondary structure, with the rooting of phylogenies developed and improved to ultimately gain insights into the deep evolution of ancient molecules. As an example, we explore the evolution of the transfer RNA (tRNA) molecules using statistical characters describing their structure. An evolutionary model of molecular accretion of tRNA modeled at the atomic level illustrates the experimental exercise.

Keywords RNA secondary structure, Phylogenetics, Molecular evolution, Tree of life, Geometrical character, Statistical character

1 Introduction

Reconstructing a Tree of Life (ToL) that summarizes the origin and evolution of organismal diversity at the planetary scale [1–4] remains a challenging proposition. Various alternative rooting hypotheses have been proposed, each of them imparting a different history onto the universal phylogeny. The root of the ToL continues to be a well-known controversial issue in the origins and evolution of early life on Earth [5–7], leading to conflicting relationships among the three superkingdoms of life (Archaea, Bacteria, and Eukarya) depending on the rooting scenarios [8–10]. With advances in next-generation DNA sequencing and the availability of numerous completely sequenced genomes, phylogenomic studies have shifted focus from individual molecules to entire molecular repertoires [4, 11–14]. For example, the very early origin of the archaeal lineage has been strongly supported, challenging the popular bacterial rooting scenario of the ToL by placing the root at various groups within Archaea [14–20]. Although ribosomal RNA (rRNA) has been the most widely used and

Gustavo Caetano-Anollés (ed.), *Evolutionary Genomics*, Methods in Molecular Biology, vol. 2981,
https://doi.org/10.1007/978-1-0716-4836-0_18,

influential evidence supporting archaeal monophyly, its validity has been questioned due to weak monophyletic support for both small [21] and large subunit rRNA [22]. Additionally, Archaea is increasingly considered paraphyletic [1, 7, 22]. Consequently, the rooting of the ToL and the origins of diversified life remain unresolved due to the lack of adequate rooting strategies. This complicates the distinction between extinct and extant basal and crown taxa of the ToL.

Molecular structure evidence has been increasingly applied in phylogenetic analysis. For instance, RNA secondary structures enhance the alignment of sequences [23–25], facilitate the exploration of molecular functions [26], and can be directly incorporated into phylogenetic analysis [18, 27–39]. The broad applications of RNA structures in phylogenetic analysis of ancient life events stem from the following advantages: (a) RNA structures exhibit higher conservation than sequences; (b) a universal tendency toward molecular order in RNA is supported by thermodynamic, statistical, and phylogenetic arguments; and (c) successfully implemented biological designs tend to be reused in nature, generally through recruitments, takeovers, or co-options [30, 31]. The structural phylogenetic methodology has been widely employed to investigate diverse biological and evolutionary questions using various RNA and protein molecules [18, 19, 28, 30–32, 34–40] and to study proteomes and molecular networks [11, 12, 41–45], including modern metabolism [46, 47].

In this chapter, we outline the experimental procedures for reconstructing global phylogenies based on RNA structural characters. This methodology highlights two interconnected aspects of evolution—patterns and processes—driving the emergence of modern biochemistry and early life. The phylogenetic approach that we here present generates not only trees of molecules (representing organisms harboring those molecules) but also trees of molecular substructures (Fig. 1). This method requires defining new types of taxa (molecular substructures) and characters (substructures in molecules) and establishing primary homology criteria related to substructural repertoires in molecular lineages. We also describe how RNA structural evidence is utilized to explore the deep evolution of ancient molecules. The supporting tree-rooting strategy that will be described has been successfully used to identify consistent or similar basal groups across various systems [14, 15, 17–19].

2 Materials

1. Access to primary RNA sequences under investigation is deposited in publicly available databases, including NCBI [48] and Rfam [49]. Sequences can also be retrieved from specialized databases such as those that hold tRNA information, including tRNAdb of Leipzig [50] and GtRNAdb of Santa Cruz [51].

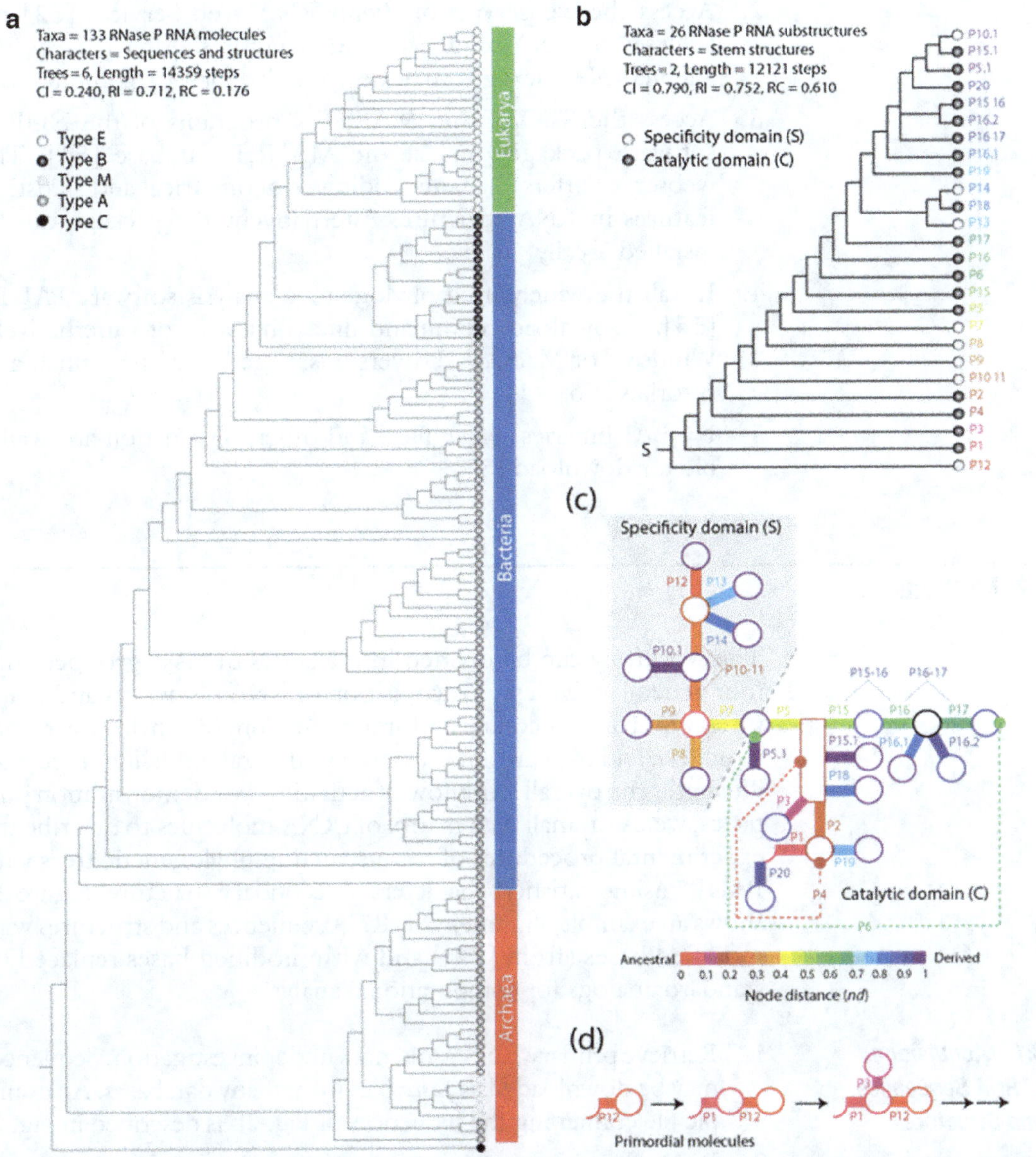

Fig. 1 Global phylogenies of RNase P RNA molecules and substructures and a model of molecular evolution [40]. (**a**) A phylogeny of molecules covering the three superkingdoms of cellular life. Note that with the exception of one Type C structure, archaeal molecules are ancestral and eukaryal molecules are derived. (**b**) A phylogeny of substructures describing the evolution of helical stems, with leaf labels colored according to their relative age (node distance; *nd*). (**c**) Schematic drawing of the secondary structure of the molecule, with stems drawn as thick lines and colored according to relative age. This evolutionary heat map allows us to unfold molecular history. (**d**) The first three steps of a model of evolutionary growth of tRNA describing how primordial molecules are transformed into fully developed molecules typical of the extant biological world. CI consistency index, RI retention index, RC rescaled consistency index

2. Access the webservers of ViennaRNA Web Services [52] for analyses of RNA sequences and structures. Alternatively, the ViennaRNA Package can be downloaded for local computation.
3. Access the STOAT and MARTEN programs of the NOBAI software package [53] at the MANET database [54]. The webserver offers character coding of geometrical and statistical features in RNA structure. Alternatively, the package can be installed locally.
4. Install the widely used phylogenetic analysis software PAUP* [55]. Download command-line binaries, or alternatively, Windows or MacOS GUI versions, together with manuals and tutorials [56].
5. NOBAI binaries, *nexus* files, and other information are available for download [57].

3 Methods

The workflow can be divided into a series of tasks grouped into four general activities: (1) acquisition of RNA sequences and structures, (2) character coding and argumentation, (3) phylogeny reconstruction, and (4) chronology-based structural modeling. Figure 2a illustrates the overall workflow of activities. For demonstration purposes, we use a small data matrix of tRNA molecules to describe the experimental procedures of reconstructing phylogenetic trees with PAUP* using statistical characters of secondary structure. Figure 2b shows an example alignment of tRNA sequences and structures with modified bases (from [37]) and with modified bases replaced by standard analogs for computational analysis.

3.1 Acquisition of RNA Sequences and Structures

1. Retrieve primary RNA sequences under investigation. Sequences may be downloaded as *fasta* files from many databases. An example file containing the sequences of Fig. 2b is described in Fig. 3.
2. RNA secondary structure is obtained by folding the primary RNA sequences and analyzing the structural landscape of the molecules using, for example, various programs that are available at ViennaRNA Web Services. *RNAfold* predicts minimum energy structures and base pair probabilities, *RNAalifold* predicts consensus secondary structure from sequence alignments, *LocARNA* does the same but with a guide tree, *RNAeval* calculates thermodynamic parameters of a sequence-structure pair, and *Barriers* unfolds RNA folding kinetics. Figure 4 shows an example structural alignment and a consensus secondary structure obtained by submitting the *fasta* file with our example sequence dataset to *LocARNA* [58].

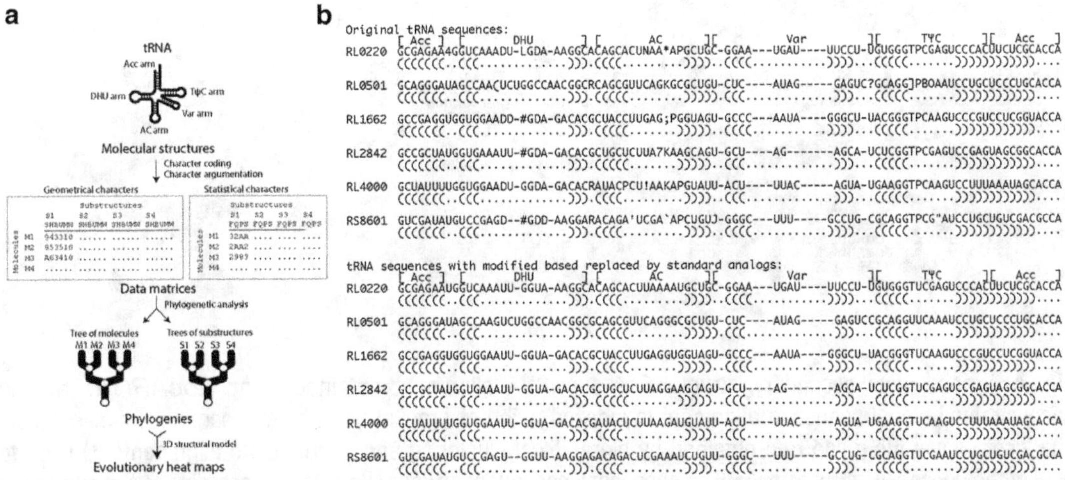

Fig. 2 General workflow of activities for the construction of trees of molecules and substructures of tRNA (**a**) and the data used to illustrate the different tasks (**b**). tRNA data were downloaded from Part 2 of the Bayreuth database (now tRNAdb at Leipzig), which provides a curated set of sequences obtained at the tRNA level with information on base modifications. Modified bases were later replaced by standard analogs for encoding with statistical characters. Secondary structure is described with dot-bracket notation. Modified bases: *, 2-methylthio-N6-isopentenyladenosine; 4, 4-thiouridine; L, N2-methylguanosine; D, dihydrouridine; P, pseudouridine; (, archaeosine; R, N2,N2-dimethylguanosine; K, 1-methylguanosine; ?, 5-methylcytidine;], 1-methylpseudouridine; B, 2'-O-methylcytidine; #, 2'-O-methylguanosine; ;, unknown modified guanosine; 7, 7-methylguanosine; K, 7-methylguanosine; !, 5-carboxymethylaminomethyluridine; ', 3-methylcytidine; ", 1-methyladenosine; and `, N6-(cis-hydroxyisopentenyl)adenosine

```
>RL0220
GCGAGAAUGGUCAAAUUGGUAAAGGCACAGCACUUAAAAUGCUGCGGAAUGAUUUCCUUGUGGGTUCGAGUCCCACUUCUCGCACCA
>RL0501
GCAGGGAUAGCCAAGUCUGGCCAACGGCGCAGCGUUCAGGGCGCUGUCUCAUAGGAGUCCGCAGGUUCAAAUCCUGCUCCCUGCACCA
>RL1662
GCCGAGGUGGUGGAAUUGGUAGACACGCUACCUUGAGGUGGUAGUGCCCAAUAGGGCUUACGGGTUCAAGUCCCGUCCUCGGUACCA
>RL2842
GCCGCUAUGGUGAAAUUGGUAGACACGCUGCUCUUAGGAAGCAGUGCUAGAGCAUCUCGGTUCGAGUCCGAGUAGCGGCACCA
>RL4000
GCUAUUUUGGUGGAAUUGGUAGACACGAUACUCUUAAGAUGUAUUACUUUACAGUAUGAAGGTUCAAGUCCUUUAAAUAGCACCA
>RS8601
GUCGAUAUGUCCGAGUGGUUAAGGAGACAGACUCGAAAUCUGUUGGGCUUUGCCUGCGCAGGTUCGAAUCCUGCUGUCGACGCCA
```

Fig. 3 File containing the *fasta* formatted tRNA sequences. All sequences are from $tRNA^{Leu}$ except for RS8601, which is for $tRNA^{Ser}$ (they are all type II tRNA with a variable loop). Origin of sequences: RL0202, Phage T4; RL501, *Haloferax volcanii*; RL1662, *Escherichia coli*; RL2842, *Glycine max* (chloroplast); RL4000, *Saccharomyces cerevisiae*; RS8601, *Nicotiana rustica*

3. Use folded RNA secondary structure information to index the sequence alignment with dots for unpaired bases and open and closed brackets for paired bases following the dot-bracket notation (as in Figs. 2b and 4). You can also use the more complex BEAR notation that indexes structural alignments for the *Web-Beagle* structural alignment server [59]. *Web-Beagle* accepts sequences in *fasta* format, folds the

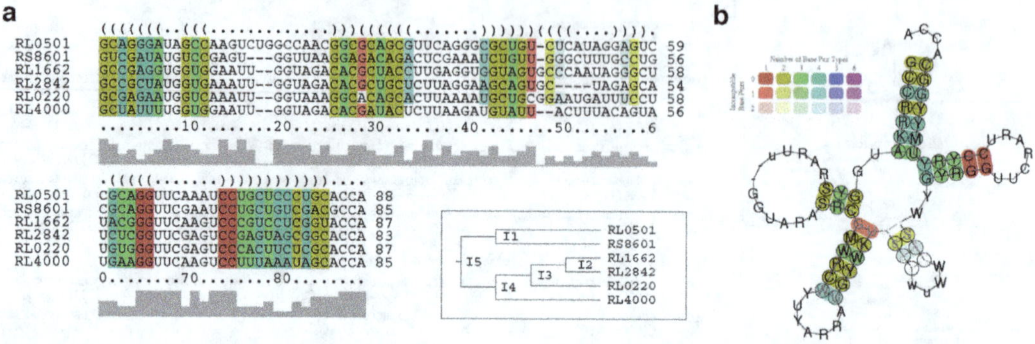

Fig. 4 A consensus secondary structure of our tRNA alignment obtained using *LocARNA*. (**a**) The unconstrained global sequence alignment is indexed with dot-bracket notation and sequence conservation. The inset shows the guide tree obtained by hierarchical clustering and intermediate alignments (I_i) used to predict the consensus structure. (**b**) The consensus secondary structure is colored according to the number of different base pair types at that location. Color saturation indicates base pair incompatibility, a measure of lack of structural conservation

molecules using *RNAfold*, and then aligns the structures, summarizing Steps 1–3, but does not produce multiple structural alignments.

3.2 Character Coding and Argumentation

Phylogenetic analysis of nucleic acid structure requires both the discovery and coding of phylogenetic characters and the establishment of an evolutionary direction of character transformation that would root phylogenetic trees. This must be done prior to the reconstruction of phylogenies. Completing these tasks permits the construction of a file containing a phylogenetic data matrix and definition of assumptions for phylogeny reconstruction.

1. *Coding:* In phylogenetic analysis, attributes are considered "characters" and the qualitative characteristics or the numerical or frequency distribution values they display are considered "character states." To be useful, however, characters must be appropriately coded so that they can render maximal phylogenetic information. Two types of molecular characters can be used based on the RNA sequences and folded RNA secondary structure, geometrical and statistical characters (Fig. 2a). Geometrical (shape) characters describe the geometry of the RNA molecules by measuring for example the length in nucleotides of each spatial component of secondary structure (e.g., double helical stems, hairpin loops, and bulges) (*see* **Note 1**). Statistical characters describe the branching, stability, and plasticity (uniqueness) of the RNA molecules (*see* **Note 2**). Both types of characters are appropriately coded into different character states in alphanumeric format (*see* **Note 3**). The maximum number of character states that can be assigned to a character

a

Taxon	123456789012345678901234
acceptor	GFGGGG000000IHIIII444444
D arm	1B6165101111A9A1ABAAAAAA
anticodon	EGGG05000001EIIIEE766677
V arm	GAH418030213KIKFIM7A7FA7
T arm	GG8G8G002020IIIIII666666

Character	States	Character	States
1 (F RL0220)	1EGM	13 (P 1RL0220)	AEIKM
2 (F RL0501)	ABFGM	14 (P 1RL0501)	9HIM
3 (F RL1662)	68GHM	15 (P 1RL166B)	AIKM
4 (F RL2842)	14GM	16 (P 1RL2842)	1FIM
5 (F RL4000)	0168GM	17 (P 1RL4000)	AEIM
6 (F RS8601)	58GM	18 (P 1RS8601)	BEIM
7 (Q RL0220)	01	19 (S 1RL0220)	0467A
8 (Q RL0501)	03	20 (S 1RL0501)	046A
9 (Q RL1662)	012	21 (S 1RL1662)	0467A
10 (Q RL2842)	012	22 (S 1RL2842)	046AF
11 (Q RL4000)	012	23 (S 1RL4000)	0467A
12 (Q RS8601)	013	24 (S 1RS8601)	0467A

b

Taxon	12345678901234567890
RL0220	G0I411AAE0E7G0K7G0I6
RL0501	F0H4B09AG0I6A3IAG0I6
RL1662	G0I461AAG0I6H0K782I6
RL2842	G0I411AAG0I642FFG0I6
RL4000	G0I461AA00E711IA82I6
RS8601	G0I451BA51E783M7G0I6

Character	States	Character	States
1 (1F acceptor)	FGM	11 (11P anticodon)	EIM
2 (2Q acceptor)	0	12 (12S anticodon)	067
3 (3P acceptor)	HIM	13 (13F V arm)	148AGHM
4 (4S acceptor)	04	14 (14Q V arm)	0123
5 (5F D arm)	156BM	15 (15P V arm)	FIKM
6 (6Q D arm)	01	16 (16S V arm)	07AF
7 (7P D arm)	9ABM	17 (17F T arm)	8GM
8 (8S D arm)	0A	18 (18Q T arm)	02
9 (9F anticodon)	05EGM	19 (19P T arm)	IM
10 (10Q anticodon)	01	20 (20S T arm)	06

Fig. 5 A matrix (**a**) and its transposed derivative (**b**) describe statistical characters of tRNA and its character states. Rows represent taxa (ordered from top to bottom), and columns represent characters (ordered from left to right) in the matrix. The matrix with rows representing substructures (e.g., acceptor) is used to reconstruct a tree of substructures, which describes how tRNA molecules grow. The matrix with rows representing molecules (e.g., RL0220) is used to reconstruct a tree of molecules, which describes how tRNA and the organisms they represent diversify. The tables in this figure describe characters and their character states

in PAUP* could be 16 (for a 16-bit machine), 32 (for a 32-bit machine), and 64 (for a 64-bit machine). Figure 5 illustrates our exercise with matrices of statistical characters F, Q, P, and S generated in [37]. Note that a transposition of a matrix that describes how molecules grow (Fig. 5a) results in a matrix that describes how molecules diversify (Fig. 5b), once matrices are used to build phylogenies. Our example exercise will only focus on building a phylogenetic tree of tRNA substructures using encodings obtained directly from the STOAT of the NOBAI package.

2. *Argumentation:* Character states must transform to other character states following a character state transformation model known as a "transformation series." These models are landscapes of possible evolutionary change, usually defined by character state graphs (CSGs) that connect vertices representing character states with edges representing allowed transformations. In our encodings, we use non-reticulated minimally connected CSGs of the linear type, which define *ordered* (Wagner) characters (Fig. 6). Ordered characters are fully reversible and do not impose by themselves an evolutionary direction of change, i.e., they do not "polarize" change and would not root a phylogenetic tree during the phylogenetic search process. They are widely used in the analysis of serial homologies that measure, for example, morphological features or quantitative traits in biology. Their definition comes standard in PAUP*.

3. Trees, however, must be rooted to convey history. This can be done *a posteriori* of phylogeny reconstruction. A branch holding the ancestor of each optimal unrooted tree is "pulled

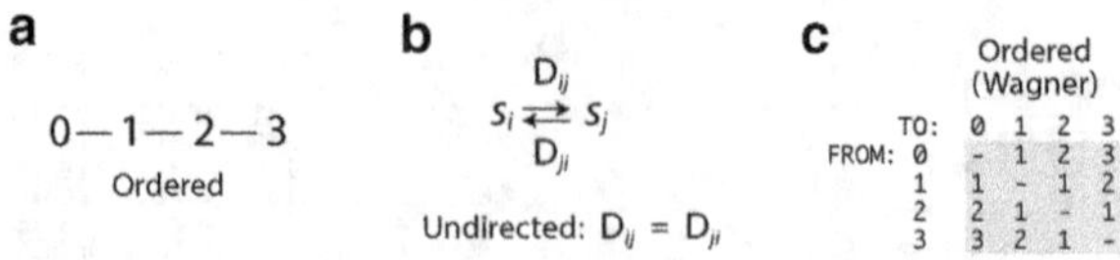

Fig. 6 Character state change in ordered characters of the Wagner type. (**a**) The character state graph (CSG) describes an equally weighted undirected character with only 4 states (0, 1, 2, and 3) that is typical of a fully ordered character. (**b**) Transformation between character states is undirected (fully reversible), in which the cost D_{ij} from character state *i* to *j* direction is equal to that from *j* to *i*. An undirected ordered character does not induce character polarization. (**c**) A character state matrix (stepmatrix) describes transformation costs (in tree lengths) from each character state to another. The more distant the character states from each other in the CSG, the more costly is their transformation. Note that the stepmatrix is symmetric and its application produces unrooted phylogenetic trees

down" to its base in a way that is most parsimonious by using Lundberg optimization [60]. This is done by using either a character state vector of a hypothetical ancestor (ancstates = ancestor) or a standard implementation (ancstates = standard) that roots the trees by nesting homologies along its branches following Weston's generality criterion [61, 62]. The definition of the ancestor complies with an evolutionary model that polarizes character transformation toward an increase in molecular order (*see* **Note 4**). Inclusion of the ancestor *a priori* in the phylogenetic search will polarize character transformations and will produce rooted trees. This option is not recommended when dealing with a large number of taxa because the solution space of rooted trees is higher in complexity than that of the unrooted trees and demands higher computational effort. However, we recommend that, when possible, all options be explored to confirm congruence between approaches.

4. *Construction of a nexus file:* An input file with a *nex* or *nxs* extension must be constructed in *nexus* format for phylogenetic analysis in PAUP*. The file must begin with the declaration #NEXUS and must contain a number of blocks describing aspects of phylogenetic analysis, including data, assumptions, trees, and PAUP*-specific directives. Blocks hold commands for execution, starting with "begin" and concluding with "end" (*see* **Note 5**). Commands and options must be followed by semicolons. Commentary can be included in brackets to avoid readout. Figure 7 describes the input file we will use to reconstruct our trees. The block holding the data matrix starts with "begin data" and is followed by the command "dimensions," which defines the number of taxa and characters of the data matrix. Command "format" is used to specify aspects of the data file, including the datatype (default is "standard"). Subcommand "symbols" defines the names of character states (in our example, 0123456789ABCDEFGHIJKLMNOPQRSTU)

```
#NEXUS

[!Data from: Sun and Caetano-Anollés, Meth Mol Biol 2025; STOAT @2005]

begin data;
dimensions  ntax=5 nchar=24;
format symbols = "0~9 A~U";
charlabels
        F_RL0220 F_RL0501 F_RL1662 F_RL2842 F_RL4000 F_RS8601
        Q_RL0220 Q_RL0501 Q_RL1662 Q_RL2842 Q_RL4000 Q_RS8601
        P_RL0220I P_RL0501 P_RL1662 P_RL2842 P_RL4000 P_RS8601
        S_RL0220 S_RL0501 S_RL1662 S_RL2842 S_RL4000 S_RS8601
        ;
matrix

[Taxon]    [Characters]

acceptor   GFGGGG000000IHIIII444444
D_arm      1B6165101111A9A1ABAAAAAA
anticodon  EGGG05000001EIIIEE766677
V_arm      GAH418030213KIKFIM7A7FA7
T_arm      GG8G8G002020IIIIII666666
;
end;

begin assumptions;
    typeset *tRNA = ord: 1-24;
        charset F = 1-6;
        charset Q = 7-12;
        charset P = 13-18;
        charset S = 19-24;
        ancstates *ancestor = M:1-6, 0:7-12, M:13-18, 0:19-24;
                                [0 for Q and S; M for F and P]
end;
```

Fig. 7 Nexus file needed to reconstruct a phylogeny of tRNA substructures. The matrix contains rows representing taxa and columns representing characters (default)

and subcommand 'transpose" flips the default format of having taxa as rows and characters as columns. Command "charlabels" identifies individual characters by name, and command "matrix" precedes the phylogenetic data matrix that will be used for reconstruction. It is often discouraged to use a data block instead of separate taxa and character blocks because it prevents the extraction of taxon information by some programs. Both formats are, however, tolerated by PAUP*. We use the data block here for simplicity.

5. The block holding assumptions defines character types, character weights, character sets, and character states of the hypothetical ancestor. Command "typeset" assigns a character type to each character. In our case, the character types are ordered (ord) for all 24 characters. However, other types are available (e.g., unordered and Dollo), including user-defined stepmatrices. Command "charsets" allows grouping characters together for quick inclusion–exclusion. This is useful when studying the contribution of sets of characters. For example, we could exclude characters F and study only the three-dimensional morphospace delimited by characters Q, P, and S [63]. Finally, the command "ancstates" defines the character state of the hypothetical ancestor used for Lundberg rooting or for inclusion in searches of tree space. Our ancestor holds character state 0 (minimum) for Q and S and character state M (maximum) for F and P. Note that S is provided by STOAT as an inverse.

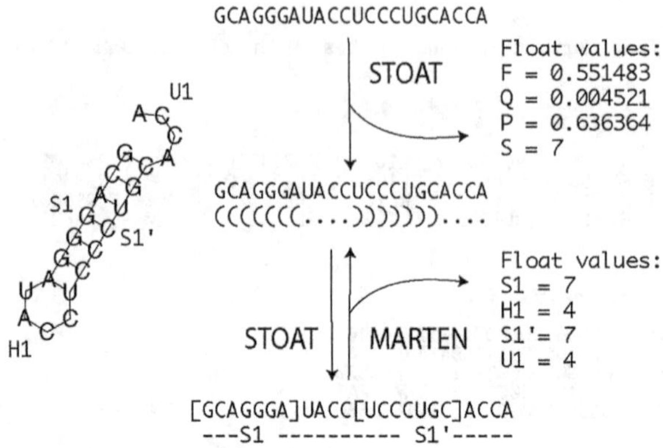

Fig. 8 Automated coding of secondary structure information. STOAT converts sequence information in *fasta* format to dot bracket notation and postscript drawing of secondary structures by invoking *RNAfold*. Float values of statistical parameters are then used to build a *nexus* file. MARTEN indexes helix strands with secondary structure information using the DCSE format, which highlights helix numbering of structured segments of the RNA molecule. Float values of geometric features are then used to build a *nexus* file

6. *Automated coding:* STOAT and MARTEN provide automated coding of secondary structure information (Fig. 8). STOAT calculates statistical values for the secondary structure of folded RNA sequences. Input files consist of RNA sequences in *fasta* format. Folded RNA secondary structures are obtained by interactions with the *RNAfold* program. Calculations are presented in two *nexus*-formatted files (local and global coding), where each row of the matrix represents statistical values of each input sequence. STOAT also translates secondary structure (provided by the user or obtained from *RNAfold*) from the dot bracket notation to the DCSA format. MARTEN counts the number of unpaired, paired, and bulged nucleotides for each helix strand in the secondary structure of folded RNA sequences. Sequence and structure information in DCSE format is provided as input. Depending on options, the program generates a *nexus*-formatted output file holding a data matrix where each row represents an input sequence and columns describe the numbers of unpaired, paired, and bulged nucleotides for each helix strand of the molecule. MARTEN also translates sequence-structure information from DCSE to dot bracket notation.

3.3 Phylogeny Reconstruction

The input *nexus* file is executed in PAUP* with the goal of generating rooted phylogenetic trees describing the evolution of either specific RNA substructures or molecules. In our exercise, we use equally weighted maximum parsimony as the optimality criterion to build a tree of tRNA substructures. A number of tasks must be completed, individually or in batches, in a process that is exploratory and iterative:

1. *Process your file:* Use the command "execute" to request processing. Commands will be executed until the end of the file is reached or a "quit" command is issued. When using batch mode, you can add all commands described in the following sections to your file under a "paup" block to complete your phylogenomic reconstruction.
2. *Include/exclude characters and taxa:* Use command "cstatus" to describe character type, weight, and informativeness. Use the command "tstatus" to confirm included taxa. Exclude uninformative characters if they are present with the "exclude uninf" command. Exclude taxa with the "delete" command if you suspect they are "rogue" and provide confounding phylogenetic signal, or when exploring selected evolutionary periods of molecular history (e.g., history of the most ancestral substructures). In our tRNA matrix, two characters are phylogenetically uninformative and are excluded from analysis.
3. *Define an optimality criterion and run your search of tree space with a tree search strategy:* Use the "set optimality" command to define the optimality criterion of maximum parsimony (default), maximum likelihood, or distance. Note that the computational efficiency and greater speed of parsimony allow a more in-depth analysis of datasets than other model-based approaches such as maximum likelihood. Run your search by selecting one of three strategies of exploration of tree solution space: (a) exhaustive search with the "alltrees" command; (b) branch-and-bound search with the "bandb" command; and (c) heuristic search with the "hsearch" command. Your choice depends on the number of taxa present in your data matrix. More than 20 taxa will force you to use heuristic approaches. Heuristic searches are the most common in a world of increasingly larger trees and datasets that is typical of phylogenomic analyses. They involve hill-climbing methods of stepwise addition and swapping of tree branches to better navigate the tree solution space. Stepwise addition adds taxa to a growing tree until all taxa are connected using, for example, the "as is," "simple," and "random" options in PAUP*. The procedure maximizes space covering and finding global rather than local optima. Standard swapping algorithms include the *nearest-neighbor interchanges* (NNI), *subtree pruning and regrafting* (SPR), and *tree bisection and reconnection* (TBR), all available options in PAUP*. These "swappers" are, however, greedy and rearrangement-costly, prompting the use of a number of strategies to make tree searches more efficient [64], including the use of ratchet, genetic, simulated annealing, and divide-and-conquer algorithms. We note that PAUP* is quite efficient in searching for optimal trees, but that tree exploration is permanently being enhanced (see benchmarking [65]

using the TNT program with phylogenomic datasets [66]). In our example, we use exhaustive searches given that the tree space for a 5-taxon statement is 15 unrooted trees or 105 rooted trees. Figure 9 illustrates a standard search of unrooted trees with a Schoenberg graph describing the space of all possible 5-taxon unrooted trees. The optimal tree that is retained shows good agreement with character state distributions, with a consistency index (CI) of 0.848, retention index (RI) of 0.575, rescaled consistency index (RC) of 0.487, homoplasy index (HI) of 0.152, and a Goloboff-fit of –17.033. Primary homologies are therefore highly corroborated with a good fit and minimum homoplasy. Note that suboptimal trees show lower CI, RI, RC, and Goloboff-fits and higher HI values, transferring suboptimal corroboration from characters to trees.

4. Use the command "describetrees /tOrder=left" to visualize trees ladderized left together with default goodness-of-fit statistics. A table of these statistics can also be obtained using the command "pscores/CI RI RC HI Gfit." Reconstructed character state vectors of internal nodes are available via the "describetrees/xout=internal" command (see Fig. 9). Finally, assess confidence by performing a bootstrap analysis with the "bootstrap" command. In Fig. 9, we used "bootstrap nreps= 2000 search=BandB" to perform a bootstrap analysis with 2000 replicates of branch-and-bound searches.

5. *Root the resulting trees:* Once one or more optimal unrooted trees are obtained, root the trees using Lundberg in its two implementations, ancestor and standard. Rooting is a necessary step to turn phylogenetic trees into bona fide phylogenies (*see* **Note 6**). Three options are available with the "root" subcommand: outgroup, Lundberg, and midpoint. As a first approximation, describe the tree with "soft" Lundberg rooting by using the command "describe 1/root=Lundberg." This will not change the length of the unrooted tree but will pinpoint a likely branch holding the root that will be brought to the base of the unrooted tree. Alternatively, you can use the "rootrees root=lundberg" command, a request that will add an additional branch at the base of the tree, reconstruct the ancestral node, and lengthen the unrooted tree by Lundberg optimization. Figure 10 compares the unrooted and rooted tree visualizations of Lundberg with the ancestor implementation in the display buffer of PAUP*. Both approaches place the branch holding the acceptor stem at the base of the trees.

6. *Include the hypothetical ancestor.* Even if not required or impossible for computational reasons, inclusion of an ancestral taxon in a search will optimize the placement of both ancestor and ingroup taxa in rooted trees and will confirm the phylogenies that we previously generated. Figure 11 shows a phylogenetic reconstruction

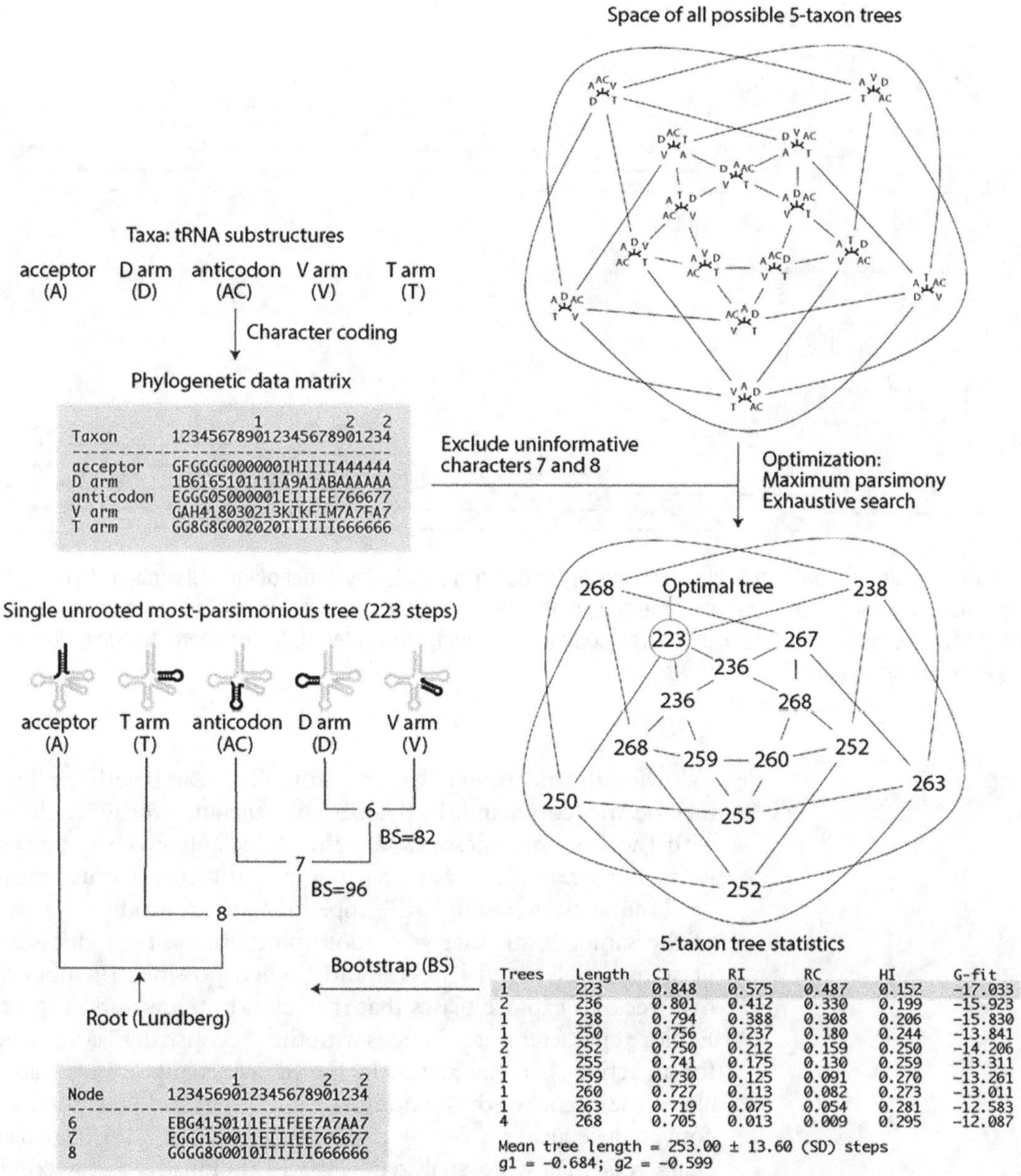

Fig. 9 Reconstructing the evolution of tRNA structure. All characters are undirected and ordered. Two phylogenetically uninformative characters are first excluded from analysis. Phylogenetic reconstruction using maximum parsimony and an exhaustive search produces one optimal unrooted tree of 223 steps in length that is selected as the optimal solution out of the 15 possible unrooted trees. The space of possible trees is depicted with a Schoenberg graph in which trees are linked by nearest-neighbor interchanges that convert one tree into another. The length of each of the 15 trees is then traced onto the graph. The left table shows goodness-of-fit statistics for all trees. Lundberg rooting (arrow) identifies the most parsimonious branch to root the tree (see below). Bootstrap support (BS) analysis with 2000 branch-and-bound replicates evaluates branch reliability, revealing significant support for individual clades of the unrooted tree. The bottom table shows character state reconstructions for internal nodes of the tree labeled 6, 7, and 8

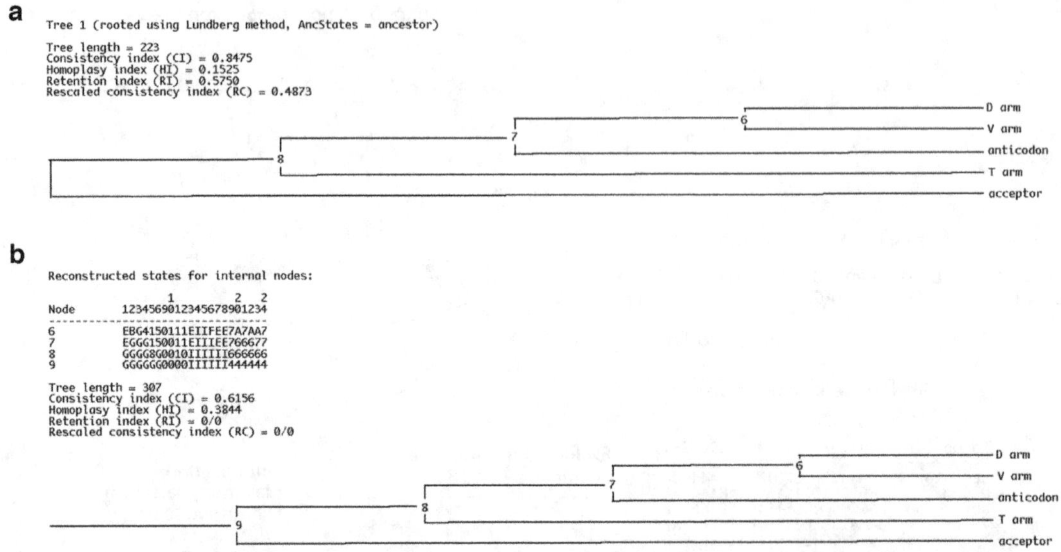

Fig. 10 Unrooted and rooted tree visualizations of Lundberg in the display buffer of the GUI version of the PAUP* program. (**a**) Unrooted tree rooted with the "describe 1/root=lundberg" command. (**b**) The same unrooted tree rooted with the "roottrees root = lundberg" command showing character states of internal nodes. Note the difference in tree lengths

following the inclusion of the ancestor in the search with the 'pset includeAnc' command. In the example, the ancestor was defined with the "assume ancstates=ancestor" (default given the asterisk in our *nexus* file). However, defining the ancestor using the "assume ancstates=standard" implementations produced a tree of the same length that was isomorphic. Request goodness-of-fit statistics of optimal tree(s) and, when possible, of suboptimal trees, to explore biases that may exist in tree solution space. Explore character state changes with the "reconstruct" command for all active character states. In the figure, changes were traced along the branches of the optimal rooted tree and illustrated for convenience using a 6 × 4 matrix describing the 24 statistical characters that were analyzed (Fig. 11a). Finally, assess confidence by performing a bootstrap analysis with the "bootstrap" command. Note the strong support for all hypothetical ancestors of the tree (nodes 6–9). Goodness-of-fit statistics provide an evaluation of how well a tree explains its supporting data (Fig. 11b).

7. *Visualize trees.* Most parsimonious trees can be visualized as cladograms or phylograms. Both options are available in PAUP*, which can produce publication-quality output files of phylogenetic trees in memory. Trees can also be saved in Newick format for visualization in graphical viewers such as FigTree [67] or iTol [68]. Figure 12 describes the most parsimonious tree that we previously reconstructed, visualized as a phylogram or as a cladogram.

a

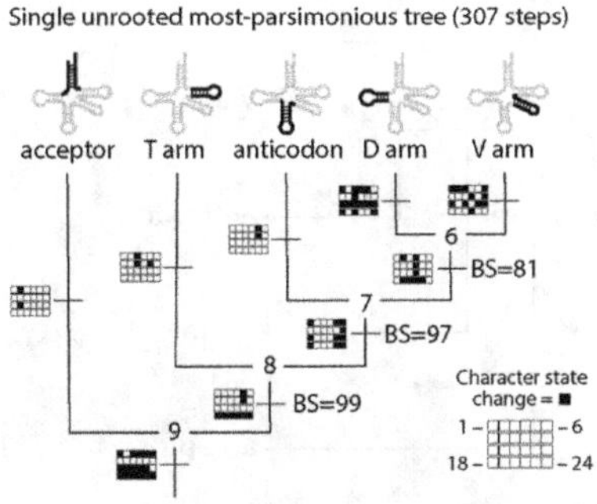

Character state reconstruction of ancestors:

Node	1234569012345678901234
6	EBG4150111EIIFEE7A7AA7
7	EGGG150011EIIIEE766677
8	GGGG8G0010IIIII666666
9	GGGGGG0000IIIII444444

b

5-taxon tree statistics (105 rooted trees)

Trees	Length	CI	RI	RC	HI	G-fit
1	307	0.860	0.656	0.564	0.140	-16.360
1	320	0.825	0.552	0.455	0.175	-15.249
1	321	0.822	0.544	0.447	0.178	-14.999
1	323	0.817	0.528	0.432	0.183	-14.394
2	327	0.807	0.496	0.400	0.193	-13.772
1	335	0.788	0.432	0.340	0.212	-13.125
2	337	0.783	0.416	0.326	0.217	-12.770
6	340	0.776	0.392	0.304	0.224	-13.116
5	341	0.774	0.384	0.297	0.226	-12.916
1	343	0.770	0.368	0.283	0.230	-12.773
2	344	0.767	0.360	0.276	0.233	-12.318
1	345	0.726	0.352	0.269	0.235	-12.068
4	349	0.756	0.320	0.242	0.244	-11.618
3	353	0.748	0.288	0.215	0.252	-11.360
1	354	0.746	0.280	0.209	0.254	-11.318
1	355	0.744	0.272	0.202	0.256	-10.924
1	356	0.742	0.264	0.196	0.258	-10.882
6	357	0.739	0.256	0.189	0.261	-11.601
1	358	0.737	0.248	0.183	0.263	-11.451
2	360	0.733	0.232	0.170	0.267	-10.963
2	366	0.721	0.184	0.133	0.279	-10.673
3	368	0.717	0.168	0.121	0.283	-10.355
1	369	0.715	0.160	0.114	0.285	-10.105
7	370	0.714	0.152	0.108	0.286	-10.063
1	372	0.710	0.136	0.097	0.290	-10.344
6	373	0.708	0.128	0.091	0.292	-10.094
8	378	0.698	0.088	0.061	0.302	-10.520
5	379	0.697	0.080	0.056	0.303	-10.478
1	385	0.686	0.032	0.022	0.314	-10.073
7	386	0.684	0.024	0.016	0.316	-9.223
5	387	0.682	0.016	0.011	0.318	-9.673
12	388	0.680	0.008	0.005	0.320	-9.728

Mean tree length = 364.00 ± 19.70 (SD) steps
g1 = -0.637; g2 = -0.490

Fig. 11 Including the hypothetical ancestor in the search. (**a**) After exclusion of phylogenetically uninformative characters 7 and 8, a phylogenetic exhaustive search that includes the "ancestor" recovers one optimal rooted tree of 307 steps. Tracing bootstrap support (BS) values assessed with 2000 branch-and-bound replicates reveals significant support for individual clades of the rooted tree. Reconstruction of character state changes along branches of the tree is given as a 6 × 4 matrix describing the characters used to reconstruct the rooted tree. The bottom table shows character state reconstructions for internal nodes of the tree labeled 6, 7, 8, and 9. (**b**) Goodness-of-fit statistics for all 105 possible 5-taxon rooted trees describe how well a tree explains its supporting data. For example, CI values tend to 1 and RC values tend to 0 indicate that characters fit the data as good or as poorly as possible, respectively. Note how statistics follow a tight correlation between suboptimality and the poor fit of data to the tree

3.4 Chronology-Based Structural Modeling

The rooted trees that are reconstructed can be used to generate evolutionary heat maps of RNA structure that color two- or three-dimensional (atomic) structural models of RNA with molecular ancestries [23, 40, 41, 54]. Figure 1c shows an example heat map of RNA secondary structure. The relative scales of the heat map describe the number of nodes from the hypothetical ancestor at the base of the tree, which is unique for trees of different substructures. Because the reconstructed trees are intrinsically rooted and are highly imbalanced, the relative age (ancestry or time of origin) of each substructure is established by measuring a distance in nodes from the hypothetical ancestor on a relative 0–1 scale; the number of nodes is counted in every lineage from the root to the terminals of the tree and this number is divided by the maximum number of nodes in a lineage [35]. The rooted trees of tRNA structures that we generated from the molecules of only six organisms were congruent with trees reconstructed from an analysis of a dataset of 571 molecules that contained modified base information [37]. These trees place the acceptor arm of tRNA secondary structure at the base of the phylogeny and provide an unambiguous statement describing the structural progression responsible for the evolutionary growth

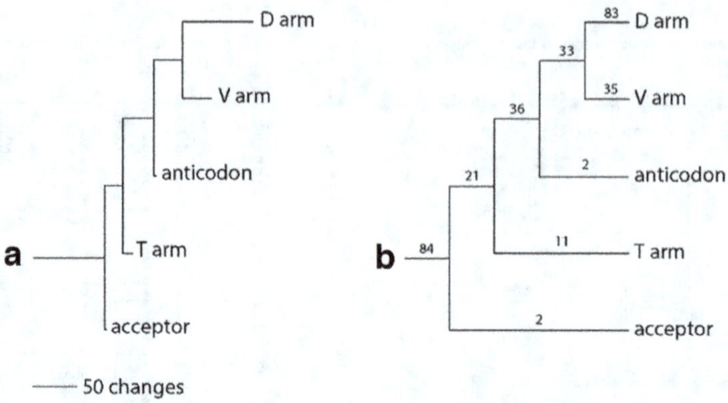

Fig. 12 The most optimal rooted tree obtained by including the hypothetical ancestor in the search is visualized as a phylogram (**a**) or cladogram (**b**) in the GUI versions of PAUP*. Bootstrap support values and character state changes are indexed in the branches of the phylogram and cladogram, respectively

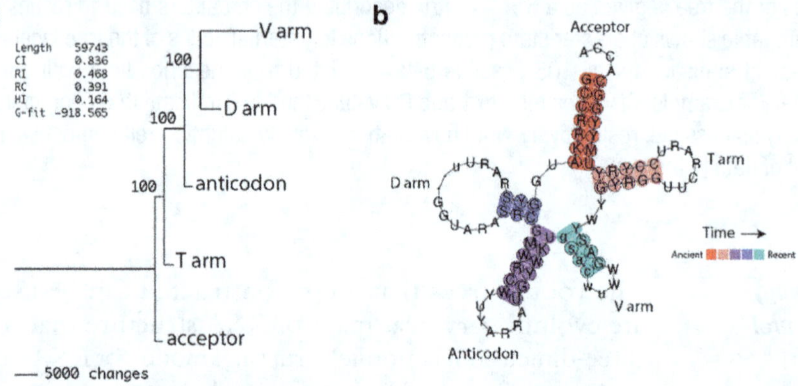

Fig. 13 Evolutionary heat map of the tRNA molecules. (**a**) Phylogeny describing the evolution of 571 tRNA structures. BS values were assessed with 2000 branch-and-bound replicates. (**b**) The consensus secondary structure obtained in Fig. 4b is colored according to times of origin derived from the trees of tRNA substructures

of tRNA. Figure 13 describes the phylogeny generated from an analysis of 2284 statistical characters and an evolutionary heat map that colors secondary structures with times of origin of the individual tRNA substructures.

The basal placement of the acceptor stem indicates an origin of the molecule in a relatively short stem loop structure (a primordial "minihelix"), which given appropriate sequence and structure information, can be modeled using AI-driven deep learning software such as trRosettaRNA [69]. Figure 14 describes the structural growth of tRNA from this primordial minihelix to the full-fledged

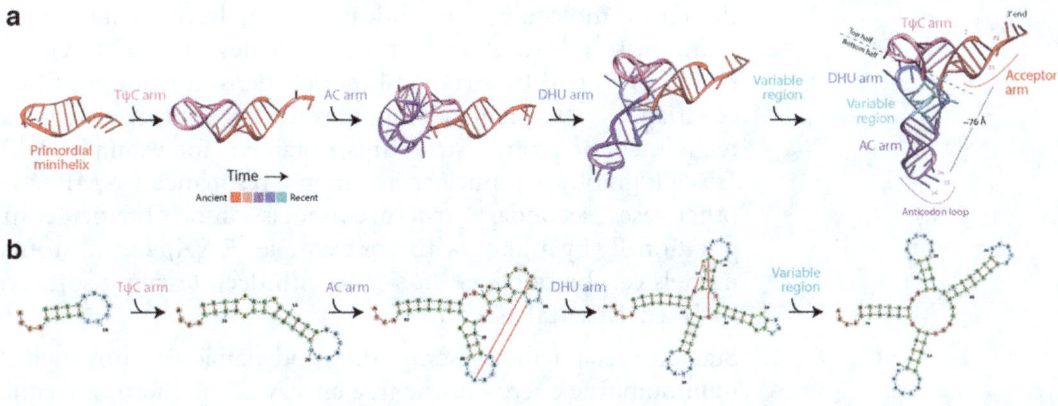

Fig. 14 The atomic structures (**a**) and secondary structures (**b**) of evolving tRNA structures modeled using deep learning trRosettaRNA from an ancestor sequence reconstructed from a tRNA dataset of 571 molecules using maximum likelihood methods

atomic structure of a universal ancestor of tRNA. The sequence of this ancestor was first reconstructed directly from the 571 tRNA sequences used to build the tree of tRNA substructures [37] using maximum likelihood methods of character state reconstruction in [33]. The ancestral sequences of the growing substructures were then subjected to analysis with the deep learning-based algorithm of trRosettaRNA, producing three-dimensional structural predictions at the atomic level. trRosettaRNA uses a transformer network to predict two-dimensional inter-nucleotide geometries that result in an optimal three-dimensional structure that can be refined by energy minimization. The per-residue pLDDT for predicted structures, which measures model confidence, was always larger than 70, with 11%–59% of modeled residues having pLDDT scores above 90. The high confidence levels of the models show accurate structural predictions at the atomic level that match those of experimentally determined crystallographic structures.

4 Notes

1. Geometrical characters, also known as "shape" characters, are attributes that measure geometrical features in molecules, such as spatial components of secondary structure describing coaxial helical segments (stems). Most of these measurements are calculated using the MARTEN program of NOBAI [52]. Geometrical characters include the length of stems (S), hairpin loops (H), bulges (B), and unpaired regions (U), and numbers of modified bases (M) and weak G:U base pairs (W) (Fig. 2a). Measuring and coding these characteristics in the context of

the entire molecule is an endeavor of molecular topography. In the case of tRNA and other molecules (e.g., rRNA), the task is facilitated by structural models derived from positional covariance in sequences or confirmed by using biochemical techniques or atomic structures obtained, for example, with crystallography or nuclear magnetic resonance (NMR). In other cases, secondary structure models can be obtained computationally by folding with, for example, *RNAfold* and atomic models can be obtained *ab initio* with deep learning software such as trRosettaRNA [69].

2. Statistical characters describe thermodynamic features such as minimum free energy or the free energy of the thermodynamic ensemble. One group of especially useful statistical features is those that define a complete structural morphospace [63, 70]. Three types of statistical characters [32, 37] are coded, including F (Frobenius norm of the probability matrix), Q (Shannon entropy of the probability matrix), P (base pair propensity), and S (mean length of helical stems), which are calculated using the STOAT program of NOBAI [53]. The following equations define the four statistical characters:

$$F=\left[\frac{1}{N}\sum_{i=1}^{N-1}\sum_{j=i+1}^{N}(P_{ij})^2\right]^{1/2}$$

$$Q=-\left(\frac{1}{Q_m}\right)\sum_{i=1}^{N-1}\sum_{j=i+1}^{N}P_{ij}\log_2 P_{ij}$$

$$P=\frac{C}{N}$$

$$S=\frac{A}{B}=\frac{1}{B}\sum_{k=1}^{B}A_k$$

with P_{ij} being the base pair probability, *Ak* the length of stem *k* (number of paired nucleotides in that stem), *C* the number of base pairs, *N* the number of nucleotides, *B* the number of stems, and $Q_m = 0.5N\log_2 N$. *F* evaluates the Euclidean spread of base pairing interactions during RNA folding, *Q* measures conflicting inter- and intra-molecular interactions, and *P* and *S* describe how extensively folded and ramified are the folded RNA molecules.

3. Character homology is determined by the relative position of substructures in the folded RNA secondary structure and coded character states are based on the length (number of bases or base pairs) and number of these substructures or by statistical features of the substructures; character states are defined in alphanumerical format with numbers from 0 to 9 and letters from A to Z and from a to z. It is noted that the uppercase and lowercase of the same letter represent two different character states. Missing substructures are given the minimum state (0). Partitioned data matrices are generated using either: (a) characters collected from the substructures which function to either stabilize (e.g., stems and G:U pairs) or destabilize (e.g., bulges, hairpin loops, and other single-stranded regions) the RNA molecules, or (b) molecules belonging to each of the three superkingdoms of life or viruses/bacteriophages.
4. The structural and statistical characters transform from one character state to another through time and along the branches of the trees and do so following a model of character transformation (a "transformation series") of linearly ordered and reversible pathways of character state change. An evolutionary direction to transformation is imposed by identifying the ancestral state in the pathway. This "polarization" of character state transformation is driven by an evolutionary search of structural order that is supported by statistical mechanics, thermodynamic, and phylogenetic arguments, ultimately distinguishing ancestral states as those thermodynamically more stable, i.e., maximum character states are defined as the ancestral states for structures stabilizing the RNA molecules, while minimum states (0) are treated as the ancestral states for structures destabilizing the RNA molecules. The validity of character argumentation and the use of maximum parsimony have been discussed in detail elsewhere [33, 34, 36], but is here confirmed with congruent reconstructions that use Lundberg with both the ancestor and standard implementations. The model of character state change and the ancestor is therefore supported by Weston's generality criterion.
5. The syntax of main individual blocks of commands is given in Fig. 15. Consult the PAUP* manual for the meaning of commands and subcommands.
6. Rooting identifies and "pulls down" the branch that holds the ancestor of the taxa being analyzed, i.e., the study group or "ingroup" [71]. Two general types of rooting approaches are used in phylogenetic reconstruction, direct and indirect methods. Direct methods focus exclusively on ingroup taxa and the most recent common ancestor of the ingroup. Direct methods include the generality criterion, optimization-based polarization, and distance and parametric-based rooting. The

a

```
BEGIN DATA [block-name] ;
   DIMENSIONS NTAX=number-of-taxa NCHAR=number-of-characters;
   [ FORMAT
         [ MISSING=missing-symbol ]
         [ LABELPOS={ LEFT | RIGHT} ]
         [ SYMBOLS="symbols-list" ]
         [ INTERLEAVE ]
         [ MATCHCHAR=match-symbol ]
         [ EQUATE="> symbol = expansion <-]" ]
         [ TRANSPOSE ]
         [ RESPECTCASE ]
         [ DATATYPE = {STANDARD | DNA | RNA | PROTEIN} ]
         [ GAP=gap-symbol ] ; ]
    [ OPTIONS
         [ IGNORE={ NONE  | INVAR  | UNINFORM } ]
         [ MSTAXA = { UNCERTAIN  | POLYMORPH } ]
         [ ZAP = "character-list" ]
         [ GAPMODE = { MISSING  | NEWSTATE} ] ; ]
    [ CHARLABELS character-name - ; ]
    [ TAXLABELS taxon-name - ; ]
    [ STATELABELS charnum-and-state-list [ , charnum-and-state-list ] ... ;
MATRIX data-matrix ;
END;
```

b

```
BEGIN ASSUMPTIONS [block-name] ;
    [ OPTIONS
         [ DEFTYPE=default-character-type ]
         [ POLYTCOUNT={ MINSTEPS | MAXSTEPS } ];
    [ USERTYPE name [ { STEPMATRIX | CSTREE } ]
         = description ; ]
    [ CHARSET character-set-name = character-list ; ]
    [ TYPESET [*] name = character-type : character-list
         [ , character-type : character-list ] - ; ]
    [ WTSET [*] weight-set-name = character-weight : character-list
         [ , character-weight : character-list ] - ; ]
    [ EXSET [*] exclusion-set-name = character-list ; ]
    [ ANCSTATES [*] ancestor-name = character-state : character-list
         [ , character-state : character-list ] - ; ]
END;
```

c

```
BEGIN TREES [block-name] ;
    [ TRANSLATE token taxon-name [ , token taxon-name ] - ; ]
    [ TREE [*] name = tree-specification; ]
END;
```

d

```
BEGIN DISTANCES;
    [ Dimensions [ NewTaxa ] NTax = number-of-taxa
         NChar = number-of-characters;]
    [Format
         [Triangle = {Lower  | Upper | Both} ]
         [ [No] Diagonal]
         [ [No] Labels]
           [Missing = Symbol]
         [Interleave] ; ]
    [ TaxLabels taxon-name taxon-name...;]
    Matrix distance-matrix;
END;
```

Fig. 15 Syntax structures of the data (**a**), assumptions (**b**), tree (**c**), and distances (**d**) blocks in PAUP*

generality criterion, which encompasses Nelson's ontogenetic criterion, Weston's rule, and the stratigraphic criterion, is the most powerful approach as it minimizes the number of assumptions that are invoked. In contrast, indirect methods make use of information from taxa external to the ingroup. They make use of outgroup comparison, which takes advantage of the distribution of character states in the ingroup and a sister group that is external to the ingroup. The application of outgroups is the most used strategy to determine the evolutionary direction in phylogenetic trees [72]. However, the appropriate selection of an outgroup ultimately impacts the optimization of phylogenetic characters and interpretation of the ingroup relationships [73–75]. The outgroups are not conveniently selected in phylogenetic studies at the superkingdom level (i.e., Archaea, Bacteria, and Eukarya), and subsequently altering the evolutionary arrow on the unrooted global phylogenetic tree [5]. In fact, it is generally difficult, if not impossible, to determine the appropriate outgroups and subsequently the evolutionary directions on the unrooted global phylogenetic trees.

Acknowledgments

Fengjie Sun was supported by the Educational and Professional Leave program at Georgia Gwinnett College during the course of this work. This research was supported by grants from the National Science Foundation (MCB-0749836 and OISE-1132791) and the United States Department of Agriculture (ILLU-802-909 and ILLU-483-625) and by Blue Waters supercomputer allocations from the National Center for Supercomputing Applications (NCSA), all to GCA.

References

1. Baldauf SL, Palmer JD, Doolittle WF (1996) The root of the universal tree and the origin of Eukaryotes based on elongation factor phylogeny. Proc Natl Acad Sci USA 93:7749–7754
2. Brown JR, Doolittle WF (1995) Root of the universal tree of life based on ancient aminoacyl-tRNA synthetase gene duplications. Proc Natl Acad Sci USA 92:2441–2445
3. Lake JA, Skophammer RG, Herbold CW, Servin JA (2009) Genome beginnings: rooting the tree of life. Phil Trans R Soc B 364:2177–2185
4. Hug L, Baker B, Anantharaman K, Brown CT, Probst AJ, Castelle CJ et al (2016) A new view of the tree of life. Nat Microbiol 1:16048
5. Forterre P, Philippe H (1999) Where is the root of the universal tree of life. Bioessays 21:871–879
6. Gouy R, Baurain D, Phillippe H (2015) Rooting the tree of life: the phylogenetic jury is still out. Phil Trans R Soc B 370:20140329
7. Doolittle WF (2020) Evolution: two domains of life or three? Curr Biol 30:R177–R179
8. Leffers H, Kjems J, Ostergaard L, Larson N, Garrett RA (1987) Evolutionary relationships amongst Archaebacteria: a comparative study of 23S ribosomal RNAs of a sulphur-dependent extreme thermophile, an extreme halophile and a thermophilic methanogen. J Mol Biol 195:43–61
9. Olsen GJ, Woese CR, Overbeek R (1994) The winds of (evolutionary) change: breathing new life into microbiology. J Bacteriol 176:1–6
10. Fournier GP, Poole AM (2018) A briefly argued case that Åsgard archaea are part of the eukaryote tree. Front Microbiol 9:1896
11. Caetano-Anollés G, Caetano-Anollés D (2003) An evolutionarily structured universe of protein architecture. Genome Res 13:1563–1571
12. Wang M, Caetano-Anollés G (2006) Global phylogeny determined by the combination of protein domains in proteomes. Mol Biol Evol 23:2444–2454
13. Chen S, Olsthoorn RCL, Yu C (2021) Structural phylogenetic analysis reveals lineage-specific RNA repetitive structural motifs in all coronaviruses and associated variations in SARS-CoV-2. Virus Evol 7:veab021
14. Xue H, Tong K-L, Marck C, Grosjean H, Wong JT-F (2003) Transfer RNA paralogs: evidence for genetic code–amino acid biosynthesis coevolution and an archaeal root of life. Gene 310:59–66

15. Xue H, Ng S-K, Tong K-L, Wong JT-F (2005) Congruence of evidence for a Methanopyrus-proximal root of life based on transfer RNA and aminoacyl-tRNA synthetase genes. Gene 360:120–130
16. Di Giulio M (2007) The tree of life might be rooted in the branch leading to Nanoarchaeota. Gene 401:108–113
17. Wong JT-F, Chen J, Mat W-K, Ng S-K, Xue H (2007) Polyphasic evidence delineating the root of life and roots of biological domains. Gene 403:39–52
18. Sun F-J, Caetano-Anollés G (2008) Evolutionary patterns in the sequence and structure of transfer RNA: early origins of Archaea and Viruses. PLoS Comput Biol 4:E1000018
19. Sun F-J, Caetano-Anollés G (2009) The evolutionary history of the structure of 5S Ribosomal RNA. J Mol Evol 69:430–443
20. Raymann K, Brochier-Armanet C, Gribaldo S (2015) The two-domain tree of life is linked to a new root for the Archaea. Proc Natl Acad Sci USA 112:6670–6675
21. Barnes SM, Delwiche CF, Palmer JD, Pace NR (1996) Perspectives on archaeal diversity, thermophily and monophyly from environmental rRNA sequences. Proc Natl Acad Sci USA 93:9188–9193
22. De Rijk P, Van de Peer Y, Van den Broeck I, De Wachter R (1995) Evolution according to large ribosomal subunit RNA. J Mol Evol 41:366–375
23. Higgs PG (2000) RNA secondary structure: physical and computational aspects. Quart Rev Biophys 33:199–253
24. Higgs PG, Jameson D, Jow H, Rattray M (2003) The evolution of tRNA-Leu genes in animal mitochondrial genomes. J Mol Evol 57:435–445
25. Akiyama M, Sakakibara Y (2022) Informative RNA base embedding for RNA structural alignment and clustering by deep representation learning. NAR Genomics Bioinf 4:lqac012
26. Spitale RC, Incarnato D (2022) Probing the dynamic RNA structurome and its functions. Nat Rev Genet 24:178–196
27. Billoud B, Guerrucci MA, Masselot M, Deutsch JS (2000) Cirripede phylogeny using a novel approach: molecular morphometrics. Mol Biol Evol 17:1435–1445
28. Collins LJ, Moulton V, Penny D (2000) Use of RNA secondary structure for studying the evolution of RNase P and RNase Mrp. J Mol Evol 51:194–204
29. Caetano-Anollés G (2001) Novel strategies to study the role of mutation and nucleic acid structure in evolution. Plant Cell Tiss Org Cult 67:115–132
30. Caetano-Anollés G (2002) Evolved RNA secondary structure and the rooting of the universal tree of life. J Mol Evol 54:333–345
31. Caetano-Anollés G (2002) Tracing the evolution of RNA structure in ribosomes. Nucleic Acids Res 30:2575–2587
32. Caetano-Anollés G (2005) Grass evolution inferred from chromosomal rearrangements and geometrical and statistical features in RNA structure. J Mol Evol 60:635–652
33. Harish A, Caetano-Anollés G (2012) Ribosomal history reveals origins of modern protein synthesis. PLoS ONE 7:e32776
34. Swain TD, Taylor DJ (2003) Structural rRNA characters support monophyly of raptorial limbs and paraphyly of limb specialization in water fleas. Proc R Soc Lond Ser B 270:887–896
35. Grajales A, Aguilar C, Sánchez JA (2007) Phylogenetic reconstruction using secondary structures of internal transcribed spacer 2 (ITS2, rDNA): finding the molecular and morphological gap in Caribbean Gorgonian corals. BMC Evol Biol 7:90
36. Sun F-J, Fleurdépine S, Bousquet-Antonelli C, Caetano-Anollés G, Deragon J-M (2007) Common evolutionary trends for SINE RNA structures. Trends Genet 23:26–33
37. Sun F-J, Caetano-Anollés G (2008) The origin and evolution of tRNA inferred from phylogenetic analysis of structure. J Mol Evol 66:21–35
38. Sun F-J, Caetano-Anollés G (2008) Evolutionary patterns in the sequence and structure of transfer RNA: a window into early translation and the genetic code. PLoS ONE 3:E2799
39. Sun F-J, Caetano-Anollés G (2008) The evolutionary significance of the long variable arm in transfer RNA. Complexity 14:26–39
40. Sun F-J, Caetano-Anollés G (2010) The ancient history of the structure of ribonuclease P and the early origins of Archaea. BMC Bioinf 11:153
41. Wang M, Caetano-Anollés G (2009) The evolutionary mechanics of domain organization in proteomes and the rise of modularity in the protein world. Structure 17:66–78
42. Wang M, Yafremava LS, Caetano-Anollés D, Mittenthal JE, Caetano-Anollés G (2007) Reductive evolution of architectural repertoires in proteomes and the birth of the tripartite world. Genome Res 17:1572–1585
43. Aziz MF, Mughal F, Caetano-Anollés G (2023) Tracing the birth of structural domains from loops during protein evolution. Sci Rep 13: 14688
44. Caetano-Anollés G (2023) Agency in evolution of biomolecular communication. Ann N Y Acad Sci 1525:88–103
45. Caetano-Anollés G, Claverie JM, Nasir A (2023) A critical analysis of the current state of virus taxonomy. Front Microbiol 14:1240993
46. Caetano-Anollés G, Kim HS, Mittenthal JE (2007) The origin of modern metabolic networks inferred from phylogenomic analysis of

protein architecture. Proc Natl Acad Sci USA 104:9358–9363
47. Caetano-Anollés G, Yafremava LS, Gee H, Caetano-Anollés D, Kim HS, Mittenthal JE (2009) The origin and evolution of modern metabolism. Int J Biochem Cell Biol 41:285–297
48. National Center for Biotechnology Information (NCBI). https://www.ncbi.nlm.nih.gov/ Accessed January 1, 2025
49. Rfam—Collection of RNA families. https://rfam.org/. Accessed January 1, 2025
50. tRNAdb—Transfer RNA database. https://www.bioinf.uni-leipzig.de/services/webservices. Accessed January 1, 2025
51. GtRNAdb—RNAscan-SE analysis of complete genomes. https://gtrnadb.ucsc.edu. Accessed January 1, 2025
52. ViennaRNA Web Services. http://rna.tbi.univie.ac.at/. Accessed January 1, 2025
53. Knudsen V, Caetano-Anollés G (2008) NOBAI: a web server for character coding of geometrical and statistical features in RNA structure. Nucleic Acids Res 36:W85–W90
54. NOBAI software package at the MANET database. https://manet.illinois.edu/nobai/nobai.php. Accessed January 1, 2025
55. Swofford DL (2003) PAUP*: Phylogenetic Analysis Using Parsimony (* and other methods), version 4.0b10. Sinauer Associates, Sunderland, MA
56. PAUP* download page (2024) https://paup.phylosolutions.com. Accessed January 1, 2025
57. Github gcalab repository. https://github.com/gcalab/files/. Accessed January 1, 2025
58. Will S, Joshi T, Hofacker IL, Stadler PF, Backofen R (2012) LocARNA-P: Accurate boundary prediction and improved detection of structural RNAs. RNA 18:900–914
59. Mattei E, Pietrosanto M, Ferrè F, Helmer-Citterich M (2015) Web-Beagle: a web server for the alignment of RNA secondary structures. Nucleic Acids Res 43:W493–W497
60. Lundberg JG (1972) Wagner networks and ancestors. Syst Zool 21:398–413
61. Weston PH (1988) Indirect and direct methods in systematics. In: Humphries CJ (ed). Ontogeny and Systematics. Columbia University Press, New York, NY, 27–56
62. Weston PH (1994) Methods for rooting cladistic trees. In: Siebert DJ, Scotland RW, Williams DM (eds). Models in Phylogeny Reconstruction, Systematics Association Special Volume No. 52. Clarendon Press, Oxford, 125–155
63. Schultes EA, Hraber PT, LaBean TH (1999) Estimating the contributions of selection and self-organization in RNA secondary structure. J Mol Evol 49:76–83
64. Giribet G (2007) Eficient tree searches with available algorithms. Evol Bioinforma 3:341–356
65. Goloboff PA, Catalano S, Torres A (2022) Parsimony analysis of phylogenomic datasets (II): evaluation of PAUP*, MEGA and MPBoot. Cladistics 38:126–146
66. Torres A, Goloboff PA, Catalano SA (2022) Parsimony analysis of phylogenomic datasets (I): scripts and guidelines for using TNT (Tree analysis using new technology). Cladistics 38:103–125
67. FigTree, version 1.4.4. http://tree.bio.ed.ac.uk/software/figtree/. Accessed January 1, 2025.
68. iTol, version 7. https://itol.embl.de. Accessed January 1, 2025
69. Wang W, Feng C, Han R, Wang Z, Ye L, Du Z, Wei H, Zhang F, Peng Z, Yang J (2023) trRosettaRNA; automated prediction of RNA 3D structure with transformer network. Nat Commun 14: 7266
70. Gardner PP, Holland BR, Moulton V, Hendy M, Penny D (2003) Optimal alphabets for an RNA world. Proc Biol Sci 270:1177–1182
71. Caetano-Anollés G, Nasir A, Kim KM, Caetano-Anollés D (2018) Rooting phylogenies and the tree of life while minimizing ad hoc and auxiliary assumptions. Evol Bioinform 14:1176934318805101
72. Maddison WP, Donoghue MJ, Maddison DR (1984) Outgroup analysis and parsimony. Syst Zool 33:83–103
73. Dalevi D, Hugenholtz P, Blackall LL (2001) A multiple-outgroup approach to resolving division-level phylogenetic relationships using 16S rDNA data. Int J Syst Evol Microbiol 51:385–391
74. O'Brien MJ, Lyman RL, Saab Y, Saab E, Darwent J, Glover DS (2002) Two issues in archaeological phylogenetics: taxon construction and outgroup selection. J Theor Biol 215:133–150
75. Rota-Stabelli O, Telford MJ (2008) A multi criterion approach for the selection of optimal outgroups in phylogeny: recovering some support for Mandibulata over Myriochelata using mitogenomics. Mol Phylogenet Evol 48:103–111

Check for updates

Chapter 19

Tracing Molecular Chronologies Onto Growing Biological Networks

M. Fayez Aziz and Gustavo Caetano-Anollés

Abstract

Chronologies derived from phylogenetic reconstruction enable the tracing of evolutionary history onto networks that model the evolution of biological systems. Here, we outline protocols for constructing and analyzing time series of evolving biological networks, integrating molecular chronologies with dynamic network models to explore the evolution of structural and functional complexity. The methodology employs Pajek for network visualization and R packages for statistical tests to examine connectivity patterns, enabling the study of preferential attachment, degree distributions, and the interplay between long-term evolutionary trends and short-term adaptive dynamics. These approaches provide a framework for uncovering fundamental principles of molecular evolution across deep timescales.

Keywords Network growth, Hierarchical modularity, Bipartite networks, Protein loop, Protein domain, Phylogenetics, Tree of Life, SCOP, SUPERFAMILY, R, PHP, Pajek

1 Introduction

Molecular chronologies derived from evolutionary genomic analyses provides a powerful lens through which to examine the evolutionary trajectories of biological systems [1–4]. In these chronologies, a series of time events describes the history of biological organization and reveals how structural and functional complexity emerges along deep temporal frameworks. For example, the reconstruction of phylogenomic trees of nucleic acid structures allows the gradual evolution of functional RNA to be traced, from the making of rather simple structures such as tRNA [5] to the evolutionary assembly of large ribonucleoprotein complexes such as the ribosome [6]. Similarly, the reconstruction of phylogenomic trees of protein structural domains makes it possible to trace the accretion of molecular components that comprise proteomes, from their embedded disordered loop structures [7–9] to modern multidomain architectures [10, 11]. These chronologies uncover distinct evolutionary phases underpinning

Gustavo Caetano-Anollés (ed.), *Evolutionary Genomics*, Methods in Molecular Biology, vol. 2981,
https://doi.org/10.1007/978-1-0716-4836-0_19,

biological organization, including the rise of common ancestors, viruses, and cellular life [4, 12–14].

Mapping these temporal patterns onto growing biological networks demonstrates how early structural innovations become organizational hubs, often through explosive diversification. This integration of chronology and network analysis has the potential to uncover fundamental principles governing life's molecular evolution, including the size-dependent ordering of proteins through loop accretion, the emergence of functional hierarchies from super secondary motifs, and the yin-yang dynamics between structural disorder and biological specificity that continue to shape proteomes [7, 8, 9].

Unlike static networks, which describe snapshots of fixed connections, time-dependent (temporal) networks reveal how systems evolve [15]. Nodes and links of these evolving networks activate or rewire across timescales ranging from microseconds (e.g., molecular conformational shifts) to billions of years (e.g., evolutionary innovations or adaptations). This dynamism creates a spectrum of structural behaviors: rapid rewiring (minutes for metabolomic networks) contrasts with gradual evolutionary reconfigurations often shaped by the rise of hierarchical modularity [16]. These extremes stem from the interplay of network dynamics (short-term processes like information flow) and network evolution (long-term structural changes). Three scenarios govern this interplay: (i) *Slow evolution versus fast dynamics*: Network structure remains stable relative to short-term processes (e.g., hardwired metabolic pathways); (ii) *Matched timescales*: Network structure and dynamics coevolve (e.g., disease spread in contact networks); (iii) *Rapid evolution*: Fast shifts in network structure necessitate time-averaged models for tracing dynamics (e.g., social contagion). Analyzing these regimes poses significant challenges. Slow evolution demands retrodiction (reconstructing past states), while rapid changes require real-time tracking—both complicating methodology due to data intensity and temporal resolution trade-offs.

Network evolution is typically visualized through contact sequences and time-event snapshots. Contact sequences track node and link activity at specific times or intervals, enabling the identification of recurring temporal motifs [17]. Time-event snapshots, inspired by discrete event simulations [18], capture global network states at critical times but obscure localized changes. These representations highlight the tension between granular temporal patterns and holistic structural insights, underscoring the need for adaptable analytical frameworks when modeling systems like metabolic networks, where deep evolutionary processes coexist with rapid adaptive responses. Figure 1 describes the growth of networks evolving at discrete temporal intervals. Time flows from one event to another as discrete "time steps." This temporal progression can be visualized with two layouts, "waterfall" and "radial." The waterfall

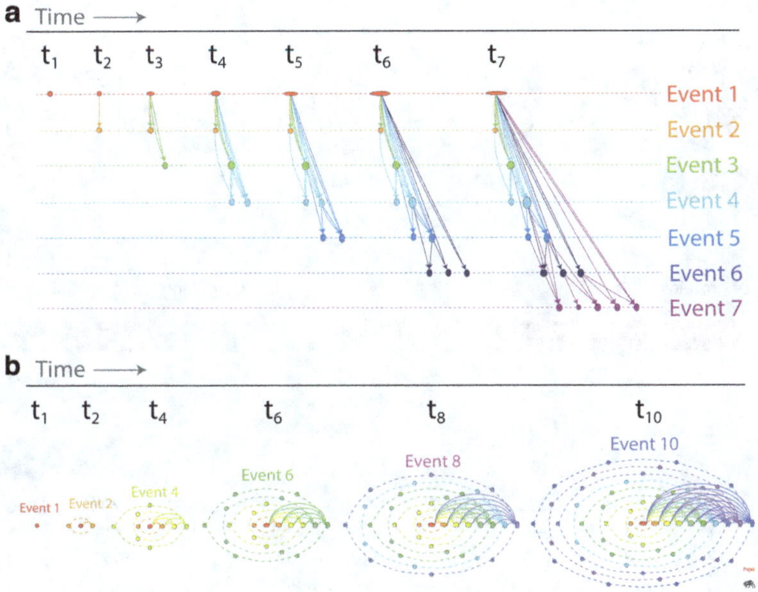

Fig. 1 Describing the evolutionary growth of systems with time-variant networks. (**a**) Network chronologies in "waterfall evolution" layout. Time progresses from left to right as "discrete events" (t_1, t_2, t_3, t_4, t_5, t_6, and t_7) unfold in network evolution. Nodes and links (time-directed arrows known as arcs) appear from top to bottom, colored according to their age. Source-sink recruitments are visualized by the width and height of symbols, which represent the outdegree and indegree of nodes, respectively. As networks evolve, the symbols of older nodes widen by outdegree accumulation, while the symbols of younger nodes grow tall by indegree accumulation. (**b**) Network chronologies in "radial evolution" layout. Networks accumulate nodes in concentric rings (orbitals), each reflecting a time event. We illustrate radial evolution with six snapshots of a growing network (time events t_1, t_2, t_4, t_6, t_8, and t_{10}), The toy networks reach a size 55 sparsely connected nodes by growing nodes (*n*) in orbitals (*r*) at an $r + 1$ rate and connecting one node per orbital to single nodes in each of the other orbitals. Diagrams modified from [11]

layout arranges nodes top-down by age. The radial layout arranges nodes in concentric rings (orbitals) with duration from origin increasing from the center to the periphery. The time series of networks describes how networks grow, making evolutionary recruitment evident as time progresses downward or outward in waterfall and radial networks, respectively.

We illustrate the radial visualization with a time series of networks of protein domain organization (Fig. 2). These networks embed evolutionary information extracted from the sequence and structure of millions of protein sequences encoded in hundreds of genomes [11]. Evolving interactions between individual domains are captured with five operative criteria defined by composition, pairwise occurrence, adjacency, splicing, and adjacent splicing of domains in a protein molecule. We only illustrate two of these criteria with the time series of spliced pairwise (SPX) and a spliced pairwise adjacency (SPAX) networks (Fig. 2a). Evolving networks revealed two ancient waves of structural novelty arising from ancient "p-loop" and "winged

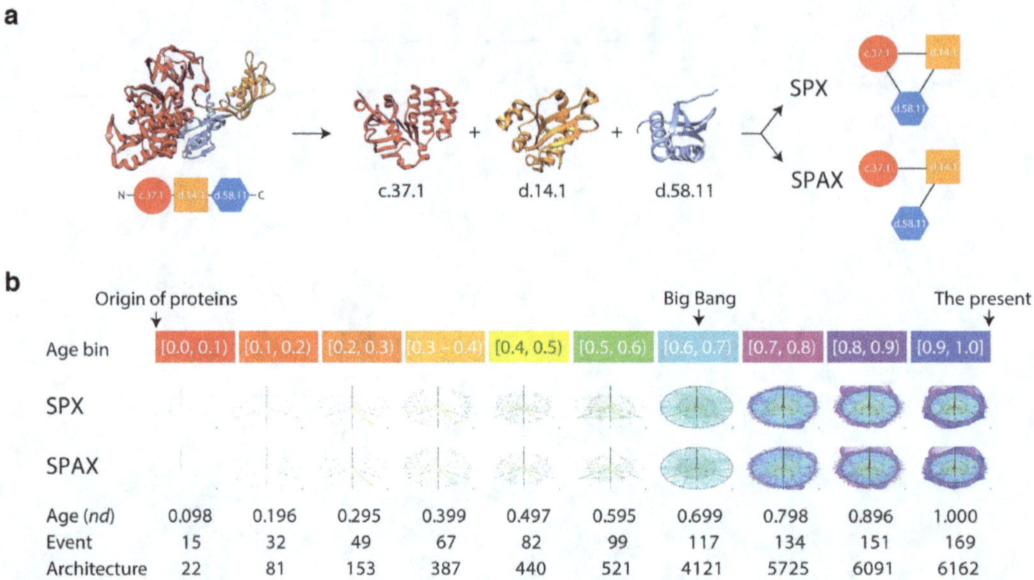

Fig. 2 Evolving networks of protein domain organization in radial evolution layout. (**a**) A census of protein domains and their combinations in the proteomes of 749 organisms was used to construct spliced pairwise (SPX) and spliced pairwise adjacency (SPAX) networks that connect domain nodes dissected from protein architectures when domain pairs or when adjacent domain pairs are present in proteins, respectively. Domains were defined at *superfamily* level with SCOP classifications strings (e.g., c.37.1). Note that the c.37.1|d.14.1|d.58.1.1 architecture is common in dehydratase enzymes and elongation factors. (**b**) Snapshots of network growth describe the evolution of 6162 domain, supradomain and multidomain architectures when 1643 domains are spliced from them. They represent 10 out of 161 time events of the evolutionary timeline, which are indexed with an evolutionary age (*nd*, ranging from 0.0 to 1.0). Age bins were custom RGB color-coded to highlight the flow of time, from top to bottom. The evolving SPX and SPAX networks reveal the gradual evolutionary accumulation of nodes and links. Diagrams modified from [11]

helix" fold structures and a massive "big bang" of protein domain organization, which marks a significant evolutionary period where architectural diversity increased dramatically [10, 11]. Key donor domains like the P-loop containing nucleoside triphosphate hydrolase, NAD(P)-binding Rossmann-fold and acetyl-CoA synthetase-like structures played crucial roles in the combinatorial processes, while significant acceptor domains like the Homeodomain-like and Acyl carrier protein (ACP)-like structures highlighted the ancient role of DNA binding and editing functions [11]. The visualization emphasizes the dynamic nature of protein networks during 3.8 billion years of evolution, with ongoing recruitment and innovation contributing to the diversity of modern protein domains.

Here, we outline the experimental procedures for building a time series of evolving networks. The methodology uses Pajek [19] to visualize networks and R [20] for statistical packages. This allowed exploration of various network qualities, including network connectivity with a focus on degree distributions, the emergence of preferential attachment [21] using power law distribution exponents and Kolmogorov–Smirnov statistical tests [22, 23], traces of randomness through von Neumann's ratio test for Randomness [24], and the rise of modularity [25].

2 Materials

2.1 Equipment

We assume the user has some experience working with the Pajek network analysis program and the R statistical analysis packages. Although not mandatory, familiarity with a scripting language like PHP would be useful to automate the generation of Pajek network files and R data sets. This can supplement manual labor with larger networks while using plain text editors and spreadsheets. We also assume the user is familiar with the specifics of each biological system that is studied. For example, when constructing networks of protein domain organization, the user must be familiar with the NCBI, SCOP, and SUPERFAMILY databases and must know how to download proteomic data for organisms of interest and how to assign SCOP classified families and superfamilies to downloaded proteomes using public tools.

1. We recommend using a computer with a 64-bit operating system possessing at least 8 GB RAM.

2.2 Software

The required software can be acquired as follows:

1. Download and install the Pajek program [19] from its download page [26]. Various versions are available for different platforms (Mac, Linux, Windows). Documentation, tutorials, and links to introductory books are also available in the same resource.
2. Download and install an interpreter for the PHP scripting language [27]. Help files on scripting and tutorials for built-in functions are available under the same link.
3. Download and install R statistical packages [20] from [28]. The link provides detailed instructions on how to set up and utilize various statistical computing, and graphing packages. A useful Integrated Development Environment (IDE) recommendation for editing R files is R-Studio [29].
4. A suggested text and source code file editor is Notepad++ [30]. It is highly recommended for quickly searching through stored files using operating system-independent search features and easily editing scripts or code files using color-coding and line numbers. It can also be used to compare multiple versions of plain text files to spot differences.

5. A standard spreadsheet editor (e.g., Microsoft Excel or Google Sheets) would understandably be required to view and edit raw data files.

3 Methods

Figure 3 outlines a workflow of main activities that starts with the generation of a chronology of parts of biological systems from a phylogeny that describes the evolution of parts. These systems are then translated into constructed networks, which describe how parts connect in some theoretical or physical framework. The networks can then be mapped or indexed with times of origin (*nd*) calculations, enabling their growth and evolution to be visualized as time events, and their structure and organization to be analyzed with statistical methods.

3.1 Phylogeny Reconstruction

Use phylogenetic methods to produce a phylogeny that describes the evolution of parts of a system (*see* **Note 1**). Using Fig. 2 as an example, a phylogeny is reconstructed that describes the evolution of protein architectures (parts) in proteomes (wholes). The phylogenetic data matrix used to search for optimal trees in tree solution space is constructed in which protein architectures (domains, supradomains, and multidomains) are treated as taxa and proteomes are treated as phylogenetic characters. Note that character states may represent occurrences, abundances, or some other feature of importance for protein domain organization. The phylogeny of parts is highly unbalanced, and consequently, it accommodates stepwise generative models of innovation [31]. This justifies counting the number of branching events along the tree and calculating a "node distance" (*nd*) that measures the time of origin (ancestry, age) of each taxon. This allows for the generation of chronologies describing the stepwise appearance of parts of an evolving system.

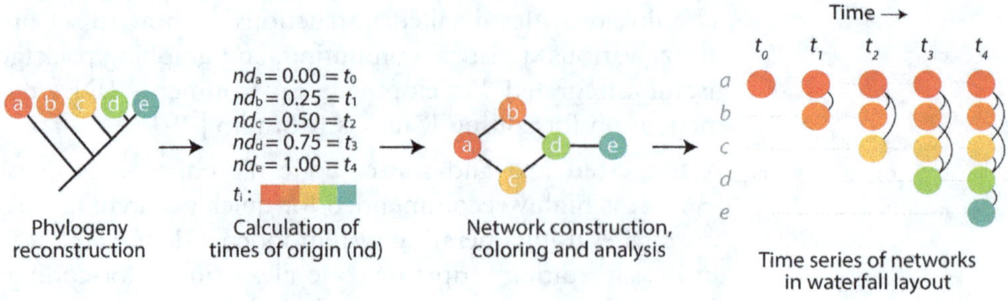

Fig. 3 Workflow of main activities

Table 1
Types of data matrices

Matrix type	Matrix objective
By "node age" (NOA)	Used for box plots and x–y line plots. Categorical columns are ordinal numbers, age bin, age and node label of nodes in the networks. Additional columns describe *cwdn* in increasing order of events. Rows are sorted by node.
By "network age" (NEA)	Used for power law distribution graphs. They are essentially transpositions of NOA data. Columns are ordinal number, age bin, and age of networks, followed by *cwdn* arranged by nodes. Rows are sorted by events.
By "degree dispersion" (DD)	Used for stacked bar charts. Column and row order are the same as the NOA data. However, columns provide the distributions of final *cwdn* (i.e., at *nd*=1) across connected node age bins.

3.2 Network Data Analyses

Visualize and analyze networks using Pajek at multiple temporal stages while they unfold along the evolutionary timeline. Encode nodes (vertices) and links (arcs/edges) of graphs (including bipartite graphs and corresponding projections) in spreadsheet or plain text files, per convenience, to be processed by the network visualization program. Generate the quantitative network property of *cumulative weighted degree per node* (*cwdn*) using custom Pajek macros. Compile *cwdn* as three types of data matrices for each network (Table 1). The data rows of matrices define parts of systems and are sorted in ascending order of node—(individual *nd*) or network—(event *nd*) age. Organize separate matrices for the "in" and "out" degree types as well as each portion of the bipartite node-set.

3.3 Network Visualization

Generate visualization attributes and layouts using the Pajek menu commands listed in Table 2. Node categories are made distinguishable using various shapes and color options. Unfold evolutionary patterns by color-coding the age of nodes. Use a color palette that represents a gradient ranging, for example, from red for the most ancient node ($nd = 0$) to orchid for the most recent node ($nd = 1$) to simulate a color-coded growth in time.

3.4 Charts, Graphs, and Benchmarking

Use graphing code constructs and packages of R [20] to visualize the *cwdn* (Table 3). Generate reference networks for comparative analysis of power law and modularity, using "barabasi" methods [32] of the R's *igraph* package [33]. Implement the ideal power law model with "barabasi.game ()". Implement an extended model with "aging. prefatt.game ()". This model simulates the scale-free evolution of random graphs by altering the probability of preference of an old vertex growing multifold, exponentially with age. Create networks of various comparable sizes to simulate reference controls for the corresponding networks of study, with properties of directionality

Table 2
Network visualization tools and commands

Pajek tool and command	Output
Network: Create Vector: Centrality: Weighted Degree: "All," "Output\Input"	The weighted degree vectors for undirected and directed networks, respectively.
Draw: "Network + Vector"	Visualization of the weighted degree of nodes as node size.
Network: Create Partition: Communities: VOS clustering: "multi-level coarsening + multi-level refinement"	The community-based layout of the networks uses the VOS clustering method. In addition, modularity indices are obtained. Default parameters are used.
Draw: "Network + Partition" and Draw: Layers: "In y Direction"	Vertical arrangement of nodes according to their age in both bipartite and waterfall layouts of the networks with age mappings.
Draw: Layout: Energy: Kamada-Kawai: "separate components" and "Optimize inside clusters only"	Visual distribution of partitions (clusters or communities) in a waterfall layout, manually separated to refine the most energetically favored network configurations.

directly transferred. Assign ages to individual nodes in incremental order to keep the age proportion per event consistent with the timelines of the test networks.

3.5 Power Law Statistics

Several power law statistics can be evaluated (*see* **Note 2**). *P(k)* vs. *k* and *log–log* plots provide preliminary insight into the scale-free behavior of a network. The gamma coefficient from *log P*(*k*) vs. *log k* scatter plots supplements the analysis of power law behavior and scale-free behavior of a network. Power law behavior usually manifests in degree distributions exhibiting long tails (like *Poisson*). Utilize R libraries and operations to run the analysis (Table 4) and color-code distributions.

3.6 Modularity Indices

Calculate modularity indices for each network using various capabilities of the *igraph* package [33] (Table 5) (*see* **Note 3**). Delete isolated vertices using a custom function and adjust corresponding partitions, as required by the modularity algorithms. Use *cwdn* as input in the calculation of all modularity indices. Generate VOS modularity indices and partitions using Pajek. Compute *NG* modularity indices [25] using two types of memberships (or partitions) as input: VOS clustering (NG_{vos}) and age (NG_{age}). Determine clustering ratios using custom functions by dividing the number of clusters by the connected node set and its size.

3.7 Randomness Quantification

Compute the randomness quality of domain networks using the rank version of von Neumann's ratio test for Randomness, i.e., the Bartels

Table 3
Graphing and charting constructs and operations

R *package*: function	Result
Reshape: *	Transform textual data tables to vector form.
Lattice: *	Generate panels of network-age and node-age bins.
LatticeExtra, grid: *	Resize panels.
LatticeExtra: "`panel.lmline ()`"	Draw linear regression model lines and equations, e.g., in *log-log* graphs.
Data.table: *	Edit data tables, e.g., to remove empty age bins.
igraph: *	Read network graph files to calculate modularity and apply the power law distribution model to calculate its fit statistics.
R base: "`bwplot ()`"	Box and whisker's plots based on NOA files, to depict measures of central tendency of *cwdn* over network age.
Lattice: "`xyplot ()`"	XY scatter and line plots based on NOA files, drawing the final *cwdn* attained at network age 1.
R color palette: *	The color-coding of data points by node age.
ggplot2: "`ggplot ()`," "`geom_bar ()`" and "`grid.draw ()`"	Produce degree dispersion stacked bar charts and graphical trend curves of power law behavior and modularity, based on DD, NEA data and VOS modularity report files, respectively.
ggplot2: `scale_fill_manual ()` `scale_color_manual ()`	Color-coded stacked bars representing the distribution of final *cwdn* over the ages of the connected node set.
ggplot2: "`layer ()`"	Stack negative-valued bars in reflection.
ggplot2: "`theme ()`"	Color-coding of multiple curves of power law statistics and modularity indices.
gridExtra: "`grid.arrange ()`" and "`arrangeGrob ()`"	Plotting together of curves for comparison.
ggplot2: "`geom_smooth ()`"	Addition of linear regression lines to some modularity plots.
ggplot2: "`geom_tile ()`," "`scale_fill_gradient ()`," "`scale_x/y_discrete ()`," etc.	Customized modularity heat maps.
ggdendro: *	Plotting of dendrograms.
R base: "`aggregate ()`"	Calculate averages of categorized data.
plotrix: "`std.error ()`"	Measure dispersion of data through standard error.

rank test of randomness. Input the network degree data from NEA files along with the parameters such as alternative = "two.sided," to test the null hypothesis of randomness against non-randomness, and p-value = beta, to compute p-value using an approximation given by

Table 4
R operations used in power law analysis

R *package*: function	Result
Lattice: "`xyplot ()`"	Plot distributions generated from NEA files.
Data.table: "`table ()`"	Compute *k* frequency tables for *P(k)* vs. *k* plots.
R base: "`length ()`"	Calculate the length of the degree table to determine *P(k)* through the quotient of *k* frequencies by the maximum degree.
R base : "`ppois()`"	Compute *Poisson* (*k*) to be plotted against *k*.
R base: "`log ()`"	Implement *log P(k)* vs. *log k* scatter plots.
Lattice: "`lm ()`" and "`panel.abline ()`'	Generate and draw linear regression models.
R base: "`summary (lm)$ coefficients`"	Retrieve the γ-slope power law coefficient and R^2 determination coefficient.
R base: "`ks.test ()`"	Calculate primary power-law stats from Kolmogorov–Smirnov tests.
igraph: "`power.law.fit ()`"	Obtain additional statistics supporting the preferential attachment principle.

Table 5
R's *igraph* package operations used in modularity analysis

igraph's function	Result
"`read.graph ()`"	Import network graph data.
"`fastgreedy.community ()`"	Calculate the *FGC* modularity index.
"`as.undirected ()`"	Collapse directed edges for *FGC*.
"`modularity ()`"	Compute *NG* modularity indices.
"`transitivity ()`"	Determine average clustering coefficient (*C*), using the "`average`" option.

the Beta distribution [24]. The algorithm removes missing values by default. Generate the random graphs to be included as controls using the Erdős–Rényi model implementation of *igraph* package [33] (Table 6). Color-coded distributions for improved visual inspection.

3.8 Heatmaps and Dendrograms

Compute the NG-pairwise modularity matrix for each network using "mod.matrix ()" [25]. Three types of memberships (or partitions) are used as input: VOS, age, and *FGC*. *FGC* partition is extracted by "membership ()." Scale each matrix by the absolute log10 value of the overall modularity score of the network, before being drawn as a heatmap with "heatmap.2 ()." Saturate pairwise scores to the

Table 6
R operations used in the calculation of randomness

R *package*: function	Result
randtests: "`bartels.rank.test ()`"	Calculate von Neumann's ratio test for Randomness (RVN) over network degree.
igraph: "`erdos.renyi.game ()`"	Generate random graphs per the Erdős–Rényi model.

range [−1, 1]. Color-code the x–y axis node labels and order them by age. In dendrograms, the x–y axis node ticks are color-coded by age, and labels are color-coded by bipartite node type. Generate dendrograms using "dist ()" by calculating squared Euclidean distance matrices that indicate dissimilarities between the cluster means [25]. Hierarchically cluster the distance (or dissimilarity) matrices using "hclust ()" with Ward''s minimum variance method, aiming at finding compact, spherical clusters [33, 34]. Order nodes within clusters that were clustered in the dendrograms by age and color-code them as such.

4 Notes

1. The history of biological systems can be recovered without the need to reconstruct phylogenetic trees using advanced tools of computational analysis. Examples include experimental retrodiction approaches such as the resurrection of computationally reconstructed or preserved macromolecules in the laboratory, the preservation of cellular organisms or viruses in time to track their evolution (e.g., SARS-CoV-2 sampling for genomic sequencing throughout the COVID-19 pandemic), and extracting ancient DNA from amber, ice cores, or other materials. History can be predicted with nomothetic methods such as algorithms or modeling approaches that mimic evolution.
2. The power law behavior of networks, including the bipartite network and its projections, can be evaluated using six differential indicators of preferential attachment, comparing these to two reference models: Barabási and Barabási-age. The metrics include the KS fit statistic, KS *p*-value, exponent of the power-law distribution, slope of the power-law linear regression model, log-likelihood of fitted parameters, and the coefficient of determination, with calculations made for each event along the evolutionary timeline.

3. Network modularity can be assessed using several indicators, showing linear regression coefficients and determination coefficients for both networks by age and network size. Two modeled control sets, Barabási and Barabási-age, can be used as references to reveal differences in modularity curves and highlight cumulative and weighted connectivity inputs over time.

Acknowledgments

Research in the Evolutionary Bioinformatics laboratory was supported by grants from the National Science Foundation (MCB-0749836 and OISE-1132791) and the United States Department of Agriculture (ILLU-802-909 and ILLU-483-625) and by Blue Waters supercomputer allocations from the National Center for Supercomputing Applications (NCSA).

References

1. Caetano-Anollés G, Wang M, Caetano-Anollés D, Mittenthal JE (2009) The origin, evolution and structure of the protein world. Biochem J 417(3):621–637
2. Caetano-Anollés G, Caetano-Anollés D (2015) Computing the origin and evolution of the ribosome from its structure—uncovering processes of macromolecular accretion benefiting synthetic biology. Comp Struct Biotechnol J 13:427–447
3. Caetano-Anollés G, Aziz MF, Mughal F, Caetano-Anollés D (2021) Tracing protein and proteome history with chronologies and networks: folding recapitulates evolution. Exp Rev Proteomics 18:863–880
4. Caetano-Anollés G, Mughal F, Aziz MF, Caetano-Anollés K (2024) Tracing the birth and intrinsic disorder of loops and domains in protein evolution. Biophys Rev 16(6):723–735
5. Sun F-J, Caetano-Anollés G (2008) The origin and evolution of tRNA inferred from phylogenetic analysis of structure. J Mol Evol 66:21–35
6. Harish A, Caetano-Anollés G (2012) Ribosomal history reveals origins of modern protein synthesis. PLoS One 7(3):e32776
7. Aziz MF, Caetano-Anollés K, Caetano-Anollés G (2016) The early history and emergence of molecular functions and modular scale-free network behavior. Sci Rep 6(1):25058
8. Aziz MF, Mughal F, Caetano-Anollés G (2023) Tracing the birth of structural domains from loops during protein evolution. Sci Rep 13(1):14688
9. Mughal F, Caetano-Anollés G (2023) Evolution of intrinsic disorder in protein loops. Life 13(10):2055
10. Wang M, Caetano-Anollés G (2009) The evolutionary mechanics of domain organization in proteomes and the rise of modularity in the protein world. Structure 17:66–78
11. Aziz MF, Caetano-Anollés G (2021) Evolution of networks of protein domain organization. Sci Rep 11: 12075
12. Wang M, Yafremava LS, Caetano-Anollés D, Mittenthal JE, Caetano-Anollés G (2007) Reductive evolution of architectural repertoires in proteomes and the birth of the tripartite world. Genome Res 17:1572–1585
13. Nasir A, Caetano-Anollés G (2015) A phylogenomic data-driven exploration of viral origins and evolution. Sci Adv 1(8):e1500527
14. Caetano-Anollés G (2024) Are viruses taxonomic units? A protein domain and loop-centric phylogenomic assessment. Viruses 16(7):1061
15. Holme P, Saramäki J (2012) Temporal networks. Phys Rep 519(3):97–125
16. Caetano-Anollés G, Aziz MF, Mughal F, Gräter F, Koç I, Caetano-Anollés K, Caetano-Anollés D (2019) Emergence of hierarchical modularity in evolving networks uncovered by phylogenomic analysis. Evol Bioinf 15:1176934319872980
17. Kovanen L, Karsai M, Kaski K, Kertész J, Saramäki J (2011) Temporal motifs in time-dependent networks. J Stat Mech P11005
18. Delaney, W, Vaccari, E, 1989. Dynamic models and discrete event simulation. Marcel Dekker Inc, New York

19. Mrvar A, Batagelj V (2016) Analysis and visualization of large networks with program package Pajek. Complex Adapt Syst Model 4:1–8
20. Ihaka R, Gentleman R (1996) R: a language for data analysis and graphics. J Comput Graph Stat 5(3):299–314
21. Li L, Alderson D, Doyle JC, Willinger W (2005) Towards a theory of scale-free graphs: definition, properties, and mplications. Internet Math 2(4):431-523.
22. Newman MEJ (2005) Power laws, Pareto distributions and Zipf's law. Contemp Phys 46:323–351
23. Clauset A, Shalizi CR, Newman MEJ (2009) Power-law distributions in empirical data. SIAM Rev 51:661–703
24. Bartels R (1982) The rank version of von Neumann's ratio test for randomness. J Am Stat Assoc 77:40–46
25. Newman ME, Girvan M (2004) Finding and evaluating community structure in networks. Phys Rev E 69(2):026113
26. Pajek download page. http://mrvar.fdv.uni-lj.si/pajek/
27. PHP scripting language interpreter. https://www.php.net/
28. R package. https://www.r-project.org/
29. R-Studio. https://rstudio.com/products/rstudio/download/
30. Notepad++ application. https://notepad-plus-plus.org/downloads/
31. Tal G, Boca SM, Mittenthal J, Caetano-Anollés G (2016) A dynamic model for the evolution of protein structure. J Mol Evol 82:230–243
32. Barabási AL, Albert R (1999) Emergence of scaling in random networks. Science 286(5439):509–512
33. Csardi G, Nepusz T (2006) The igraph software package for complex network research. Intl J Complex Syst 1695(5):1–9
34. Borg I, Groenen P (2003) Modern multidimensional scaling: theory and applications. J Educ Meas 40(3):277–280

Subject Index

A

B

C

D

E

F

G

H

Gustavo Caetano-Anollés (ed.), *Evolutionary Genomics*, Methods in Molecular Biology, vol. 2961,
https://doi.org/10.1007/978-1-0716-4836-0,

www.ingramcontent.com/pod-product-compliance
Lightning Source LLC
Chambersburg PA
CBHW081021070526
PP18256900001BA/31
* 9 7 8 1 0 7 1 6 4 8 3 5 3 *